AF388937

From Laboratory Studies to Court Evidence: Challenges in Forensic Entomology

From Laboratory Studies to Court Evidence: Challenges in Forensic Entomology

Editors

Damien Charabidze
Daniel Martín-Vega

MDPI • Basel • Beijing • Wuhan • Barcelona • Belgrade • Manchester • Tokyo • Cluj • Tianjin

Editors
Damien Charabidze
UMR 8025, Centre dHistoire
Judiciaire
Université de Lille,
Lille
France

Daniel Martín-Vega
Departamento de Ciencias de la
Vida
Universidad de Alcalá
Madrid
Spain

Editorial Office
MDPI
St. Alban-Anlage 66
4052 Basel, Switzerland

This is a reprint of articles from the Special Issue published online in the open access journal *Insects* (ISSN 2075-4450) (available at: www.mdpi.com/journal/insects/special_issues/insects_forensic_entomology).

For citation purposes, cite each article independently as indicated on the article page online and as indicated below:

LastName, A.A.; LastName, B.B.; LastName, C.C. Article Title. *Journal Name* **Year**, *Volume Number*, Page Range.

ISBN 978-3-0365-1708-7 (Hbk)
ISBN 978-3-0365-1707-0 (PDF)

Contents

About the Editors

Damien Charabidze

Damien Charabidze is a doctor of biology and a specialist of necrophagous insects. He is a lecturer at the University of Lille (France) and a scientific associate at Bruxelles University (ULB, Belgium). His research focuses on the behavioural and social adaptations of carrion insects and their consequences on larval development. He is a forensic expert in entomology and is currently investigating the articulation of science and law in the French legal system.

Daniel Martín-Vega

Daniel Martín-Vega is a lecturer at the University of Alcalá and a scientific associate at the Natural History Museum, London. His research focuses on the morphological and developmental aspects of insects of medical, forensic and veterinary importance. He is currently applying micro-computed tomographic methods to document the developmental changes that take place during the metamorphosis of blow flies (Calliphoridae) and bot flies (Oestridae) and to explore and visualise the host–parasite interface in different fly vectors of diseases.

Editorial

Looking Back to Move Forward: How Review Articles Could Boost Forensic Entomology

Damien Charabidze [1,2,*] and Daniel Martín-Vega [3,*]

1   UMR 8025, Centre d'Histoire Judiciaire, Université de Lille, F-59000 Lille, France
2   Unit of Social Ecology (USE), Université Libre de Bruxelles (ULB), 1050 Bruxelles, Belgium
3   Departamento de Ciencias de la Vida, Universidad de Alcalá, 28805 Alcalá de Henares, Madrid, Spain
*   Correspondence: damien.charabidze@univ-lille.fr (D.C.); daniel.martinve@uah.es (D.M.-V.)

check for **updates**

Citation: Charabidze, D.; Martín-Vega, D. Looking Back to Move Forward: How Review Articles Could Boost Forensic Entomology. *Insects* 2021, *12*, 648. https://doi.org/10.3390/insects12070648

Received: 2 July 2021
Accepted: 13 July 2021
Published: 15 July 2021

Publisher's Note: MDPI stays neutral with regard to jurisdictional claims in published maps and institutional affiliations.

The Locard's exchange principle (1930) holds that the perpetrator of a crime leaves traces behind that can later be sampled and used as forensic evidence. On the contrary, insects are autonomous cues: they appear, grow and move away from a crime scene without any human action. Forensic entomologists rely on these living cues to reconstruct a posteriori the post-mortem chronology.

In the Editorial of the first Forensic Entomology Special Issue (2001), Benecke [1] stated *"There are three things that we need in forensic entomology: more young researchers (. . .), open discussion about possible flaws in our methods and PMI calculations, and access to information for scientists all over the world (. . .)"*. Thanks to research, forensic entomology has since investigated several possible flaws and evolved from a rather empirical clue into a mature evidence-based forensic science. This is the consequence of the deep involvement of many students and researchers worldwide, who performed extensive studies and published hundreds of articles, case reports and technical refinements. However, this may be too much: the above-mentioned easy access to information has on the way been blurred by an excessive amount of publication. Revealingly, the current top 3 cited "forensic entomology" documents are rather old manuals from 1992, 1991 and 2007 and many forensic pathologists and crime scene investigators still think in terms of "squads", a hypothesis yet abandoned by forensic entomologists for decades. Thus, while several factors may explain the slow rise of forensic entomology, its complexity and inability to simply answer some practical questions cannot be ignored. In this context, we believe review articles may be the next big step for forensic entomology.

Although their main (and obvious) disadvantage is that they will eventually become outdated, review articles integrate and synthesize relevant information that otherwise would be dispersed, so they can be both the perfect introduction to a novel topic [2] and a new starting point that boosts research within a particular area of knowledge. Either if they are written from a purely "narrative" perspective or with a systematic and/or meta-analytical approach, review articles are potentially useful tools for forensic entomology practitioners, researchers and students. Narrative reviews can find their place either in academic journals or in academic collaborative books (e.g., [3,4]), whereas systematic reviews that are developed around a research question and often include a meta-analysis of published data are more often published among other scientific papers in academic journals.

"Oh my god. You still think forensic entomology is all blowflies and screw-worms, don't you?"

Having trouble finding trace amounts of blood at a crime scene, police inspectors of the TV show *Brooklyn 99* call a superstar forensic entomologist, Dr. Yee (Season 6, Episode 10, March 2019). Supposedly the scientist has bred a species of flies that can detect small traces of blood, even when it has been cleaned. Unfortunately, Dr. Yee was a fraud. But while forensic entomology is still about blowflies and maggots, and not mutant flies, some practices of forensic entomology have greatly evolved during the last 40 years. In the first article of this Special Issue, UK forensic entomologist M. Hall shares his long experience and his point of view on the links between fundamental research and its application in forensic cases [5]. Review articles are one of these bridges, as they are useful for all, can focus on any step of the forensic entomology analyses and can be written by both academics and case-workers (contrary to books that require long-time and deep knowledge). Unfortunately, such valuable sources of knowledge currently seem under-represented compared to classical research (Figure 1).

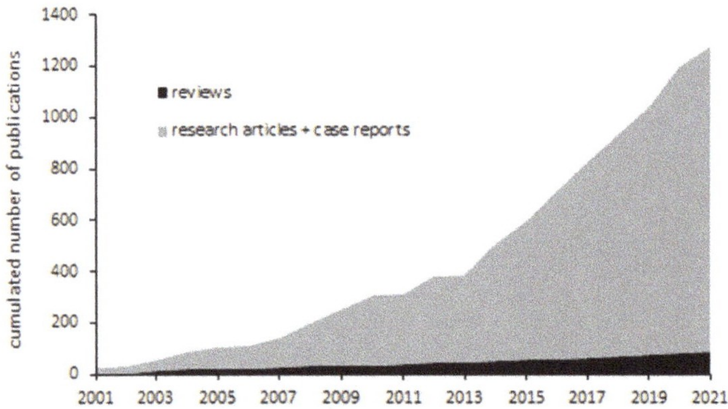

Figure 1. Cumulated number of reviews (black) and research (grey) articles published during the last 20 years (PubMed research performed in March 2021 using "forensic entomology" as keyword).

Due to their low number, forensic entomologists cannot move to all crime scenes, and sampling is thus often performed by non-specialists [6]. Subsequent transport, conservation and temperature control are also extremely important for forensic entomology analysis. To avoid bias, crime scene technicians or forensic pathologists who sample insects must follow clear, concise and efficient sampling, fixation and conservation protocols. While the *Best practice in forensic entomology—standards and guidelines* paper published by Amendt et al. in 2007 [7] remains one of the most cited papers in forensic entomology, it provides general guidelines rather than a clear sampling protocol and is now out of date. In practice, each country or even laboratory currently has its own procedures, and there is a lack of widely accepted international standards. Furthermore, researchers are still working on some key points. As an example, up-to-date protocols are still needed for those eventual cases where eggs could be the only entomological evidence recovered in a death scene. While Amendt et al. [7] recommended to kill and preserve egg specimens—if maintaining living samples was not possible—by directly placing them in 70–95% ethanol, it was recently shown that this method results in a marked decomposition of tissues, thus making it impossible to reliably use age-specific morphological markers [8]. Similarly, the fixation of fly larvae using very hot water prior to storage in 70–95% ethanol requires anticipation and material and is most of the time difficult or impossible under harsh crime scene conditions. Finally, the guidelines by Amendt et al. [7] did not provide any recommendations on how to fix and preserve the intra-puparial stages. However, this period can be particularly crucial in minimum post-mortem interval estimations as it can account for more than half of the developmental duration of the blow fly life cycle. Accordingly, in the last decade, a number of studies have tested the suitability of different fixation and preservation methods: at least eight different articles on sample storage have been published in the last ten years, including one in this Special Issue [9,10]. In this context, it is obviously difficult for a non-specialist to stay informed and make the right methodological choices. There is no doubt that an up-to-date review published under a large scientific consensus and establishing a step-by-step sampling protocol would be a great milestone for the whole forensic entomology community.

The next step in the forensic entomology process, species identification, is also crucial. While the adults of the most common necrophagous species can be easily identified, this task is more challenging for some incidental species or early instars. Whereas identification keys to both the larval and adult stages of the most common blow fly species occurring on cadavers have been published, only spare morphological descriptions might be available for other forensically relevant taxa, with the immature stages of many species remaining unknown. Molecular biology, and especially DNA-based methods, are thus widely investigated. In this Special Issue, Gzywacz et al. [11] provide a good example of the value of molecular methods when morphological identification is not possible. However, while powerful and increasingly accessible, these methods remain more expensive and time-consuming than traditional morphological identification. On the other hand, the idea of traditional morphological identification being a simple and straightforward method can be misleading. First, it requires a solid background in insect taxonomy and many forensic practitioners might not be properly trained in the observation of certain diagnostic characters. Furthermore, some diagnostic morphological characters may be obscured or altered if the sample preservation is not optimal or if the practitioner lacks expertise in the handling, processing and/or mounting of insect specimens. In this Special Issue, Pradelli et al. [12] present simple methods to clean puparial samples in order to facilitate their morphological identification. Some practitioners may also "feel tempted" to use keys from a different biogeographic region, sometimes leading to misidentifications. Thus, reviews focusing on simple and efficient DNA-based methods and straightforward taxonomic keys dedicated to carrion insects, clearly delimited to particular biogeographical regions and supported by unambiguous, high-resolution images of the most useful diagnostic characters, should be encouraged.

Beside molecular species identification, significant changes occurred in forensic entomology during the last decades. In the early 1990s, the "squads" method, first formalized by J.P. Megnin a century ago [13], was progressively abandoned. As summarized in the Crime Scene Investigation TV show, this idea was simple and appealing: "*Insects arrive at a corpse in a specific order. Like summer follows the spring. And you can pinpoint time of death, based on the type and age of insects present on the body*" (Grissom, CSI Las Vegas 1.10). However, what happens in the field is rarely simple enough to be summarized in a punchline, and Megnin's necrophagous squads were not reliable enough for forensic purposes [14]. Research indeed evidenced that using the chronological faunal succession for post-mortem interval estimations actually requires large-scale local studies involving numerous cadavers, repetitions over years and extensive statistical analysis [15]. In this Special Issue, LeBlanc et al. [16] show how the occurrence and assemblage composition of necrophagous flies in small bait traps may differ from that of cadavers, thus warning about the potential issues of extrapolating trap collection data to courtroom proceedings. Thus, the step between field experiments and forensic cases application is high: Moreau [15] reviewed these pitfalls, but also acknowledged the interest of ecological decomposition experiments. He suggest that "the authors should explicitly recognize that the (successional field) study is descriptive, thus not allowing for transposition of the results to other situations or use in court".

On the contrary, development–time datasets, which are quite simple to perform and an indispensable prerequisite for aging larvae, have received little attention during the last years. In this Special Issue, Matuszewski [17] highlights the need for development datasets for potentially useful species that regularly breed on cadavers, but for which no reference developmental data are available. For those species with wide geographical distributions, local developmental studies are required as there might be differences in the developmental rates between populations. In this context, Shin et al. [18] present baseline developmental data for the blow fly species *Lucilia sericata* in South Korea, commonly used as a forensic indicator and for which development–time datasets from other geographical regions are available [17]. Regrettably, many researchers may be discouraged to perform basic developmental studies as they could sometimes be seen as "not innovative enough" for being published in some high-profile academic journals. In addition, as in the case of sample collection at the forensic scene, an updated review providing detailed and standardized rearing and sampling protocols for developmental studies would be highly beneficial and enable the comparison of different datasets [17]. Laboratory analyses have indeed demonstrated that numerous biotic and abiotic parameters such as food type, larval behavior or bacterial load, to name a few, could significantly affect larval development [19]. However, most of these studies lack concrete applied methodologies to take account of the effect observed experimentally. In such a situation, forensics readers may wonder how they are supposed to proceed, and the most pessimistic observers would question the reliability of forensic entomology analyses. Thus, while there is currently almost no review dedicated to synthesizing the biology and development of key species (e.g., *Lucilia sericata*), such reviews could distinguish incidental effects from clear trends and suggest appropriate case-related methodology.

Even if there are some key areas where much research still needs to be carried out [17], forensic entomology is today a dynamic science, with the above-mentioned increasing number of scientific publications and active researchers and practitioners across the world. However, many other researchers and practitioners can face significant challenges if they approach this discipline for the first time. Again, updated and comprehensive review articles can be a valuable tool for those who aim to start forensic entomology in a particular country or region. In this Special Issue, Wang et al. [20] review the recent progress in forensic entomology research and application made in China, providing an inventory of the different types of baseline studies conducted, as well as highlighting the need for proper entomological training for forensic practitioners.

Collaboration between forensic practitioners and researchers is indeed of mutual benefit as it enables access to large forensic cases databases, providing new insights into the biology, ecology and forensic relevance of the insect species associated with human cadavers. Such meta-analyses of forensic cases can identify, for example, patterns in the seasonal activity, the colonization of cadavers or the co-occurrence with other species, thus improving the interpretation of further forensic entomological evidence and stimulating additional research on forensically relevant species (e.g., [21,22]). They can also shed some light on the forensic implications of traditionally neglected aspects of the insect activity on cadavers (e.g., [23]). However, these types of *a posteriori* analyses of forensic databases are still scarce. More frequent is the publication of case reports addressing one or a few particular cases (e.g., [24]). The publication of those particular case reports can be useful if, for example, they describe the occurrence of a species which had not previously been considered as forensically relevant [25], highlight how an entomological analysis can provide new and significant insights into a case [26] or show other utilities and applications of the evidence [27]. Nonetheless, meta-analyses of large forensic cases datasets are desirable as they provide strong reference data for the most usual and required type of entomological analyses in case reports. In this sense, the preparation of a solid case report is pivotal, as it must show the reliability and significance of the performed entomological analyses. In this Special Issue, Kotzé et al. [28] propose an overview of sections to be considered when drafting a forensic entomology case report, which could certainly contribute to the standardization of these reports among both experienced and novel practitioners.

As a very specific area of study, forensic entomology is more often requested in unusual cases that require additional and/or alternative approaches, rather than routinely applied in forensic casework. Even if a case requiring an entomological analysis is not particularly "unusual", it will always be unique, differing spatially and temporally from a priori similar cases [5]. Therefore, the conclusions drawn from that entomological analysis and stated in a forensic report will be opinions based on science rather than undisputable scientific facts. That those opinions should be based on science means that they should rely both on a solid experience in the area and on related and published peer-reviewed studies.

This emphasizes the importance of forensic entomology research, either if it consists of baseline studies stimulated by general needs (e.g., development–time datasets) or on ad hoc experiments simulating the conditions of a specific case [5,17]. On the other hand, this also highlights the need for self-criticism and awareness among researchers and practitioners of the limitations of the application of forensic entomology [29,30].

As pointed out by Hall [5] in the opening of this Special Issue, *"Opinion evidence is clearly important, but the opinions expressed in an expert witness statement come from an interpretation of the facts of the case (e.g., the presence of larvae or the temperatures recorded) in the light of the expert witness' experience and their knowledge of data (e.g., rates of insect development) that are generated through research focused on supporting casework"*. Accordingly, this Special Issue presents a series of remarkable research studies, forensic case reports and reviews, all of them highlighting the complex and challenging implications of applying forensic entomology studies to court. It is our opinion that it might be the right time for critical, disruptive reviews that clearly identify the weaknesses and limitations of our field. Acknowledging the real nature of forensic entomology and its inherent or current limitations will avoid misinterpretations, overstatements and troubles in court. It may be the basis for a more probabilistic/Bayesian approach to the forensic entomology analysis, transforming what might appear to be weaknesses into a strength.

Funding: This research received no external funding.

Institutional Review Board Statement: Not applicable.

Conflicts of Interest: The authors declare no conflict of interest.

References

1. Benecke, M. Forensic entomology: The next step. *Forensic Sci. Int.* **2001**, *120*, 1. [CrossRef]
2. Amendt, J. Insect Decline—A Forensic Issue? *Insects* **2021**, *12*, 324. [CrossRef]
3. Byrd, J.H.; Castner, J.L. *Forensic Entomology: The Utility of Arthropods in Legal Investigations*, 2nd ed.; CRC Press: Boca Raton, FL, USA, 2009.
4. Amendt, J.; Campobasso, C.P.; Goff, M.L.; Grassberger, M. (Eds.) *Current Concepts in Forensic Entomology*; Springer: Dordrecht, The Netherland, 2010.
5. Hall, M.J.R. The relationship between research and casework in forensic entomology. *Insects* **2021**, *12*, 174. [CrossRef] [PubMed]
6. Lutz, L.; Verhoff, M.A.; Amendt, J. To be there or not to be there, that is the question—On the problem of delayed sampling of entomological evidence. *Insects* **2021**, *12*, 148. [CrossRef]
7. Amendt, J.; Campobasso, C.; Gaudry, E.; Reiter, C.; LeBlanc, H.; Hall, M.J.R. Best practice in forensic entomology—Standards and guidelines. *Int. J. Legal Med.* **2007**, *121*, 90–104. [CrossRef] [PubMed]
8. Martín-Vega, D.; Hall, M.J.R. Estimating the age of *Calliphora vicina* eggs (Diptera: Calliphoridae): Determination of embryonic morphological landmarks and preservation of egg samples. *Int. J. Legal Med.* **2016**, *130*, 845–854. [CrossRef] [PubMed]
9. Niederegger, S. Effects of makeshift storage in different liquors on larvae of the blowflies *Calliphora vicina* and *Lucilia sericata* (Diptera: Calliphoridae). *Insects* **2021**, *12*, 312. [CrossRef] [PubMed]
10. Brown, K.; Thorne, A.; Harvey, M. Preservation of *Calliphora vicina* (Diptera: Calliphoridae) pupae for use in post-mortem interval estimation. *Forensic Sci. Int.* **2012**, *223*, 176–183. [CrossRef]
11. Grzywacz, A.; Jarmusz, M.; Walczak, K.; Skowronek, R.; Johnston, N.P.; Szpila, K. DNA barcoding identifies unknown females and larvae of *Fannia* R.-D. (Diptera: Fanniidae) from carrion succession experiment and case report. *Insects* **2021**, *12*, 381. [CrossRef]
12. Pradelli, J.; Tuccia, F.; Giordiani, G.; Vanin, S. Puparia cleaning techniques for forensic and archaeo-funerary studies. *Insects* **2021**, *12*, 104. [CrossRef]
13. Mégnin, P. *La faune des cadavres. Application de l' entomologie à la médecine légale, Encyclopédie scientifique des Aides-Mémoire*; Masson: Paris, France, 1894.
14. Franceschetti, L.; Pradelli, J.; Tuccia, F.; Giordiani, G.; Cattaneo, C.; Vanin, S. Comparison of accumulated degree-days and entomological approaches in post mortem interval estimation. *Insects* **2021**, *12*, 264. [CrossRef] [PubMed]
15. Moreau, G. The pitfalls in the path of probabilistic inference in forensic entomology: A review. *Insects* **2021**, *12*, 240. [CrossRef] [PubMed]
16. LeBlanc, K.; Boudreau, D.R.; Moreau, G. Small bait traps may not accurately reflect the composition of necrophagous Diptera associated to remains. *Insects* **2021**, *12*, 261. [CrossRef] [PubMed]
17. Matuszewski, S. Post-mortem interval estimation based on insect evidence: Current challenges. *Insects* **2021**, *12*, 314. [CrossRef] [PubMed]

18. Shin, S.E.; Park, J.H.; Jeong, S.J.; Park, S.H. The growth model of forensically important *Lucilia sericata* (Meigen) (Diptera: Calliphoridae) in South Korea. *Insects* **2021**, *12*, 323. [CrossRef]
19. Hans, K.; Vanlaerhoven, S.L. Impact of comingled heterospecific assemblages on developmentally based estimates of the post-mortem interval—A study with *Lucilia sericata* (Meigen), *Phormia regina* (Meigen) and *Calliphora vicina* Robineau-Desvoidy (Diptera: Calliphoridae). *Insects* **2021**, *12*, 280. [CrossRef] [PubMed]
20. Wang, Y.; Wang, Y.; Wang, M.; Xu, W.; Zhang, Y.; Wang, J. Forensic entomology in China and its challenges. *Insects* **2021**, *12*, 230. [CrossRef]
21. Dekeirsschieter, J.; Frederickx, C.; Verheggen, F.J.; Boxho, P.; Haubruge, E. Forensic entomology investigations from Doctor Marcel Leclercq (1924-2008): A review of cases from 1969 to 2005. *J. Med. Entomol.* **2013**, *50*, 935–954. [CrossRef] [PubMed]
22. Charabidze, D.; Vincent, B.; Pasquerault, T.; Hédouin, V. The biology and ecology of *Necrodes littoralis*, a species of forensic interest in Europe. *Int. J. Legal Med.* **2016**, *130*, 273–280. [CrossRef]
23. Viero, A.; Montisci, M.; Pelleti, G.; Vanin, S. Crime scene and body alterations caused by arthropods: Implications in death investigation. *Int. J. Legal Med.* **2019**, *133*, 307–316. [CrossRef]
24. Arnaldos, M.I.; García, M.D. Entomological contributions to the legal system in southeastern Spain. *Insects* **2021**, *12*, 429. [CrossRef] [PubMed]
25. Michalski, M.; Gadawski, P.; Klemm, J.; Szpila, K. New species of soldier fly—*Sargus bipunctatus* (Scopoli, 1763) (Diptera: Stratiomyidae), recorded from a human corpse in Europe—A case report. *Insects* **2021**, *12*, 302. [CrossRef] [PubMed]
26. Niederegger, S.; Mall, G. Flies do not jump to conclusions: Estimation of the minimum post-mortem interval for a partly skeletonized body based on larvae of *Phormia regina* (Diptera: Calliphoridae). *Insects* **2021**, *12*, 294. [CrossRef] [PubMed]
27. Introna, F.; Cattaneo, C.; Mazzarelli, D.; De Micco, F.; Campobasso, C.P. Unusual application of insect-related evidence in two European unsolved murders. *Insects* **2021**, *12*, 444. [CrossRef] [PubMed]
28. Kotzé, Z.; Aimar, S.; Amendt, J.; Anderson, G.S.; Bourguignon, L.; Hall, M.J.R.; Tomberlin, J.K. The forensic entomology case report—A global perspective. *Insects* **2021**, *12*, 283. [CrossRef] [PubMed]
29. Charabidze, D.; Gosselin, M.; Hédouin, V. Use of necrophagous insects as evidence of cadaver relocation: Myth or reality? *PeerJ* **2017**, *5*, e3506. [CrossRef]
30. Charabidze, D.; Hédouin, V. Temperature: The weak point of forensic entomology. *Int. J. Legal Med.* **2019**, *133*, 633–639. [CrossRef]

Review

The Relationship between Research and Casework in Forensic Entomology

Martin J. R. Hall

Department of Life Sciences, Natural History Museum, London SW7 5BD, UK; m.hall@nhm.ac.uk

Simple Summary: Forensic entomology concerns the use of insects as evidence in legal investigations. Many sorts of investigation can benefit from an interpretation of insects associated with the crime scene, but insect evidence is most frequently used in investigations of death. The interpretation of insect evidence in casework is guided by the data supplied through research. Such data are essential to improve the casework interpretation of insect evidence, thereby improving the robustness of the legal systems in which it operates. This paper explores the mutually beneficial relationship between research and casework in forensic entomology, contrasting the different challenges that each presents and giving examples of how each can support the other in delivering results of real societal benefit. It is written from the perspective of the Criminal Justice System of England and Wales, but many of the points raised are relevant to legal systems worldwide.

Citation: Hall, M.J.R. The Relationship between Research and Casework in Forensic Entomology. *Insects* **2021**, *12*, 174. https://doi.org/10.3390/insects12020174

Academic Editors: Damien Charabidze and Daniel Martín-Vega

Received: 26 January 2021
Accepted: 10 February 2021
Published: 17 February 2021

Publisher's Note: MDPI stays neutral with regard to jurisdictional claims in published maps and institutional affiliations.

Abstract: Research is a vital component of all forensic sciences and is often stimulated by casework, which identifies gaps in our knowledge. In such a niche area of forensic science as entomology there should be a close and mutually beneficial relationship between research and casework: to some extent there is a continuum between the two and many forensic entomologists are involved in both to a greater or lesser degree. However, research and casework involve quite differing challenges, from the replicated, highly controlled, sometimes esoteric aspects of research to the very individual, sometimes chaotic and disruptive, but highly applied aspects of casework. Ideally casework will include the full involvement of a forensic entomologist, who will collect the insect and climate evidence at the scene and produce a robust expert witness statement based on a full analysis of this data. Unfortunately, it can also include situations where samples, if collected at all, are poorly preserved, not representative of the full cadaver fauna available and presented to the entomologist months or years after the event, without local temperature data. While research is recognised through publications and their citation indices, casework and its associated expert witness statements often receive no credit in an academic workplace, although they do have a positive societal impact and many other benefits of teaching and public engagement value. This manuscript examines the relationship between research and casework from a UK perspective, to raise awareness of the need to create an environment that values the contribution of both, for future generations to flourish in both areas.

Keywords: casework; court; criminal justice systems; expert witness; forensic entomology; insect evidence; research

1. Introduction

Before the 1980s, research publications in the field of forensic entomology were relatively infrequent [1] (p. 418), and therefore the research support to investigations of insect evidence during casework was limited. The first case in which forensic entomology evidence was used in the UK was the "Ruxton" case of 1935 [2]. Analysis of the insect evidence in this case was based on records of insect development that were unpublished at the time, nevertheless it was important as corroboration of other evidence even if not used in court [3,4].

Publication of Ken Smith's Manual of Forensic Entomology in 1986 [5] was a major catalyst to research, bringing together the widely scattered forensic entomology literature available at that time and combining it with relevant taxonomic and ecological literature recording the biology of carrion fauna. I believe that it is no coincidence that its publication was in the decade that saw the start of an almost exponential rise in scientific publications in the field of forensic entomology [1] (p. 418), which continues to increase year-on-year [6]. A substantial portion of Smith's book was devoted to insect identification, highlighting the critical importance of assigning the correct name to insect evidence at the start of any casework. This landmark book clearly demonstrated the potential value of insect evidence in criminal investigations, summarizing 19 case histories and suggesting areas for future research of value to forensic entomology casework, thereby being of importance in highlighting the relationship between research and casework.

In 2009, Magni et al. [7] published results of a survey of forensic entomologists worldwide. Of some 300 individuals contacted, 70 responded and by far the largest group (60%) were employed in universities. However, 67% of all respondents considered forensic entomology to be only their secondary occupation. Some 79% of respondents published in the area of forensic entomology or taxonomy while a similar proportion, 74%, undertook casework. Nevertheless, only 25% of the caseworkers had worked on >100 cases, most (64%) had worked on <50 cases and half of those on <10 cases. In my opinion, it seems likely that there are many forensic entomologists who spend far more of their time on research (and teaching) than on casework, in part because of the difficulty of gaining casework experience due: (1) to the relatively low number of cases in which it is used; and (2) the generally pragmatic nature of those who require a forensic entomology input to their investigations, minimizing risk and/or honouring service contracts by selecting those they have previously used with success.

The following comparison of research and casework in forensic entomology is very much a personal perspective developed from working within a UK context. The policing and legal systems of the UK are complex and differ between those of Northern Ireland, Scotland and England and Wales—the majority of my experience is within the Criminal Justice System of England and Wales (abbreviated here as CJS-EW). I look at some of the challenges of working in casework within an academic or other scientific institute background, e.g., universities and museums, which Magni et al.'s [7] survey shows to be the dominant workplace for forensic entomologists.

2. Research and Casework—Similarities and Differences

Table 1 summarises some of the differences between undertaking research and undertaking casework in forensic entomology, indeed in virtually all branches of forensic science. In addition to the likely requirement for caseworkers of a research degree, such as a PhD, and subsequently acquired practical experience of the subject that qualifies them to provide an expert opinion on specific, insect related aspects of a criminal investigation, it is of great benefit to the forensic entomologist to acquire training in expert witness skills: if only to make them aware of the potentially stressful and intimidating environment of a court room [8], to help in writing a witness statement and to prepare for the likely style of questioning.

Table 1. Some differences between research and casework in forensic entomology.

Areas of Difference	Research	Casework
Qualifications needed	PhD required	PhD and expert witness skills
Nature of study	Experimental replication	One-off, unique scenario
	Proactive	Reactive
	Planned schedule	Often highly disruptive
	Self-managed deadlines	Imposed, strict deadlines
Productions from study	Research publications	Expert witness statements
Rewards for production	Citation indices, academic credit	Often no academic credit but knowledge of the societal benefit

Most practicing forensic entomologists can expect to appear in court at some time to be questioned on the opinions and conclusions in their expert witness statements. Of course, court rooms and the legal systems practiced within them vary enormously from country to country. My background is in the adversarial system practiced in the UK and, during my career, I appeared in court in just 12.3% (18/146) of the cases for which I submitted one or more expert witness statements. In some cases that did not result in a court appearance the entomology evidence may either have been submitted, but was not contested, or it may not have had evidential value. Without the training I received from an expert witness training provider, which had been assessed, quality assured and certified by a university school of law [9], I would have been much less comfortable in the court room. In addition to courses, there are a number of published articles and numerous books, e.g., [10,11], to help prepare an expert witness for court, in addition to the online aids available from the judiciaries of England and Wales [12,13] and Scotland [14] and from the office of the Forensic Science Regulator [15,16]. There are also some invaluable aids written specifically for forensic entomologists [17,18]. An interesting and personal perspective on the challenges of presenting forensic entomology evidence in court is given in an opinion paper by Disney [19], discussed by reference to a number of specific cases, which together illustrate the wide variety of ways in which insect evidence can be used.

Major differences between research and casework appear in the actual nature of the study. All research relies on adequate replication [20]. However, each case is a one-off, unique scenario—even apparently similar cases usually vary spatially and temporally. This can present real challenges in interpretation. Sometimes valuable information towards an understanding of the biology of a case can be gathered by simulation of important aspects of the original case, e.g., use of a buried pig to simulate the decomposition of a victim in a shallow grave [21]. However, such simulation is often not possible due to time constraints imposed by the investigation or difficulties in working at the crime scene, especially if indoors, e.g., due to limited access.

While, by and large, a research study can be conducted within a planned schedule to self-managed deadlines, casework is often highly disruptive to the timetables of research and teaching that most forensic entomologists also undertake and it runs to strict deadlines, imposed both by the needs of the investigation and of any subsequent trial. Research is much more amenable to control of its timetable than is casework.

Other than products such as patents, the final products from most research work are peer-reviewed publications in the scientific literature. For these the researcher receives academic credit based on some form of citation indicator [22]. To my knowledge and experience, no academic credit is usually given for the expert witness statements which are the product of a forensic investigation. However, many factors compensate for the lack of standard academic credit and the caseworker can take credit from the knowledge that their evidence has benefited more than science, it has made a positive contribution to society, contributing to the criminal justice system generally as well as to specific individuals, such as relatives of victims who, from personal experience, can benefit from the closure given through insect evidence [23]. Feedback is sometimes received by caseworkers from someone in the judicial system that also demonstrates the societal value of the casework. Three examples of feedback that I have received are given below and illustrate the value of forensic entomology, not only in the court room but earlier in the investigation, for example as a guide to what time period for the investigators to focus on, leading to savings in increasingly limited resources:

- "The opinions you expressed in your statement went a considerable way to support our hypothesis, based on a number of other known facts. Your statement was accepted in evidence by the defence. I was very pleased with the accuracy of the opinions you expressed, which I am sure went someway to causing the defendant to change his account. Had he not changed his account at the eleventh hour we would have relied considerably upon your evidence to convince the jury (of our hypothesis).

> I am therefore satisfied that the evidence you provided was useful to the case and represented good value for money."

- " ... the investigation team were very happy to be told that death occurred on the day [the victim] went missing as this hugely reduced the amount of CCTV they had to view." (NB: investigators still had to view >11,000 h of CCTV in this case.).
- "Your statement was crucial in securing guilty pleas. The two accused were pleading not guilty up until the moment of the trial beginning. The trial, had it run, would have lasted a week or so at great expense, so your statement led to those guilty pleas. So, value for money it was worth it."

In addition to the knowledge that casework is of societal value, engagement in casework provides researchers with practical experience that they can share with students, adding value to their teaching, it provides ideas for projects (e.g., see Section 3 below), it can provide income for the institute and it provides experience for use in public engagement activities.

Writing an expert witness statement can be as challenging as writing a scientific paper, if not more so. While each witness statement is usually founded on a smaller scientific study than a research paper, the implications for the conclusions and, therefore, the consequences of errors, are profound. For forensic entomology cases investigating neglect or death, each statement usually involves a factual account of when and where specimens were collected, how they were processed, how they were identified and how they were aged based on estimated temperatures for the period of development. However, it is not just about facts, but about how these facts are then interpreted based on experience to assist the court. This summation of facts and opinion results in an estimation of the time that the earliest female insect laid her eggs/larvae on the body of the victim, living or dead. In an investigation of death, this equates to the minimum post-mortem interval, minPMI [24,25]. The witness statement can also include a testing of any hypotheses proposed by either prosecution (H_p) or defence (H_d) [26–28].

Of course, with some cases it is possible to prepare not only an expert witness statement but also a scientific publication, with details of cases suitably anonymized [29–31]. Case reports are widely accepted in medical science, with several journals dedicated to their publication, e.g., *BMJ Case Reports*, *Oxford Medical Case Reports* and *Journal of Medical Case Reports*. Such publications are also of value in forensic science and can be of enormous benefit in describing unusual cases (e.g., [32,33]), providing a comparison of similar cases (e.g., [34–36]), identifying local fauna (e.g., [37,38]), and in discussing the application and validation of methods (e.g., [39–43]), especially if there is an independent and accurate verification of time of death to compare with the entomology estimates.

With forensic scientists being under greater scrutiny now than ever before, there is a great deal of responsibility and legal obligations on any expert witness, as detailed in a document produced by the Forensic Science Regulator regarding the CJS-EW [44]. I believe that the pressures on an expert witness have always been there, but they have increased over the last four decades with, for example, greater scrutiny, greater accreditation, shorter time scales and tighter budgets. From the first statement I produced in 1992, the only way that anyone might know that I had any knowledge of entomology was the fact that my address included "Department of Entomology at the Natural History Museum, London". In contrast, in addition to a detailed statement of my qualifications and experience, my final statement produced in 2019 included a one page declaration with more than 20 bullet points based on guidance from the Criminal Practice Directions [45] (19B, pp. 35–37), essentially declaring awareness of my duties as an expert witness, including that "I am likely to be the subject of public adverse criticism by the judge if the Court concludes that I have not taken reasonable care in trying to meet the standards set out above". I think this declaration is entirely fair and reasonable but, in its clear and unambiguous statement, it is still a stark reminder of the onus on an expert witness. As mentioned above, the pressure on expert witnesses has always been there, evidenced by the autobiography of Professor Keith Simpson [46] in which he says, "My insistence on the timing of death had become pretty well known, to the police, the Director of Public Prosecutions, the lawyers—and the Press, who would have scented a public disgrace for me if I'd been wrong." [46] (p. 310). However, in this case, the so-called "Lydney murder", Professor Simpson had opened himself up to unnecessary pressure by declaring a period of death, rather than a minPMI, at the scene, before he had a chance to confirm the larval identification and take into account the ambient climate conditions [46] (pp. 300–310). His estimate was not unreasonable, but I would not recommend making such an unequivocal statement at a crime scene without caveats.

The lack of control of many factors in casework, mentioned above, can make it very difficult to produce a robust expert witness statement. In an ideal world, a forensic entomologist would visit the scene and/or the post-mortem to collect the insect evidence. My own experience is of visiting the scene in just 18.5% (27/146) of the cases for which I produced a statement. Not visiting a scene does not mean that it is impossible to produce a robust statement, but the confidence intervals might have to be widened and more caveats introduced. Additionally, the forensic entomologist can give trained crime scene personnel timely additional advice and instruction by mobile telephone as they collect insect evidence at the scene, and the entomologist can visit the scene virtually when they later view crime scene photographs and videos, including those produced by 360° imaging techniques [47,48]. However, non-attendance at the scene by an entomologist can introduce the potential for complications, illustrated in Table 2: examples of those I have experienced include:

- Provision of just six larvae for analysis when many thousands are evident on scene photographs.
- Puparia (apparent on scene photographs) were overlooked in collections because they were not on the body and did not move, and so were just not considered part of the insect evidence, although they were most likely the oldest stages present.
- All larvae collected alive for rearing were dead on arrival at the laboratory as they were transported in sealed plastic pots inside air-tight evidence bags (Figure 1).
- Being asked to identify and determine the age of dried, flattened larvae 3.5 years post-collection, following their storage without preservative in a sealed glass jar kept in a fridge (remarkably, immersion in potassium hydroxide (KOH) [5] enabled them to be identified to species and assigned to a life stage) (Figure 2).

Table 2. Examples of differences between collection and analysis of forensic entomology evidence in an ideal world and in reality.

In an Ideal World the Forensic Entomologist:	In Reality There Are Cases Where:
visits the scene;	the forensic entomologist does not visit the scene;
collects a good range of insect evidence;	insect evidence collected is suboptimal in range;
preserves the insects appropriately;	insect evidence is not preserved properly;
retains some specimens alive for rearing;	specimens collected alive die before delivery;
collects scene meteorological data to compare with the nearest weather station;	no meteorological data is collected at the scene;
collects evidence at the start of the investigation;	evidence is only made available to the forensic entomologist months/years after the case;
prepares a robust expert witness statement.	it is extremely difficult to prepare a robust expert witness statement.

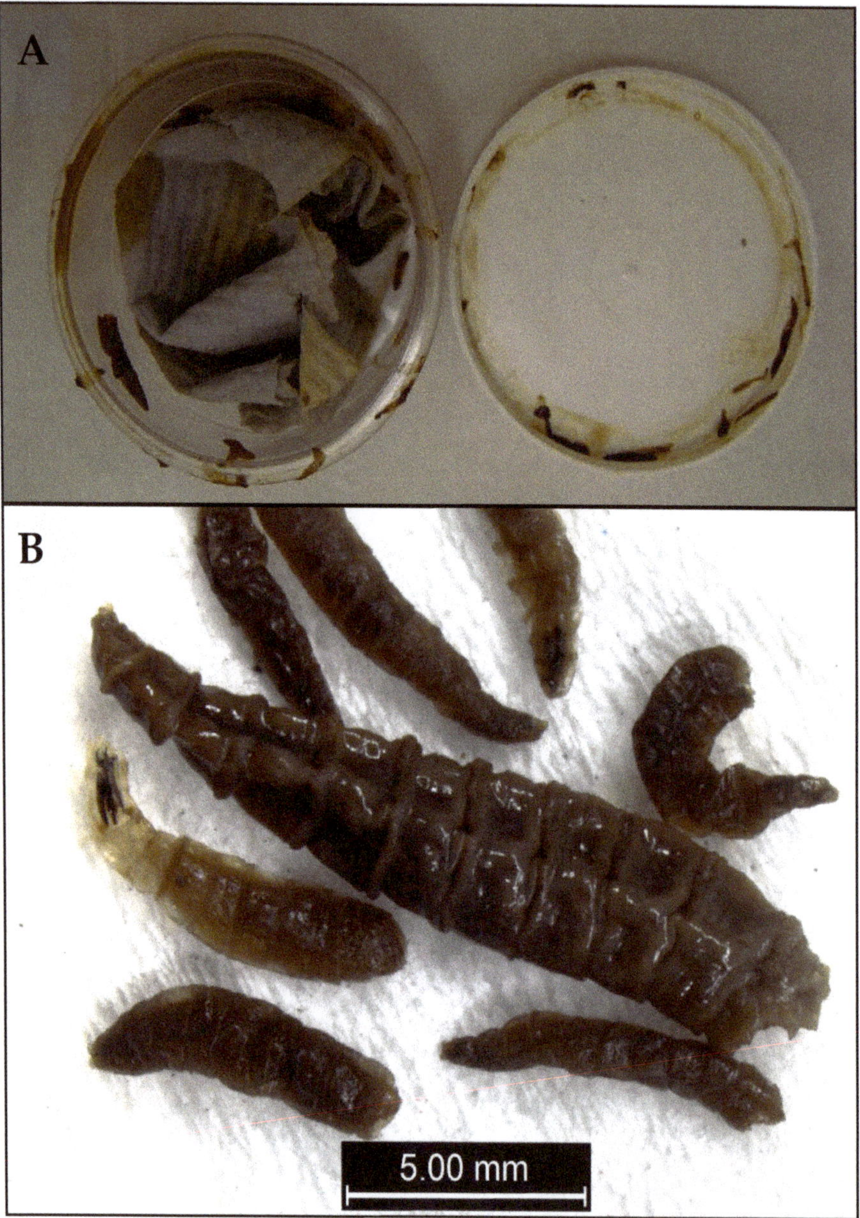

Figure 1. Example of forensic exhibit containing larvae of *Calliphora vomitoria* (Diptera: Calliphoridae) collected alive, but then sealed inside an airtight plastic pot (**A**) inside a plastic evidence bag so that all had died (**B**) by the time of their arrival at the laboratory.

Figure 2. Example of suboptimal maintenance of insect evidence, stored without preservative for 3.5 years in a sealed glass jar in a fridge. (**A,B**) dried flattened fragments of fly larvae, resembling flakes of rusted metal; (**C**) oldest specimens revealed to be at 2nd–3rd instar moult and early 3rd instars of *Calliphora vomitoria* (Diptera: Calliphoridae) following immersion in 10% aqueous potassium hydroxide (KOH) [5].

Lutz et al. [49] highlighted the importance of collecting insect evidence at the scene, especially the oldest stages, as in the second example above: in an analysis of 29 death investigations, they found that in 21 cases where puparia were collected at the scene, only in 14 (67%) of the subsequent autopsies were puparia collected. A published case that mirrors the last example given above is that of the re-examination of specimens in a "cold case" review nine years after the initial investigation [50]. Effective training of crime scene personnel with regular updating and use of recommended entomological collection equipment (e.g., [25]) are obvious ways to combat the illustrated complications of non-attendance of an entomologist at the scene.

3. Research and Casework—A Mutually Beneficial Relationship

Casework continuously identifies lacunae in our knowledge that research can address. Casework therefore poses questions and research can provide the solutions. Those solutions must be realistic, practical, validated and, ultimately, accepted in Court. However, researchers must not be limited at an early stage by what might seem impractical at the current time. For example, I sometimes questioned how practical for casework our own research using micro-computed tomography (micro-CT) to explore the intra-puparial development of the blow fly (Diptera: Calliphoridae) would be [51]. This research was stimulated by knowledge that the intra-puparial stages occupied at least 50% of the developmental period of blow flies, yet there were no adequate means available to estimate the stage reached by puparia collected from a scene. Micro-CT offers a means to virtually and non-destructively examine puparia, a bonus when this is forensic evidence, and can provide physiological aging at 10% intervals, equivalent to about one day at UK summer temperatures [52]. While a micro-CT is still an expensive capital item, it is increasingly possible to buy time on them, to use even remotely, and this trend is only likely to increase, as a survey of the use of CT in the field of forensic science reveals, showing a huge increase in publications in that area since 2000 [53]. Micro-CT in forensic entomology has great future potential and the research also develops our understanding of biology at a more fundamental level, e.g., of metamorphosis [54,55].

Examples of casework from my own experience that directly stimulated research include the following:

- Andrew Hart and I worked on two cases in quick succession in central and southern England in which the body of the victim was concealed inside a suitcase. In both cases there were fly larvae on the bodies and the question was, "can adult flies deposit their eggs on a suitcase in such a way that the larvae can develop on the body inside?" The subsequent research showed that not only can first instar larvae penetrate through suitcase zippers, but also that female flies can insert their ovipositors through the zippers and lay eggs inside the suitcase, enabling the hatching larvae to colonise the body [56].
- The oldest insect stages in a case in northern England in late Autumn were newly hatched first instar larvae from egg batches laid around a neck ligature and in the facial orifices. Most of the available information on blow fly egg stages gives a period from egg-laying to egg-hatching, therefore, aging of the specimens in this case was possible as they had just hatched: it was around six days due to the cool temperatures. However, if they had not hatched we would have struggled to age them and, at the low temperatures of this case or even lower, this embryonic period could be quite lengthy. Therefore, for use in similar future cases, we developed a simple method to estimate the age of *Calliphora vicina* (Diptera: Calliphoridae) eggs by morphological characters [57].
- An indoor case I attended in southern UK featured a large number of dead adult flies on the floor together with many dispersing larvae and puparia, but none of the latter in our collections were empty. Some adults were just emerging from puparia at the time of collection, so could we have overlooked empty puparia and was there any way of distinguishing if adults flies found at a scene had developed on the body or had, instead, flown in from outside? Developing the age-grading technique of wing fray, that was first used for tsetse flies (Diptera: Glossinidae) [58], we found that there were indeed major differences between the wing fray of populations of flies that emerged and died in a room, after developing as larvae on a body, compared with those that flew into the room from outdoors [59].

Much research is not stimulated by specific cases but arises from a general need recognized in all casework. For example, the provision of insect developmental data gathered in the laboratory and validated in the field [60], development of techniques for killing and preserving insect specimens in casework [61,62] and of methods for collecting temperature data at scenes of body recovery [63,64]. In most of these papers recommendations for practitioner protocols were included, such that in addition to the science there was guidance for those involved in casework.

New methods need to be validated before presentation in Court and peer review through scientific publication is one method of third-party examination of data, although not infallible. Validity of methods is a requirement highlighted in the Criminal Practice Directions for England and Wales which state, "Therefore factors which the court may take into account in determining the reliability of expert opinion, and especially of expert scientific opinion, include: (a) the extent and quality of the data on which the expert's opinion is based, and the validity of the methods by which they were obtained;" [45] (19A.5, p. 33). With regard to new techniques, the Crown Prosecution Service of England and Wales sets out that, "Caution should always be exercised in assessing whether a new technique or novel science is accredited or is sufficiently sound to be admissible as evidence at trial". One of the four factors to be considered in assessment was, "Whether the theory or technique has been subject to peer review and publication;" [13] (Part 1—Guidance: Admissibility of Expert Evidence, New or Novel Techniques).

At the time of writing, accreditation of individuals or organisations in England and Wales is only a requirement for DNA and fingerprint evidence [13] (Part 1—Guidance: Accreditation) [65]. Forensic science providers are assessed against ISO standards (especially ISO17020 and ISO17025) and the Codes of Practice and Conduct of the Forensic Science Regulator [44]. Some of the challenges around the accreditation of forensic entomology techniques and practitioners in the UK—e.g., validation of techniques, demonstration of practitioner competence and evaluating the strength of evidence—have been raised by Hall et al. [2]. However, it will surely only be a matter of time before some form of accreditation of individuals for casework in forensic entomology, and for their subsequent submission of evidence, is required, as appendices on entomology and other niche areas of forensic science have been drafted for the Forensic Science Regulator's Code and will be considered in due course (G. Tully, personal communication). Indeed, the groundwork for this was laid in Europe with accreditation of the Department of Forensic Entomology of the Institut de Recherche Criminelle de la Gendarmerie Nationale, France, in 2007, the process for which is described by Gaudry and Dourel [66], focussing on preparation of documentation on good working standards and introduction of a quality assurance system. A thorough discussion of mandatory certification of forensic scientists in general from a United States perspective was provided by Melbourn et al. [67].

4. Challenges for the Future

Forensic science has faced a challenging last decade or more in the UK, due mainly to changes in the way forensic services are provided and reduced funding levels, prompting media headlines, such as "Forensics in Crisis" and "Police forensic science at 'breaking point'", e.g., [68,69]. Discussion of this dynamic situation has also been stimulated in the scientific literature, with commentaries that consider the impacts of closure of the UK Government's Forensic Science Service in 2012 and the pressures that force forensic scientists in the commercial market to focus on chargeable casework at the expense of the broader aspects of science, including research and mentoring the next generation, e.g., [70,71].

To address the concerns highlighted above, a thorough review of the provision of forensic services in England and Wales was led by the Home Office, the Association of Police and Crime Commissioners and the National Police Chiefs' Council [72]. One of its recommendations was to, "Ensure policing and the CJS [Criminal Justice System] benefits from advances in science and technology by developing and implementing new forensic techniques more coherently. Change is needed to bring about structured engagement across CJS partners, industry, science and academia in the testing, evaluation and development of new forensic techniques, improving the case for investment and helping forensic science providers to bring new innovation to market." This essentially endorses the close links that should exist between research and casework.

The UK's House of Lords conducted a similar review through its Science and Technology Select Committee [73]. Among its conclusions were recommendations to increase levels of funding for both technological advances and foundational research in forensic science and to create a National Institute for Forensic Science, to set priorities and oversee research on forensic science. One response to these recommendations was the formation of the Forensic Capability Network, designed and developed under the Transforming Forensic Programme [74], with the challenging yet exciting task of delivering high quality forensic science in England and Wales.

It is my opinion that there has never been a better time for more casework-related research, especially in areas which are not considered sufficiently "blue skies" to attract regular science funding but are essential "coal face" studies, for example, developmental studies, e.g., [75,76] of insect species that are frequently encountered in casework but are currently unavailable for use as evidence because of a lack of data. Areas for research could be provided by caseworkers in a valuable synergy, as exemplified by that reported in the social sciences some 70 years ago [77]. We also need to provide opportunities for young researchers to gain experience in forensic entomology casework and to receive appropriate recognition from academic employers of the value of casework. The relationship between research and casework seems clear, but the ability of personnel to move from the research end of the continuum to the casework end needs to be made more available and straightforward. One avenue that could be explored would be to copy the system for professional certification of forensic anthropologists adopted by the Royal Anthropological Institute, whereby aspiring, but inexperienced, forensic anthropologists are mentored by senior forensic anthropologists, enabling them to gain casework experience under expert tutelage [78]. Mentoring of students and early-career researchers by experienced forensic entomologists could improve the quality of entomology casework and, at an early stage, it could also be a highly influential factor in encouraging interest in the subject. This is demonstrated by the positive influence of mentors in stimulating the interest of medical students in a career in forensic pathology [79]. An article that discusses some of the challenges and rewards of a career as an expert witness in the forensic sciences concluded with a statement by Owen Jones, Professor of Law and Biological Sciences at the Vanderbilt Law School in Nashville, USA, saying, in effect, that legal systems will never be better informed than when those in science spend their time helping those systems advance along a more constructive and accurate path [80]. I agree strongly with that sentiment, but scientists, especially early-career researchers, need to be helped to do this in both their research and casework.

A useful and informative sheet summarising what an expert witness is describes an expert witness as one who, "may give opinion evidence within their expertise and in addition evidence of facts" [81]. Opinion evidence is clearly important, but the opinions expressed in an expert witness statement come from an interpretation of the facts of the case (e.g., the presence of larvae or the temperatures recorded) in the light of the expert witness' experience and their knowledge of data (e.g., rates of insect development) that are generated through research focussed on supporting casework. This meld of experience and data is crucial so that the opinions are based on science rather than anecdote. Research and casework always work best together and the more the casework opinions can be based on and informed by research data, the better the value of the evidence delivered in court.

Funding: This work received no external funding.

Institutional Review Board Statement: Not applicable.

Acknowledgments: This manuscript was developed from a presentation delivered at the Annual Meeting of the European Association for Forensic Entomology in Bordeaux, France, in 2019 [82]. I am grateful to all those students and post-docs who have undertaken forensic entomology research projects with me over many years and have been a great inspiration and encouragement to me, both in my research and casework: Meghan Beutler, Poulomi Bhadra, Liz Bromley, Abigail Chong, Sarah Donovan, Juliet Dukes, Rebecca Grimsley, Maike Hartmann, Ines Hofer, Hélène LeBlanc, Molly Mactaggart, Daniel Martín-Vega, Matteo Mondani, Samantha Pickles, Olga Retka, Cameron Richards, Amoret Whitaker, and Runchao Zhang. I am also very grateful to Andrew Hart for reading and suggesting improvements to an early version of this manuscript, to Gillian Tully for her personal communication, to four anonymous reviewers for their helpful comments and to the police forces and solicitors who gave permission to share their feedback and/or use images of insect evidence.

Conflicts of Interest: The author declares no conflict of interest.

References

1. Tomberlin, J.K.; Benbow, M.E. Current global trends and frontiers. In *Forensic Entomology: International Dimensions and Frontiers*; Tomberlin, J.K., Benbow, M.E., Eds.; CRC Press, Taylor and Francis Group: Boca Raton, FL, USA, 2015; pp. 417–422.
2. Hall, M.J.R.; Whitaker, A.P.; Hart, A.J. United Kingdom. In *Forensic Entomology: International Dimensions and Frontiers*; Tomberlin, J.K., Benbow, M.E., Eds.; CRC Press, Taylor and Francis Group: Boca Raton, FL, USA, 2015; pp. 89–99.
3. Glaister, J.; Brash, J.C. *Medico-Legal Aspects of the Ruxton Case*; E. and S. Livingstone: Edinburgh, Scotland, UK, 1937.
4. Mearns, A.G. Larval infestation and putrefaction. In *Recent Advances in Forensic Medicine*, 2nd ed.; Smith, S., Ed.; Churchill: London, UK, 1939; pp. 250–255.
5. Smith, K.G.V. *A Manual of Forensic Entomology*; The Trustees of the British Museum (Natural History): London, UK, 1986.
6. Lei, G.; Liu, F.; Liu, P.; Zhou, Y.; Jiao, T.; Dang, Y.-H. A bibliometric analysis of forensic entomology trends and perspectives worldwide over the last two decades (1998–2017). *Forensic Sci. Int.* **2019**, *295*, 77–82. [CrossRef]
7. Magni, P.; Guercini, S.; Leighton, A.; Dadour, I. Forensic entomologists: An evaluation of their status. *J. Insect Sci.* **2013**, *13*. [CrossRef]
8. Shepherd, J.P. Presenting expert evidence in criminal proceedings. *Br. Med. J.* **1993**, *307*, 817–818. [CrossRef] [PubMed]
9. Bond Solon—Expert Witness—Overview. Available online: https://www.bondsolon.com/expert-witness/ (accessed on 19 January 2021).
10. Wall, W.J. *Forensic Science in Court: The Role of the Expert Witness*; John Wiley & Sons: Chichester, UK, 2009.
11. Hackman, S.; Raitt, F.; Black, S. *The Expert Witness, Forensic Science and the Criminal Justice Systems of the UK*; CRC Press, Taylor and Francis Group: Boca Raton, FL, USA, 2019.
12. The Criminal Procedure Rules. Expert Evidence. In *Criminal Procedure Rules and Practice Directions 2020*. Available online: https://www.gov.uk/guidance/rules-and-practice-directions-2020#evidence (accessed on 18 December 2020).
13. Crown Prosecution Service, Code for Crown Prosecutors, Expert Evidence (updated 09/10/2019). Available online: https://www.cps.gov.uk/legal-guidance/expert-evidence (accessed on 1 December 2020).
14. Crown Office and Procurator Fiscal Service. Guidance Booklet for Expert Witnesses; The Role of the Expert Witness and Disclosure. Available online: https://www.copfs.gov.uk/images/Documents/Prosecution_Policy_Guidance/Guidelines_and_Policy/Guidance%20booklet%20for%20expert%20witnesses.PDF (accessed on 1 December 2020).
15. Forensic Science Regulator. In *Guidance, Expert Report Guidance*; FSR-G-200, Issue 4; The Forensic Science Regulator: Birmingham, UK, 2020. Available online: https://assets.publishing.service.gov.uk/government/uploads/system/uploads/attachment_data/file/881550/Statement_and_Report_Guidance_Appendix_Issue_4.pdf (accessed on 2 December 2020).
16. Forensic Science Regulator. *Information, Legal Obligations*; FSR-I-400; Issue 8, The Forensic Science Regulator: Birmingham, UK, 2020. Available online: https://assets.publishing.service.gov.uk/government/uploads/system/uploads/attachment_data/file/882074/FSR_Legal_Obligations_-_Issue_8.pdf (accessed on 2 December 2020).
17. Hall, R.D. Chapter 14: The forensic entomologist as expert witness. In *Forensic Entomology: The Utility of Arthropods in Legal Investigations*; Byrd, J.H., Castner, J.L., Eds.; CRC Press: Boca Raton, FL, USA, 2001; pp. 379–400.
18. Greenberg, B.; Kunich, J.C. *Entomology and the Law: Flies as Forensic Indicators*; Cambridge University Press: Cambridge, UK, 2002.
19. Disney, R.H.L. Forensic science is not a game. *Pest Technol.* **2011**, *5*, 16–22.
20. Moreau, G.; Michaud, J.-P.; Schoenly, K.G. Chapter 19: Experimental design, inferential statistics and computer modelling. In *Forensic Entomology: International Dimensions and Frontiers*; Tomberlin, J.K., Benbow, M.E., Eds.; CRC Press, Taylor and Francis Group: Boca Raton, FL, USA, 2015; pp. 205–229.
21. Turner, B.D.; Wiltshire, P. Experimental validation of forensic evidence: A study of the decomposition of buried pigs in a heavy clay soil. *Forensic Sci. Int.* **1999**, *101*, 113–122. [CrossRef]
22. Aksnes, D.W.; Langfeldt, L.; Wouters, P. Citations, citation indicators, and research quality: An overview of basic concepts and theories. *SAGE Open* **2019**. [CrossRef]
23. Michigan State University AgBioResearch. Fall/Winter 2015, "Mystery Solved, Thanks to Insects", Interview with Eric Benbow. Futures Mag. Available online: https://www.canr.msu.edu/news/mystery_solved_thanks_to_insects (accessed on 19 January 2021).
24. Amendt, J.; Campobasso, C.; Gaudry, E.; Reiter, C.; LeBlanc, H.; Hall, M.J.R. Best practice in forensic entomology—Standards and guidelines. *Int. J. Legal Med.* **2007**, *121*, 90–104. [CrossRef]

25. Hall, M.J.R.; Whitaker, A.P.; Richards, C. Chapter 8: Forensic entomology. In *Forensic Ecology Handbook: From Crime Scene to Court*; Márquez-Grant, N., Roberts, J., Eds.; John Wiley and Sons Limited: Chichester, UK, 2012; pp. 111–140.

26. Cook, R.; Evett, I.W.; Jackson, G.; Jones, P.J.; Lambert, J.A. A model for case assessment and interpretation. *Sci. Justice* **1998**, *38*, 151–156. [CrossRef]

27. Triggs, C.M.; Buckleton, J.S. Comment on: Why the effect of prior odds should accompany the likelihood ratio when reporting DNA evidence. *Law Probab. Risk* **2004**, *3*, 73–82. [CrossRef]

28. Fenton, N. Assessing evidence and testing appropriate hypotheses. *Sci. Justice* **2014**, *54*, 502–504. [CrossRef]

29. Greenberg, B. Forensic entomology: Case studies. *Bull. Ent. Soc. Am.* **1985**, *31*, 25–28. [CrossRef]

30. Hart, A.J.; Whitaker, A.P.; Hall, M.J.R. The use of forensic entomology in criminal investigations: How it can be of benefit to SIOs. *J. Homicide Major Incid. Investig.* **2008**, *4*, 37–47.

31. Wang, M.; Chu, J.; Wang, Y.; Liao, M.; Shi, H.; Zhang, Y.; Hu, G.; Wang, J. Forensic entomology application in China: Four case reports. *J. Forensic Leg. Med.* **2019**, *63*, 40–47. [CrossRef]

32. Benecke, M. Six forensic entomology cases: Description and commentary. *J. Forensic Sci.* **1998**, *43*, 797–805. [CrossRef]

33. Sukontason, K.L.; Narongchai, P.; Sukontason, K.; Methanitikorn, R.; Piangjai, S. Forensically important fly maggots in a floating corpse: The first case report in Thailand. *J. Med. Assoc. Thai.* **2005**, *88*, 1458–1461.

34. Pohjoismäki, J.L.O.; Karhunen, P.J.; Goebeler, S.; Saukko, P.; Sääksjärvi, I.E. Indoors forensic entomology: Colonization of human remains in closed environments by specific species of sarcosaprophagous flies. *Forensic Sci. Int.* **2010**, *199*, 38–42. [CrossRef]

35. Charabidze, D.; Colard, T.; Vincent, B.; Pasquerault, T.; Hedouin, V. Involvement of larder beetles (Coleoptera: Dermestidae) on human cadavers: A review of 81 forensic cases. *Int. J. Legal Med.* **2013**, *128*, 1021–1030. [CrossRef]

36. Syamsa, R.A.; Omar, B.; Zuha, R.M.; Faridah, M.N.; Swarhib, M.S.; Hidayatulfathi, O.; Shahrom, A.W. Forensic entomology of high-rise buildings in Malaysia: Three case reports. *Trop. Biomed.* **2015**, *32*, 291–299.

37. Sandford, M.R. Insects and associated arthropods analyzed during medicolegal death investigations in Harris County, Texas, USA: January 2013–April 2016. *PLoS ONE* **2017**, *12*, e0179404. [CrossRef]

38. Sharma, A.; Bala, M.; Singh, N. Five case studies associated with forensically important entomofauna recovered from human corpses from Punjab, India. *J. Forensic Sci. Crim. Inves.* **2018**, *7*. [CrossRef]

39. Introna, F., Jr.; Campobasso, C.P.; Di Fazio, A. Three case studies in forensic entomology from southern Italy. *J. Forensic Sci.* **1998**, *43*, 210–214. [CrossRef]

40. Arnaldos, M.I.; Garcia, M.D.; Romera, E.; Presa, J.J.; Luna, A. Estimation of postmortem interval in real cases based on experimentally obtained entomological evidence. *Forensic Sci. Int.* **2005**, *149*, 57–65. [CrossRef]

41. Lindgren, N.K.; Sisson, M.S.; Archambeault, A.D.; Rahlwes, B.C.; Willett, J.R.; Bucheli, S.R. Four forensic entomology case studies: Records and behavioural observations on seldom reported cadaver fauna with notes on relevant previous occurrences and ecology. *J. Med. Ent.* **2015**, *52*, 143–150. [CrossRef]

42. Thyssen, P.J.; Aquino, M.F.K.; Purgato, N.C.S.; Martins, E.; Costa, A.A.; Lima, C.G.P.; Dias, C.R. Implications of entomological evidence during the investigation of five cases of violent death in Southern Brazil. *J. Forensic Sci. Res.* **2018**, *2*, 1–8. [CrossRef]

43. Sanford, M.R. Comparing species composition of passive trapping of adult flies with larval collections from the body during scene-based medicolegal death investigations. *Insects* **2017**, *8*, 36. [CrossRef]

44. Forensic Science Regulator. Codes of Practice and Conduct: For Forensic Science Providers and Practitioners in the Criminal Justice System. FSR-C-100; Issue 5, 2020. Available online: https://assets.publishing.service.gov.uk/government/uploads/system/uploads/attachment_data/file/880708/Codes_of_Practice_and_Conduct_-_Issue_5.pdf (accessed on 20 January 2021).

45. Criminal Practice Directions 2015 Division V, Evidence. 19A: Expert evidence. 19B: Statements of Understanding and Declarations of Truth in Expert Reports. Available online: https://www.justice.gov.uk/courts/procedure-rules/criminal/docs/2015/crim-practice-directions-V-evidence-2015.pdf (accessed on 19 January 2021).

46. Simpson, K. *Forty Years of Murder*; Granada Publishing Limited: London, UK, 1980.

47. Sheppard, K.; Cassella, J.P.; Fieldhouse, S. A comparative study of photogrammetric methods using panoramic photography in a forensic context. *Forensic Sci. Int.* **2017**, *273*, 29–39. [CrossRef] [PubMed]

48. Raneri, D. Enhancing forensic investigation through the use of modern three-dimensional (3D) imaging technologies for crime scene reconstruction. *Aust. J. Forensic Sci.* **2018**, *50*, 697–707. [CrossRef]

49. Lutz, L.; Verhoff, M.A.; Amendt, J. To be there or not to be there, that is the question—On the problem of delayed sampling of entomological evidence. *Insects* **2021**, *12*, 148. [CrossRef]

50. Magni, P.A.; Harvey, M.L.; Saravo, L.; Dadour, I.R. Entomological evidence: Lessons to be learnt from a cold case review. *Forensic Sci. Int.* **2012**, *223*, e31–e34. [CrossRef]

51. Richards, C.S.; Simonsen, T.J.; Abel, R.L.; Hall, M.J.R.; Schywn, D.A.; Wicklein, M. Virtual forensic entomology: Improving estimates of minimum post-mortem interval with 3D micro-computed tomography. *Forensic Sci. Int.* **2012**, *220*, 251–264. [CrossRef]

52. Martín-Vega, D.; Simonsen, T.J.; Wicklein, M.; Hall, M.J.R. Age estimation during the blow fly intra-puparial period: A qualitative and quantitative approach using micro-computed tomography. *Int. J. Legal Med.* **2017**, *131*, 1429–1448. [CrossRef] [PubMed]

53. Jeong, Y.; Woo, E.J.; Lee, S. Bibliometric analysis on the trend of the Computed Tomography (CT)-related studies in the field of forensic science. *Appl. Sci.* **2020**, *10*, 8133. [CrossRef]

54. Martín-Vega, D.; Simonsen, T.J.; Hall, M.J.R. Looking into the puparium: Micro-CT visualisation of the internal morphological changes during metamorphosis of the blow fly, *Calliphora vicina*, with the first quantitative analysis of organ development in cyclorrhaphous Diptera. *J. Morphol.* **2017**, *278*, 629–651. [CrossRef]

55. Hall, M.J.R.; Martín-Vega, D. Visualization of insect metamorphosis. *Philos. Trans. R. Soc. Lond. B Biol. Sci.* **2019**, *374*, 20190071. [CrossRef]

56. Bhadra, P.; Hart, A.; Hall, M.J.R. Factors affecting accessibility of bodies disposed in suitcases to blowflies. *Forensic Sci. Int.* **2014**, *239*, 62–72. [CrossRef]

57. Martín-Vega, D.; Hall, M.J.R. Estimating the age of *Calliphora vicina* eggs (Diptera: Calliphoridae): Determination of embryonic morphological landmarks and preservation of egg samples. *Int. J. Legal Med.* **2016**, *130*, 845–854. [CrossRef]

58. Jackson, C.H.N. An artificially isolated generation of tsetse flies (Diptera). *Bull. Entomol. Res.* **1946**, *37*, 291–299. [CrossRef]

59. Beutler, M.; Hart, A.J.; Hall, M.J.R. The use of wing fray and sex ratios to determine the origin of flies at an indoor crime scene. *Forensic Sci. Int.* **2020**, *307*, 110104. [CrossRef]

60. Donovan, S.E.; Hall, M.J.R.; Turner, B.D.; Moncrieff, C.B. Larval growth rates of the blowfly, *Calliphora vicina*, over a range of temperatures. *Med. Vet. Entomol.* **2006**, *20*, 1–9. [CrossRef]

61. Adams, Z.J.O.; Hall, M.J.R. Methods used for the killing and preservation of blowfly larvae, and their effect on post-mortem larval length. *Forensic Sci. Int.* **2003**, *138*, 50–61. [CrossRef]

62. Richards, C.S.; Rowlinson, C.C.; Hall, M.J.R. Effects of storage temperature on the change in size of *Calliphora vicina* larvae, during preservation in 80% ethanol. *Int. J. Legal Med.* **2013**, *127*, 231–241. [CrossRef]

63. Hofer, I.M.J.; Hart, A.J.; Martín-Vega, D.; Hall, M.J.R. Optimising crime scene temperature collection for forensic entomology casework. *Forensic Sci. Int.* **2017**, *270*, 129–138. [CrossRef]

64. Hofer, I.M.J.; Hart, A.J.; Martín-Vega, D.; Hall, M.J.R. Estimating crime scene temperatures from nearby meteorological station data. *Forensic Sci. Int.* **2020**, *306*, 110028. [CrossRef] [PubMed]

65. The Accreditation of Forensic Service Providers Regulations 2018, UK Statutory Instruments 2018 No. 1276, Regulation 3, Scope. Available online: https://www.legislation.gov.uk/uksi/2018/1276/regulation/3/made (accessed on 20 January 2021).

66. Gaudry, E.; Dourel, L. Forensic entomology: Implementing quality assurance for expertise work. *Int. J. Legal Med.* **2013**, *127*, 1031–1037. [CrossRef] [PubMed]

67. Melbourn, H.; Smith, G.; McFarland, J.; Rogers, M.; Wieland, K.; DeWilde, D.; Lighthart, S.; Quinn, M.; Baxter, A.; Quarino, L. Mandatory certification of forensic science practitioners in the United States: A supportive perspective. *Forensic Sci. Int. Synerg.* **2019**, *1*, 161–169. [CrossRef]

68. New Scientist, Insight. Austerity Has Put UK Forensic Labs in Crisis and Justice at Risk. Available online: https://www.newscientist.com/article/mg23831754-400-austerity-has-put-uk-forensic-labs-in-crisis-and-justice-at-risk/ (accessed on 21 December 2020).

69. Police Forensic Science at 'Breaking Point', Warn peers. Financial Times. 1 May 2019. Available online: https://www.ft.com/content/f3336774-6b4c-11e9-a9a5-351eeaef6d84 (accessed on 19 January 2021).

70. Flanagan, R.J. Cut costs at all costs! *Forensic Sci. Int.* **2018**, *290*, e26–e28. [CrossRef]

71. Tully, G. Forensic science in England & Wales, a commentary. *Forensic Sci. Int.* **2018**, *290*, e29–e31. [CrossRef]

72. Home Office, Association of Police and Crime Commissioners and National Police Chiefs' Council. Forensics Review: Review of the Provision of Forensic Science to the Criminal Justice System in England and Wales. April 2019. Available online: https://assets.publishing.service.gov.uk/government/uploads/system/uploads/attachment_data/file/911660/Joint_review_of_forensics_and_implementation_plan__accessible_.pdf (accessed on 21 December 2020).

73. House of Lords, Science and Technology Select Committee. 3rd Report of Session 2017-2019, HL Paper 333. Forensic Science and the Criminal Justice System: A Blueprint for Change. May 2019. Available online: https://publications.parliament.uk/pa/ld201719/ldselect/ldsctech/333/33302.htm (accessed on 21 December 2020).

74. Forensic Capability Network. Available online: https://www.fcn.police.uk (accessed on 20 January 2021).

75. Grzywacz, A. Thermal requirements for the development of immature stages of *Fannia canicularis* (Linnaeus) (Diptera: Fanniidae). *Forensic Sci. Int.* **2019**, *297*, 16–26. [CrossRef] [PubMed]

76. Wang, Y.; Zhang, Y.; Hu, G.; Fu, Y.; Zhi, R.; Wang, J. Development of *Hydrotaea spinigera* (Diptera: Muscidae) at constant temperatures and its significance for estimating postmortem interval. *J. Med. Entomol.* **2020**, ljaa162. [CrossRef]

77. Herzog, E. What social casework wants of social science research. *Am. Sociol. Rev.* **1951**, *16*, 68–73. [CrossRef]

78. The Royal Anthropological Institute, Forensic Anthropology. Available online: https://www.therai.org.uk/forensic-anthropology (accessed on 22 December 2020).

79. Hanzlick, R.; Prahlow, J.A.; Denton, S.; Jentzen, J.; Quinton, R.; Sathyavagiswaran, L.; Utley, S. Selecting forensic pathology as a career: A survey of the past with an eye on the future. *Am. J. Forensic Med. Pathol.* **2008**, *29*, 114–122. [CrossRef]

80. Gewin, V. Careers. Science in court—Courage of conviction. *Nature* **2015**, *526*, 463–465. [CrossRef]

81. Academy of Experts. What is an Expert Witness? Information Sheet 2015/01. Available online: https://academyofexperts.org/wpcms/wp-content/uploads/2020/06/Information-Sheet-What-is-an-Expert-2015-01.pdf (accessed on 21 December 2020).

82. Hall, M.J.R. The Relationship between Research and Casework in Forensic Entomology: How Do We Encourage the Next Generation of Caseworkers from Research? In Proceedings of the 16th Meeting of the European Association of Forensic Entomology, Pôle Juridique et Judiciaire, Bordeaux, France, 6–7 June 2019; p. 17. Available online: https://issuu.com/giefitalia/docs/eafe-2019-programme (accessed on 1 December 2020).

Review

Post-Mortem Interval Estimation Based on Insect Evidence: Current Challenges

Szymon Matuszewski [1,2]

1   Laboratory of Criminalistics, Adam Mickiewicz University, Święty Marcin 90, 61-809 Poznań, Poland; szymmat@amu.edu.pl
2   Wielkopolska Centre for Advanced Technologies, Adam Mickiewicz University, Uniwersytetu Poznańskiego 10, 61-614 Poznań, Poland

Simple Summary: The post-mortem interval of human cadavers may be estimated based on insect evidence. In order to identify scientific challenges that pertain to these estimations, I review forensic entomology literature and conclude that research on the development and succession of carrion insects, thermogenesis on cadavers and the accuracy of PMI estimates are of primary importance to advance this field.

Abstract: During death investigations insects are used mostly to estimate the post-mortem interval (PMI). These estimates are only as good as they are close to the true PMI. Therefore, the major challenge for forensic entomology is to reduce the estimation inaccuracy. Here, I review literature in this field to identify research areas that may contribute to the increase in the accuracy of PMI estimation. I conclude that research on the development and succession of carrion insects, thermogenesis in aggregations of their larvae and error rates of the PMI estimation protocols should be prioritized. Challenges of educational and promotional nature are discussed as well, particularly in relation to the collection of insect evidence.

Keywords: forensic entomology; carrion insects; development; succession; validation

Citation: Matuszewski, S. Post-Mortem Interval Estimation Based on Insect Evidence: Current Challenges. *Insects* **2021**, *12*, 314. https://doi.org/10.3390/insects12040314

Academic Editor: Daniel Martín-Vega

Received: 28 February 2021
Accepted: 29 March 2021
Published: 1 April 2021

Publisher's Note: MDPI stays neutral with regard to jurisdictional claims in published maps and institutional affiliations.

1. Introduction

Carrion insects living in human cadavers can be highly useful for the estimation of the post-mortem interval (PMI) [1,2]. Methods for PMI estimation based on insect evidence are developed, validated, improved and applied by forensic entomologists. This field is growing with a constant increase in the number of scientific publications and countries where entomology-based estimation of PMI is regularly used in death investigations [3,4]. As a maturing field, forensic entomology contains several weaknesses and under-researched areas. These challenges are the focus of this article.

A PMI estimate is only as good as it is close to the true PMI. The accuracy of estimation is most important, particularly for the end users of insect evidence. Therefore, the major general challenge for the field is to reduce the estimation inaccuracy. Its sources are related to both the collection and analysis of insect evidence (Figure 1). I divided this paper into sections devoted to the collection of insect evidence, research on insect development and succession, reconstructing temperature conditions, analysis of challenging evidence and validation of the protocols for PMI estimation.

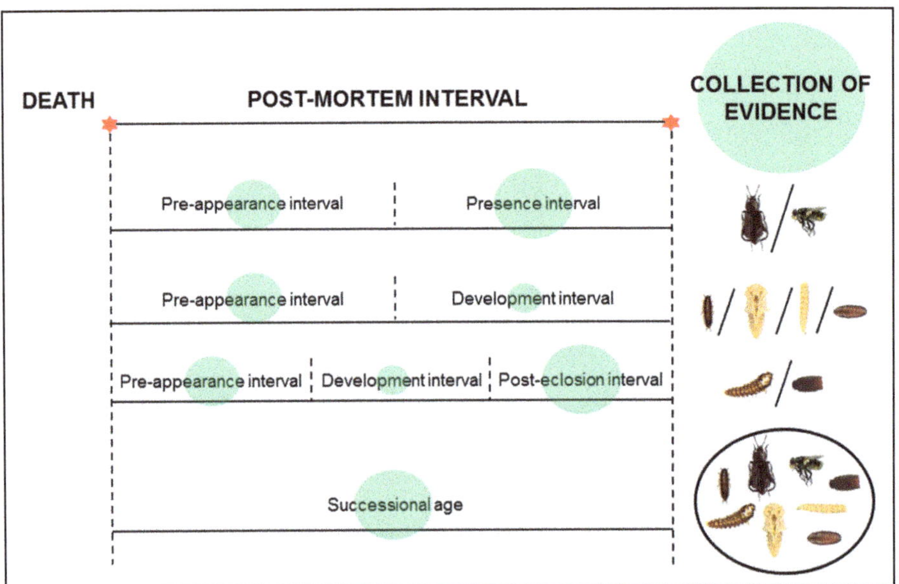

Figure 1. Sources of inaccuracy in the estimation of the post-mortem interval (PMI) based on insect evidence. Green circles represent the sources, their size represents importance of the sources. Pictures of insects were made by Anna Mądra-Bielewicz (Poznań, Poland).

2. Collection of Insect Evidence

Errors in the collection of insect evidence are certainly among the most important sources of the inaccuracy in PMI estimation. Death scene samples frequently misrepresent cadaver entomofauna. However, it is difficult to discern how bad these samples usually are and what the consequences of sampling errors are for the estimation of PMI. In most cases insects are collected by law enforcement officers or medical examiners, and rarely by entomologists. In a recent case, insects were sampled by police officers with the medical examiner and independently by entomologists, which enabled—in this paper— the comparison of samples taken by non-experts and experts [5]. The sample taken by non-experts was distinctly less diverse and did not contain insect evidence, based on which PMI has finally been estimated (see Tables 1 and 2 in [5]). If PMI was estimated in this case using only the non-expert sample, no meaningful maximum PMI would be derived, although the minimum PMI would be similar to the one estimated based on the expert sample (unpublished data). Another kind of error in the collection of evidence is the error of preservation. Insects may be preserved improperly, for example using an unsuitable preservative or a leaking container [6,7]. Such errors may limit the scope of possible analyses and in extreme cases may even destroy the evidence. Although there are no surveys of errors in the collection of insect evidence, I think that most experts share the opinion that insect samples frequently misrepresent cadaver entomofauna or are preserved improperly. We should therefore discuss whether our guidelines for the collection of evidence are truly fit-for-purpose.

Guidelines for the collection of insect evidence state that death scene samples should accurately represent cadaver entomofauna, i.e., all life stages of each important species that inhabit a cadaver should be represented in the sample [8–10]. Cadaver entomofauna may be very diverse and abundant, consisting of many life stages from many species, some in very large numbers. However, to estimate PMI only a small part of it is necessary. Usually, we choose the most developmentally advanced life stage of the most successionally advanced species, and even if the PMI estimate is based on a larger number of taxa, this is usually no more than two or three [11]. Therefore, in most cases a representative sample is redundant, and for this reason we should reshape guidelines for the collection of insect evidence and abandon our commitment to the true representativeness of death scene samples. I believe it is possible to develop guidelines that are user-friendly, quick to implement and that yield more fit-for-purpose samples, i.e., the most developmentally and successionally advanced insects only (Table 1). Insects are usually collected by the law enforcement officers with basic skills in entomology, whereas guidelines for the collection of insect evidence are usually addressed to entomologists. Therefore, we should provide guidelines for non-entomologists that specify what insect evidence they should look for and where it can be found, with pictures of the evidence and related preservation protocols.

Table 1. A sketch of guidelines for the collection of insect evidence on a death scene. To make them useful to non-entomologists, they should be combined with pictures of insect evidence and protocols suitable for the preservation of particular pieces of evidence.

State of a Cadaver	Insect Evidence	
	What to Look For	**Where to Look For**
Relatively fresh	Eggs or larvae of flies	Natural orifices (particularly of the head), wounds
Signs of putrefaction (bloating, marbling, etc.)	Larvae of flies	Natural orifices, wounds, interface cadaver/ground
Signs of active decay (large masses of insect larvae, stench of decay, leakage of decomposition fluids, etc.)	1. Larvae (particularly post-feeding) of flies 2. Larvae of beetles	1. Larval masses, the surface of soil (outdoor scenarios) or the floor (indoor scenarios) in the vicinity of a cadaver 2. Larval masses, clothes and cadaver surface, the soil in the vicinity of a cadaver (outdoor scenarios, soil samples are recommended), the floor in the vicinity of a cadaver (indoor scenarios)
Signs of advanced decay (exposed bones; greasy by-products of active decay, darkening of the remaining skin, etc.)	1. Puparia (full and empty) of flies 2. Larvae and pupae of beetles 3. Larvae of late-colonizing flies (e.g., skipper flies)	1. The soil in the vicinity of a cadaver (outdoor scenarios, soil samples are recommended), the floor (under carpets or furniture) in the vicinity of a cadaver (indoor scenarios), pockets and foldings of clothes, cadaver surface (all scenarios) 2. Larval masses, clothes and cadaver surface, the soil in the vicinity of a cadaver (outdoor scenarios, soil samples are recommended), the floor in the vicinity of a cadaver (indoor scenarios) 3. Larval masses, the surface of soil (outdoor scenarios) or the floor (indoor scenarios) in the vicinity of a cadaver
Signs of minimal insect infestation (e.g., massive putrefaction or mummification)	All types of insect evidence	Natural orifices, wounds, clothes and cadaver surface, the soil (outdoor scenarios) or the floor (indoor scenarios) in the vicinity of a cadaver

3. Insect Development

Most frequently, forensic entomologists estimate the age of immature insects collected on a death scene and use this information as the minimum PMI [12]. The reference developmental data for the species that was collected on a death scene is necessary for such estimation. Because developmental data may vary between geographical populations of insects, it is recommended to use reference data from the closest population [10,13–15]. Although there is constant progress in this field, with new species and populations gaining developmental data, still much research needs to be done. A review of developmental datasets available for insect species colonizing cadavers in central Europe reveals that among the most extensively researched are cosmopolitan species that colonize cadavers shortly after death and that were frequently reported from indoor cases (Table 2). Although several important species have many datasets (e.g., *Lucilia sericata* or *Calliphora vicina*), there are still species that regularly breed in cadavers but for which no dataset has been published (e.g., *Lucilia caesar*, *Hydrotaea ignava* or *Necrobia violacea*) or only single datasets are available (e.g., *Stearibia nigriceps*, *Necrodes littoralis*, *Omosita colon* or *Necrobia rufipes*). These species should become the hot taxa for forensic entomology research in Europe. The understudied species ought to be identified also for other geographical regions.

Another point that needs our attention is the lack of standards and guidelines for developmental studies in forensic entomology. Several elements of the protocol for such studies were found to affect the quality of the resultant developmental data [16–19]. In addition, there is unnecessary variation in the type of development data provided and the way they are presented in publications. Standard research protocols emerge in mature sciences and I feel it is time to start this discussion in forensic entomology.

Table 2. Developmental datasets available for the species that breed in large vertebrate cadavers in central Europe (species list compiled based on [20–25]).

Family	Species	Number of Published Datasets	Country of a Population's Origin	References
Calliphoridae	*Calliphora vicina*	18	US,AT,GB,RU,CA,DE,IT,EG	[26–43]
	Calliphora vomitoria	7	US,GB,RU,DE	[26,29,31,33,37,44,45]
	Chrysomya albiceps	9	BR,RU,AT,ZA,CO,IR,EG	[14,26,39,46–51]
	Lucilia caesar	-	-	-
	Lucilia sericata	27	US,FI,IT,GB,RU,CA,AT,CO,IR,EG,TR,FR,EC,KR,CN	[26–28,30,31,33,36,48,52–70]
	Phormia regina	7	US,RU,CA,MX	[26–28,31,71–73]
	Protophormia terraenovae	7	US,GB,RU,AT,CA	[26,31,33,44,74–76]
Sarcophagidae	*Sarcophaga argyrostoma*	3	AT,DE,TR	[29,77,78]
	Sarcophaga caerulescens	-	-	-
Muscidae	*Hydrotaea dentipes*	-	-	-
	Hydrotaea ignava	-	-	-
	Hydrotaea pilipes	-	-	-
Fanniidae	*Fannia canicularis*	2	US,PL	[79,80]
	Fannia scalaris	-	-	-
	Fannia leucosticta	-	-	-
Piophilidae	*Stearibia nigriceps*	1	RU	[26]
Silphidae	*Necrodes littoralis*	2	PL	[81,82]
	Thanatophilus rugosus	1	DZ	[83]
	Thanatophilus sinuatus	1	CZ	[84]
Histeridae	*Margarinotus brunneus*	-	-	-
	Saprinus planiusculus	-	-	-
	Saprinus semistriatus	-	-	-
Staphylinidae	*Aleochara curtula*	-	-	-
	Creophilus maxillosus	5	US,CN,PL	[85–89]
	Philonthus politus	-	-	-
Dermestidae	*Dermestes frischii*	3	GB,ES,IT	[90–92]
	Dermestes lardarius	2	GB	[93,94]
	Dermestes murinus	-	-	-
Nitidulidae	*Omosita colon*	1	CN	[95]
Cleridae	*Necrobia rufipes*	1	CN	[96]
	Necrobia violacea	-	-	-
Pteromalidae	*Nasonia vitripennis*	5	AT,US,AU,CN,BR	[97–101]

AT—Austria, AU—Australia, BR—Brazil, CA—Canada, CN—China, CO—Colombia, CZ—Czech Republic, DE—Germany, DZ—Algeria, EC—Ecuador, EG—Egypt, ES—Spain, FI—Finland, FR—France, GB—United Kingdom, IR—Iran, IT—Italy, KR—Republic of Korea, MX—Mexico, PL—Poland, RU—Russian Federation, TR—Turkey, US—United States and ZA—South Africa.

4. Insect Succession

There are several forensic reasons to study insect succession on cadavers. First, these studies yield inventories of carrion insects for habitats and geographical locations that form a starting point for any further research in forensic entomology. Such inventories were published for many habitats and locations around the world (recently reviewed in [102]), but there are still white spots on this map.

Second, succession studies provide reference data on the pre-appearance interval (PAI) and the presence interval (PI) of particular insect taxa. Such data are essential to use insects that colonize cadavers late in decomposition, as their PAI may be longer than the development interval, and to get meaningful PMI it may be necessary to combine insect age with the PAI [5,11,103]. PAI may also support estimates of maximum PMI when insect evidence is absent [104,105]. PAI may be estimated using the temperature models for PAI [106]. However, such models are available only for some taxa, and for several important taxa (e.g., blow flies) PAI may not be estimated using the temperature data [107,108]. In such cases, insect succession studies with animal cadavers (preferably large pigs [102]) yield the best PAI reference data (e.g., average seasonal PAIs). As for the PI, it has a more complex causal background than PAI, its predictions are inherently related with larger inaccuracy (Figure 1) and currently it may be approximated only based on the reference data from succession studies. Although in some habitats and locations robust PAI or PI datasets are available for many taxa, usually there is shortage of such data (Table 3). In particular, indoor habitats need more attention. Therefore, pig decomposition studies to yield PAI and PI data of forensically relevant insects should be one of the priority research areas in forensic entomology.

Third, decomposition experiments using pig cadavers may be useful to validate the PMI estimation protocols [109]. Such experiments are especially suitable as proof-of-concept studies or initial validation studies [102]. Although validation of new methods is a priority in forensic sciences [110], datasets on the performance of insect-based methods for PMI are very limited. Validation using pig cadavers (ultimately also human cadavers [102]) should be another primary research area in forensic entomology (Section 6 of this article).

There are guidelines for decomposition studies in forensic entomology [102,111–115]. Still, however, more standardization is necessary, particularly in terms of the sampling frequency, insect identifications and the presentation of the results (summarized in [116]). We need to remember that PAI and PI data for particular taxa are necessary when results of the study are to be used for the estimation of PMI. Therefore, the data for immature insects should be prioritized. When only a few cadavers were used, a daily occurrence matrix may be the best choice to present the results in a forensically useful way [117,118]. When more cadavers were investigated, it may be necessary to present insect occurrences in a synthetic way, but still raw data from individual cadavers (or seasonal averages) should be given on the PAI and PI of particular insect taxa (e.g., [119]).

Table 3. Datasets on the pre-appearance interval (PAI) and the presence interval (PI) of the species that breed in large vertebrate cadavers in central Europe (species list compiled based on [20–25]). I reviewed datasets derived from experiments performed in Europe and on pig cadavers only.

Family	Species	PAI		PI—Seasonal Data (Country/Habitat/Season/Stage)	References
		Temperature Model	Seasonal Data (Country/Habitat/Season/Stage)		
Calliphoridae	*Calliphora vicina*	-	PL/F/S/A,L1 IT/Ou/u/Au,W/A AT/Ou/u/S,Su/A PT/Ou/u/S,Su,Au,W/A PT/Ou/u/S,Au,W/O,L1,P ES/I/S,Su,Au,W/O	PL/F/S/A,L IT/Ou/u/Au,W/A AT/Ou/u/S,Su/A PT/Ou/u/S,Su,Au,W/A PT/Ou/u/S,Au,W/E,L,P ES/I/S,Su,Au,W/E	[22,119–122]

Table 3. *Cont.*

Family	Species	PAI		PI—Seasonal Data (Country/Habitat/Season/Stage)	References
		Temperature Model	Seasonal Data (Country/Habitat/Season/Stage)		
	Calliphora vomitoria	-	PL/F/S,Su,Au/A,L1,L3 IT/Ou/u/Au,W/A AT/Ou/u/S/A PT/Ou/u/S,Su,Au,W/A PT/Ou/u/S,W/O,L1,P ES/I/S/O	PL/F/S,Su,Au/A,L IT/Ou/u/Au,W/A AT/Ou/u/S/A PT/Ou/u/S,Su,Au,W/A PT/Ou/u/S,W/E,L,P ES/I/S/E	[22,25,119–123]
	Chrysomya albiceps	-	IT/Ou/u/Su,Au/A AT/Ou/u/Su/A PT/Ou/u/Su,Au/A PT/Ou/u/Su,Au/O,L1,P ES/I/S,Su,Au/O	IT/Ou/u/Su,Au/A AT/Ou/u/Su/A PT/Ou/u/Su,Au/A PT/Ou/u/Su,Au/E,L,P ES/I/S,Su,Au/E	[22,120–122]
	Lucilia caesar	-	PL/Ou/r/S,Su/A,L3 PL/F/S,Su,Au/A,L1,L3 IT/Ou/u/Su,Au/W/A PT/Ou/u/S,Su,Au/A PT/Ou/u/S,Su,Au/O,L1,P	PL/F/S,Su,Au/A,L IT/Ou/u/Su,Au,W/A PT/Ou/u/S,Su,Au/A PT/Ou/u/S,Su,Au/E,L,P	[21,25,119–121,123]
	Lucilia sericata	-	PL/Ou/r/S,Su/A IT/Ou/u/Su,Au,W/A PT/Ou/u/S,Su,Au/A PT/Ou/u/Au/O,L1,P ES/I/S,Su,Au/O	IT/Ou/u/Su,Au,W/A PT/Ou/u/S,Su,Au/A PT/Ou/u/Au/E,L,P ES/I/S,Su,Au/E	[21,120–122]
	Phormia regina	A	PL/Ou/r/S,Su/A PL/F/S,Su,Au/A,L1 AT/Ou/u/S,Su/A PL/F/S,Su/L3	PL/F/S,Su,Au/A,L AT/Ou/u/S,Su/A	[21,22,25,107,119,123]
	Protophormia terraenovae	-	AT/Ou/u/S,Su/A	AT/Ou/u/S,Su/A	[22]
Sarcophagidae	*Sarcophaga argyrostoma*	-	-	-	-
	Sarcophaga caerulescens	-	-	-	-
Muscidae	*Hydrotaea dentipes*	A	PL/F/S,Su,Au/A PT/Ou/u/S/A PL/F/S,Su/L1	PL/F/S,Su,Au/A PT/Ou/u/S/A PL/F/S,Su/L	[25,107,119,121,123]
	Hydrotaea ignava	A	PL/Ou/r/S,Su/A,L3 PL/F/S,Su,Au/A PT/Ou/u/S,Su,Au/A PL/F/S,Su/L1 PL/F/Su/L3	PL/F/S,Su,Au/A PT/Ou/u/S,Su,Au/A PL/F/S,Su/L	[21,25,107,119,121]
	Hydrotaea pilipes	-	PL/Ou/r/S,Su/A PL/F/S,Su,Au/A	PL/F/Su,Au/A	[21,25,123]
Fanniidae	*Fannia canicularis*	-	IT/Ou/u/Au,W/A	IT/Ou/u/Au,W/A	[120]
	Fannia scalaris	-	-	-	-
	Fannia leucosticta	-	-	-	-
Piophilidae	*Stearibia nigriceps*	A,O	PL/Ou/r/S,Su/A,L3 PL/F/S,Su,Au/A,L1 IT/Ou/u/Au/A PT/Ou/u/S,Su,Au/A PT/Ou/u/Su/E PT/Ou/u/S,Su,Au/L1,P PL/F/S,Su/L3	PL/F/S,Su,Au/A,L IT/Ou/u/Au/A PT/Ou/u/S,Su,Au/A PT/Ou/u/Su/E PT/Ou/u/S,Su,Au/L,P	[21,25,107,119–121,123]
Silphidae	*Necrodes littoralis*	A,L1	PL/Ou/r/S,Su/A,L1 PL/F/S,Su,Au/A,L1	PL/F/S,Su,Au/A,L	[21,25,103,119,123,124]
	Thanatophilus rugosus	-	PL/F/S,Su,Au/A IT/F/W/A PL/Ou/r/S/L3	PL/F/Su,Au/A IT/F/W/A PL/Ou/r/S/L3	[25,120,123,125]
	Thanatophilus sinuatus	A,L1	PL/Ou/r/S,Su/a PL/F/S,Su,Au/A IT/Ou/u/W/A PT/Ou/u/S,Au,W/A PL/Ou/r/S/L3	PL/F/S,Su,Au/A IT/Ou/u/W/A PT/Ou/u/S,Au,W/A PL/Ou/r/S/L3	[21,25,119,120,124–126]
Histeridae	*Margarinotus brunneus*	A	PL/Ou/r/S,Su/A PL/F/S,Su/A PT/Ou/u/S,Su,Au,W/A	PL/F/S,Su/A PT/Ou/u/S,Su,Au,W/A	[21,25,119,124,126]
	Saprinus planiusculus	A	PL/F/S/A	PL/F/S/A	[119,124]
	Saprinus semistriatus	A	PL/Ou/r/S,Su/A PL/F/S,Su,Au/A	PL/F/S,Su,Au/A	[21,25,119,123,124]

Table 3. *Cont.*

Family	Species	PAI		PI—Seasonal Data (Country/Habitat/Season/Stage)	References
		Temperature Model	Seasonal Data (Country/Habitat/Season/Stage)		
Staphylinidae	*Aleochara curtula*	-	PL/F/S,Su/A IT/Ou/u/W/A	IT/Ou/u/W/A	[25,120]
	Creophilus maxillosus	A,L1	PL/Ou/r/S,Su/A,L1 PL/F/S,Su,Au/A,L1 IT/Ou/u/Au,W/A PT/Ou/u/S,Su,Au,W/A	PL/F/S,Su,Au/A,L IT/Ou/u/Au,W/A PT/Ou/u/S,Su,Au,W/A	[21,25,119,120,123,124,126,127]
	Philonthus politus	A	PL/F/S,Su,Au/A IT/Ou/u/Au/A	IT/Ou/u/Au/A	[25,120,124]
Dermestidae	*Dermestes frischii*	-	PL/Ou/r/S,Su/A,L1 PT/Ou/u/S,Su,Au/A	PT/Ou/u/S,Su,Au/A	[21,126]
	Dermestes lardarius	-	-	-	-
	Dermestes murinus	-	PL/F/S,Su,Au/A PL/F/S/Lm	PL/F/Su,Au/A	[25,123]
Nitidulidae	*Omosita colon*	-	-	-	-
Cleridae	*Necrobia rufipes*	A	IT/Ou/u/Su,W/A PT/Ou/u/Su,Au/A	IT/Ou/u/Su,W/A PT/Ou/u/Su,Au/A	[120,124,126]
	Necrobia violacea	A	PL/Ou/r/S,Su/A PL/F/S,Su/A PT/Ou/u/S,Su,Au,W/A PL/F/S/L3	PT/Ou/u/S,Su,Au,W/A PL/F/S/L3	[21,25,119,124,126]
Pteromalidae	*Nasonia vitripennis*	-	AT/Ou/u/S/A	AT/Ou/u/S/A	[22]

A—adult stage PAI or PI, O—oviposition PAI, E—egg PI, L1—first instar larvae PAI, L—larval PI, L3—third instar larvae PAI or PI, Lm—mature larvae PAI or PI, P—puparial/pupal PAI or PI. S—spring, Su—summer, Au—autumn, W—winter. I—indoor habitats, Ou/u—outdoor, urban habitats, Ou/r—outdoor, rural habitats, F—forests. AT—Austria, ES—Spain, IT—Italy, PL—Poland, and PT—Portugal.

5. Temperature Conditions

The succession and development of insects on cadavers is largely dependent on the temperature [124,128,129]. When estimating PMI from insect succession or development, it is necessary to reconstruct temperature conditions. The accuracy of the PMI estimation depends largely on the accuracy of the reconstructed temperature conditions. This source of error is one of the most important.

Forensic entomologists frequently use temperature data from the local weather stations. Weather station temperatures can be corrected to adjust them to the peculiarities of a death scene [130–134]. Such corrections are based on the regression analysis between recordings made on a death scene and recordings from the station and for this reason they may be unfeasible [135]. Moreover, some authors indicate that the correction protocol has uncertain benefits for the accuracy of PMI estimation [135,136]. From the other side, there are robust experimental data indicating that the protocol improves the death scene temperatures [130,132–134]. It was found beneficial in casework, as well [5,132], although it was used infrequently [134]. The protocol may be impractical and its use may have a minor risk of deteriorating the death scene temperatures; however, it is the best tool we have and we should try to use it more frequently, particularly on outdoor death scenes. We need to remember that the protocol makes the weather station temperatures closer to the cadaver's ambient temperatures only. Therefore, the corrected temperatures may still be far from the true temperatures experienced by the insects, because the protocol accounts for peculiarities of a death scene in terms of the factors that affect ambient air temperature only. In order to take into account other important factors a different approach is needed.

Some authors modelled temperature conditions in parked vehicles [137], containers [138] or specific urban and semi-natural habitats (e.g., cellars, attics or trailers) [139]. A model was also derived to extract heat profiles representing the temperatures experienced by insect populations growing on cadavers [140]. Charabidze and Hedouin [135] developed an algorithm to correct temperatures through a qualitative analysis of thermal-specific aspects of the case. The analysis consisted of six stages, starting from the conditions on a cadaver and moving towards the outside of the body. This research area is growing and both quantitative and qualitative approaches may be useful here.

The last factor that needs much more of our attention is the insect-driven thermogenesis. It has been discovered and extensively studied in aggregations of blow fly larvae [141–152]. Recently, insect-driven thermogenesis has been also reported for carrion beetles *Necrodes littoralis* L. (Silphidae), with evidence that heat is produced within the feeding matrix, which is formed by adult and larval beetles through spreading their exudates over the cadaver surface [153]. Thermogenesis in larval aggregations may be more common among carrion insects, and because it may substantially shorten the development interval, it should be factored when reconstructing temperature conditions [151,153].

When large aggregations of fly larvae (with elevated temperature) are present on a cadaver, Charabidze and Hedouin [135] suggest to use the minimum development time for the feeding stage of each species. Unfortunately, minimum times needed to reach the post-feeding phase in large aggregations of larvae are not available for any species. Accordingly, it may be tempting to use minimum development times from the laboratory development studies, as there are many such datasets (Table 2). However, in such studies minimum development times are recorded at high and constant temperatures that may be suboptimal for the insects. Experiments using the tracking of blow fly larvae within aggregations indicated that they have a strong preference for the hottest part of the aggregation [154]. This finding prompted the authors to state that the maximum temperatures of the aggregation represent the actual temperatures experienced by the larvae [154]. More recent data demonstrated that larvae continuously move between the periphery and the inside of the aggregation, with individual larvae spending from 16 to 68% (mean 43%) of their time at the aggregation periphery [155]. The periphery has a lower temperature than the inside of the aggregation [154]. For this reason, the heat gain of individual larvae may be smaller than if they were feeding for the whole time in the hottest part of the aggregation. The maximum temperature of the aggregation probably overestimates the true temperatures experienced by the larvae. Perhaps the heat benefit of the aggregated larvae is somehow related to the temperatures selected by the larvae along a thermal gradient [143,154,156]. Further research using tracking techniques to monitor heat benefits and the development time of individual larvae within large aggregations will be necessary to find the minimum development times or the optimal temperatures for the larvae that develop in an aggregation.

6. Challenging Evidence

All insect evidence can be challenging in some cases, and some types are always challenging. The puparia of flies and pupae of beetles are informative pieces of insect evidence, particularly on decomposed cadavers [5,11]. However, they are difficult to identify and it is difficult to estimate their age.

The identification of insect evidence is a necessary first step in any analysis. Significant progress has been recently made in this field, with several forensically important fly taxa gaining excellent identification keys for adult insects [13,157–161] and larvae [162–164]. Puparia should be the next step. Although some groups of carrion flies have useful descriptions of puparia [165], there is no forensically useful key for this type of insect evidence in any family of flies. This area is much less developed in the case of the forensically important beetles. There is just a single identification key for the larvae of beetles that colonize cadavers [166] and a single key for the adult carrion beetles (Silphidae) that frequent cadavers [167]. Although some descriptions of larval identification features have been published for forensically important species [168,169], this group needs more attention. Otherwise, we will still have to base our identifications on the taxonomic references that may be inaccessible to forensic entomologists with no experience in beetle taxonomy.

It is difficult to estimate the age of the fixed puparia of flies and pupae of beetles. The most promising techniques for aging such evidence consist of the qualitative morphological analyses of the intra-puparial forms of the flies [30,35,36]. Intra-puparial development has been documented for many forensically important species [35,65,78,170–175]. Although these techniques have obvious advantages (e.g., they cover most of the intra-puparial development, they are generally non-destructive, low-cost and they need a stereomicroscope only), they have also important disadvantages (e.g., they are qualitative in nature and therefore less accurate and they are also impractical due to the need to have an expert knowledge in the intra-puparial morphology) [82]. Recently, a simple-to-use technique has been developed for aging pupae of the carrion beetle *Necrodes littoralis* by means of the quantification of the eye-background contrast, with very encouraging results of the initial validation [82]. Similar quantitative techniques should be developed in other forensically important insects.

Empty puparia (i.e., hardened outer shells that remain upon the completion of immature development of some flies) are frequently collected on cadavers with long PMI, and their examination may provide an estimate of the minimum PMI [176]. This type of insect evidence poses specific difficulties. When estimating minimum PMI based on the empty puparium, it is necessary to take the post-eclosion interval (PEI) into account. The interval starts when an adult fly emerges from the puparium and ends when the empty puparium is being collected. PEI may be longer than the minimum PMI estimated based on the puparium. Although techniques to estimate PEI are being developed [177–179], they are far from the implementation to forensic casework. In a recent PMI simulation study, seasonal patterns of changes in PMI following various PEIs were revealed for the empty puparia of two species of flies, demonstrating that the simulation studies may guide estimation of the minimum PMI based on such challenging evidence [176].

7. Validation of the PMI Estimation Protocols

Forensic entomologists developed several methods for the estimation of PMI based on insect development [26,41,180–182] or succession [103,105,115,117,118,128,183]. As there are contemporary reviews of these methods [1,184], in this article I focus only on their validation (Table 4). Validation of a protocol for the estimation of PMI is of key importance, as it may demonstrate that the protocol provides robust evidence when used in a forensic context. Validation studies may also provide PMI errors that could be used to present a PMI estimate as a meaningful interval. Sometimes entomologists provide point estimates for PMI (e.g., [185,186]). The inaccuracy of the PMI estimate that usually has many and diverse sources, but which is inherently related to every analysis of insect evidence, should be explicit in casework. This may be accomplished by providing ranges for PMI. Therefore, an interval estimate for PMI should be a standard way to present the results of the insect evidence analysis. If we knew robust errors, they could be used to transform any PMI estimate (a point or a range) into a highly informative interval that takes into account all sources of inaccuracy. The error of estimation is the difference between the estimated and true PMI, expressed as a percentage of the true or estimated PMI (hereafter error I and II). If such errors were calculated for a reliable sample of PMI estimations, i.e., a large sample of forensic case reports with known true PMI or a large sample of PMI estimations for experimentally used human cadavers, they might robustly approximate the accuracy of the PMI estimation protocol in a forensic context. I believe that such errors could also yield a truly informative evaluation of the uncertainty in PMI estimates in casework.

Most of the validation studies in forensic entomology were proof-of-assumptions or proof-of-concept studies (Table 4). Experiments fully validating the estimation protocols were rare. Only a few such datasets have been published; most used pig cadavers and were replicated moderately, at most. Human cadavers in anthropology research facilities (i.e., body farms) could be used more extensively for that purpose. The estimation of PMI for such cadavers using mock crime scenarios could provide robust validation data. This research design is surprisingly underutilised at body farms.

Similarly, validations using casework data were infrequent (Table 4). In order to use the casework data for the validation, a true PMI needs to be specified based on a confession or a witness statement about when the victim was last seen alive, or other non-insect evidence. Although the non-insect evidence only approximates the true PMI sensu stricto, this is the only way to use casework data for the validation. However, published case reports rarely provide information on the true PMI sensu largo. In order to calculate the errors of insect-based protocols for PMI, I analyzed relevant case reports where the PMI was estimated based on insect development (Table 5) and separately based on insect succession (Table 6). Due to the imperfections of the data used, resultant errors need to be treated with caution. They are only rough approximations of the true errors of the insect-based protocols for PMI.

Table 4. Validation of the protocols for the estimation of PMI based on insect evidence.

Type of the Validation	Aims	Development-Based Protocols		Succession-Based Protocols	
		Number of Studies	References	Number of Studies	References
Proof-of-assumptions study [1]	Testing validity of the assumptions that are at the root of the protocol	26	[14,17–19,27,29,30,34,41,55–57,61,144,145,176,187–196]	56	[20,21,24,25,103,107,113,119,122,124,127,128,197–240]
Proof-of-concept study [1]	Testing validity of the protocol as used in a simplified setting	12	[30,73,81,82,87–89,241–245]	3	[103,127,128]
Experimental validation with non-human cadavers	Testing validity of the protocol as used for non-human cadavers in an experimental setting	6	[109,241,246–249]	6	[105,106,183,249–251]
Experimental validation using human cadavers	Testing validity of the protocol as used for human cadavers in an experimental setting	0		1	[183]
Validation using casework data [1]	Testing validity of the protocol as used in forensic casework	7	[46,118,182,185,252–254]	6	[5,108,118,186,255,256]

[1] Selected studies were referenced.

There were surprisingly large differences between the cases. Errors I (differences between the true and estimated PMI expressed as the percentage of the true PMI) ranged from 0 to 83% for the development-based estimates (Table 5) and from 2 to 43% for the succession-based estimates (Table 6). Surprisingly, average errors were larger for the development-based estimates (22.3%) than the succession-based estimates (13.4%). Although the average difference between the true and estimated PMI was almost four times lower for the development-based estimation than the succession-based estimation (1.5 and 5.6 days respectively, Tables 5 and 6), the latter type of the estimation was usually used for cadavers with a much larger PMI (Table 6); therefore, it had lower errors, which are relative values. Differences between the true and estimated PMI increased with the increase in the true PMI, and this relationship was particularly apparent when plotted for the development-based estimates (Figure 2).

Summarizing, the protocols for the estimation of PMI based on insect evidence usually lack errors and their validity has been rather poorly demonstrated in a true forensic context. Therefore, validation studies using pig or human cadavers and casework data should be prioritized in forensic entomology. I think this is our greatest challenge.

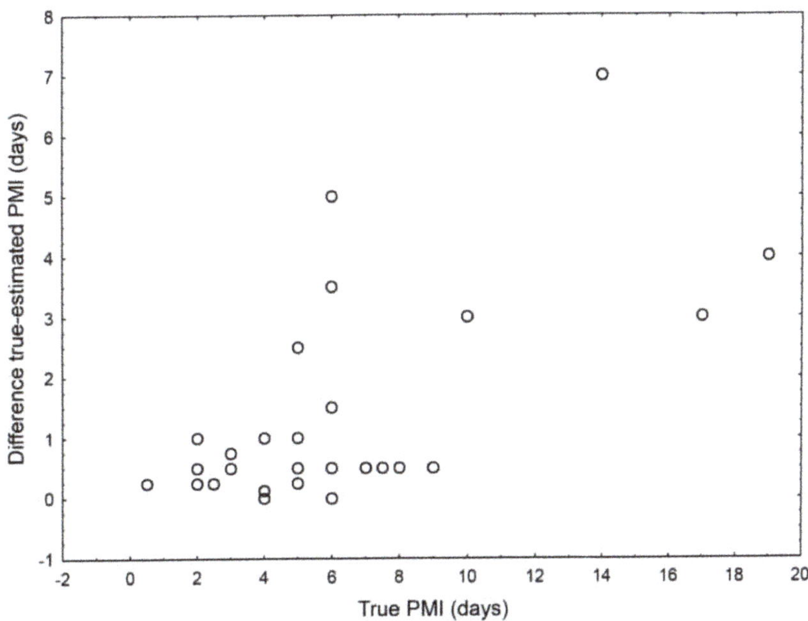

Figure 2. Differences between the true and estimated PMI plotted against the true PMI for the estimations based on insect development that were referenced in Table 5.

Table 5. Errors of the protocols for the estimation of PMI based on insect development.

Reference	N	True PMI [1] (days)		Difference True-Estimated PMI [2] (days)		Error I [3] (%)		Error II [4] (%)		Remarks
		Mean	Range	Mean	Range	Mean	Range	Mean	Range	
Goff et al., 1988 [252]	2	5.5	5–6	0.375	0.25–0.5	6.65	5–8.3	6.95	4.8–9.1	-
Kashyap, Pillay, 1989 [185]	16	4.9	0.5–9	0.438	0–1	13.74	0–50	11.65	0–33.3	No mention of temperature data
Grassberger et al., 2003 [46]	1	17	-	3	-	17.65	-	20	-	-
Reibe et al., 2010 [182]	1	4	-	0.125	-	3.125	-	3.03	-	-
Pohjoismäki et al., 2010 [253]	7	10.6	5–19	4.57	2.5–7	48.96	21.1–83.3	144.2	26.7–500	Single average temperature assumed in all cases (24 °C)
Bugelli et al., 2015 [254]	4	4.0	2–6	0.94	0.5–1.5	23.75	20–25	31.25	25–33	-

N—a number of PMI estimations in a dataset. [1] PMI determined based on non-insect evidence (a confession, a witness statement about when the victim was last seen alive, etc.). [2] An absolute difference between the true PMI and the PMI estimated based on insect development. When the estimated PMI was presented as an interval, I calculated absolute differences between the true PMI and the lower and upper limit of the estimated interval and then averaged them to get the difference between the true and estimated PMI. [3] Error I = (the difference between the true and estimated PMI/true PMI) × 100. [4] Error II = (the difference between the true and estimated PMI/estimated PMI) × 100. When the estimated PMI was presented as an interval, a midpoint of the interval was used in denominator.

Table 6. Errors of the protocols for the estimation of PMI based on insect succession.

Reference	N	True PMI [1] (days)	Estimated PMI (days)	Difference True-Estimated PMI [2] (days)	Error I [3] (%)	Error II [4] (%)	Remarks
Goff et al., 1986 [256]	1	20	19–20	0.5	2.5	2.6	-
Goff and Odom, 1987 [186]	1	53	≥52	1	1.9	1.9	-
Goff and Flynn, 1991 [255]	1	38	34–39	2.5	6.6	6.8	-
Schoenly et al., 1996 [118]	2	11	10.5–11	0.25	2.3	2.3	-
		36	34–36	1	2.8	2.9	
Archer, 2014 [108]	1	21	16–34	9	42.9	36	-
Matuszewski and Mądra-Bielewicz, 2019 [5]	1	72	30–64	25	34.7	53.2	Less reliable true PMI

N—a number of PMI estimations in a dataset. [1] PMI determined based on non-insect evidence (a confession, a witness statement about when the victim was last seen alive, etc.). [2] An absolute difference between the true PMI and the PMI estimated based on insect succession. When the estimated PMI was presented as an interval, I calculated absolute differences between the true PMI and the lower and upper limit of the estimated interval and then averaged them to get the difference between the true and estimated PMI. [3] Error I = (the difference between the true and estimated PMI/true PMI) × 100. [4] Error II = (the difference between the true and estimated PMI/estimated PMI) × 100. When the estimated PMI was presented as an interval, a midpoint of the interval was used in denominator.

8. Conclusions

Although the set of challenges elucidated in this article is somehow subjective, I believe that most forensic entomologists would construct similar sets. Some challenges should focus more of our attention, with priority for the resultant research. This applies, in particular, to the validation research, as well as to development and succession research. Studies on thermogenesis in larval aggregations on cadavers should be prioritized as well. There are also highly important challenges of educational and promotional nature. Although we should look for more optimal guidelines for insect sampling on a death scene, and this is a scientific task, improvement in the samples taken by a law enforcement personnel depends equally or even more on the promotion of forensic entomology among its end-users and on the education of the officers or medical examiners that collect insect evidence on death scenes.

Funding: This research received no external funding.

Acknowledgments: I thank the anonymous reviewers for their comments that helped to improve the manuscript. Thanks are also extended to Anna Mądra-Bielewicz (Poznań, Poland) for pictures of insects that were used in Figure 1.

Conflicts of Interest: The author declares no conflict of interest.

References

1. Villet, M.H.; Amendt, J. Advances in entomological methods for death time estimation. In *Forensic Pathology Reviews*; Turk, E.E., Ed.; Springer: Berlin/Heidelberg, Germany, 2011; pp. 213–237.
2. Catts, E.P.; Goff, M.L. Forensic entomology in criminal investigations. *Annu. Rev. Entomol.* **1992**, *37*, 253–272. [CrossRef]
3. Lei, G.; Liu, F.; Liu, P.; Zhou, Y.; Jiao, T.; Dang, Y.-H. A bibliometric analysis of forensic entomology trends and perspectives worldwide over the last two decades (1998–2017). *Forensic Sci. Int.* **2019**, *295*, 72–82. [CrossRef]
4. Tomberlin, J.K.; Benbow, M.E. *Forensic Entomology: International Dimensions and Frontiers*; CRC Press: Boca Raton, FL, USA, 2015.
5. Matuszewski, S.; Mądra-Bielewicz, A. Post-mortem interval estimation based on insect evidence in a quasi-indoor habitat. *Sci. Justice* **2019**, *59*, 109–115. [CrossRef]
6. Sanford, M.R.; Pechal, J.L.; Tomberlin, J.K. Rehydration of forensically important larval Diptera specimens. *J. Med. Entomol.* **2011**, *48*, 118–125. [CrossRef]
7. Mądra-Bielewicz, A.; Frątczak-Łagiewska, K.; Matuszewski, S. Blowfly puparia in a hermetic container: Survival under decreasing oxygen conditions. *Forensic Sci. Med. Pathol.* **2017**, *13*, 328–335. [CrossRef]
8. Sanford, M.R.; Byrd, J.H.; Tomberlin, J.K.; Wallace, J.R. Entomological evidence collection methods: American Board of Forensic Entomology approved protocols. In *Forensic Entomology Utility of Arthropods in Legal Investigations*; Byrd, J.H., Tomberlin, J.K., Eds.; CRC Press: Boca Raton, FL, USA, 2019; pp. 63–86.

9. Lord, W.D.; Burger, J.F. Collection and preservation of forensically important entomological materials. *J. Forensic Sci.* **1983**, *28*, 936–944. [CrossRef]

10. Amendt, J.; Campobasso, C.P.; Gaudry, E.; Reiter, C.; LeBlanc, H.N.; Hall, M.J. Best practice in forensic entomology—Standards and guidelines. *Int. J. Leg. Med.* **2007**, *121*, 90–104. [CrossRef]

11. Bajerlein, D.; Taberski, D.; Matuszewski, S. Estimation of postmortem interval (PMI) based on empty puparia of *Phormia regina* (Meigen) (Diptera: Calliphoridae) and third larval stage of *Necrodes littoralis* (L.) (Coleoptera: Silphidae)—Advantages of using different PMI indicators. *J. Forensic Leg. Med.* **2018**, *55*, 95–98. [CrossRef]

12. Amendt, J.; Richards, C.S.; Campobasso, C.P.; Zehner, R.; Hall, M.J. Forensic entomology: Applications and limitations. *Forensic Sci. Med. Pathol.* **2011**, *7*, 379–392. [CrossRef]

13. Greenberg, B.; Kunich, J.C. *Entomology and the Law: Flies as Forensic Indicators*; Cambridge University Press: Cambridge, UK, 2002; p. 306.

14. Richards, C.S.; Paterson, I.D.; Villet, M.H. Estimating the age of immature *Chrysomya albiceps* (Diptera: Calliphoridae), correcting for temperature and geographical latitude. *Int. J. Leg. Med.* **2008**, *122*, 271–279. [CrossRef]

15. Owings, C.G.; Spiegelman, C.; Tarone, A.M.; Tomberlin, J.K. Developmental variation among *Cochliomyia macellaria* Fabricius (Diptera: Calliphoridae) populations from three ecoregions of Texas, USA. *Int. J. Leg. Med.* **2014**, *128*, 709–717. [CrossRef]

16. Midgley, J.M.; Villet, M.H. Effect of the killing method on post-mortem change in length of larvae of *Thanatophilus micans* (Fabricius 1794) (Coleoptera: Silphidae) stored in 70% ethanol. *Int. J. Leg. Med.* **2009**, *123*, 103–108. [CrossRef]

17. Richards, C.S.; Villet, M.H. Factors affecting accuracy and precision of thermal summation models of insect development used to estimate post-mortem intervals. *Int. J. Leg. Med.* **2008**, *122*, 401–408. [CrossRef]

18. Richards, C.S.; Villet, M.H. Data quality in thermal summation development models for forensically important blowflies. *Med. Vet. Entomol.* **2009**, *23*, 269–276. [CrossRef]

19. Frątczak-Łagiewska, K.; Matuszewski, S. The quality of developmental reference data in forensic entomology: Detrimental effects of multiple, in vivo measurements in *Creophilus maxillosus* L.(Coleoptera: Staphylinidae). *Forensic Sci. Int.* **2019**, *298*, 316–322. [CrossRef]

20. Matuszewski, S.; Bajerlein, D.; Konwerski, S.; Szpila, K. Insect succession and carrion decomposition in selected forests of Central Europe. Part 2: Composition and residency patterns of carrion fauna. *Forensic Sci. Int.* **2010**, *195*, 42–51. [CrossRef]

21. Matuszewski, S.; Frątczak, K.; Konwerski, S.; Bajerlein, D.; Szpila, K.; Jarmusz, M.; Szafałowicz, M.; Grzywacz, A.; Mądra, A. Effect of body mass and clothing on carrion entomofauna. *Int. J. Leg. Med.* **2016**, *130*, 221–232. [CrossRef]

22. Grassberger, M.; Frank, C. Initial study of arthropod succession on pig carrion in a central European urban habitat. *J. Med. Entomol.* **2004**, *41*, 511–523. [CrossRef]

23. Anton, E.; Niederegger, S.; Beutel, R.G. Beetles and flies collected on pig carrion in an experimental setting in Thuringia and their forensic implications. *Med. Vet. Entomol.* **2011**, *25*, 353–364. [CrossRef]

24. Feddern, N.; Mitchell, E.A.; Amendt, J.; Szelecz, I.; Seppey, C.V. Decomposition and insect colonization patterns of pig cadavers lying on forest soil and suspended above ground. *Forensic Sci. Med. Pathol.* **2019**, *15*, 342–351. [CrossRef]

25. Jarmusz, M.; Grzywacz, A.; Bajerlein, D. A comparative study of the entomofauna (Coleoptera, Diptera) associated with hanging and ground pig carcasses in a forest habitat of Poland. *Forensic Sci. Int.* **2020**, *309*, 110212. [CrossRef] [PubMed]

26. Marchenko, M.I. Medicolegal relevance of cadaver entomofauna for the determination of the time of death. *Forensic Sci. Int.* **2001**, *120*, 89–109. [CrossRef]

27. Greenberg, B. Flies as forensic indicators. *J. Med. Entomol.* **1991**, *28*, 565–577. [CrossRef] [PubMed]

28. Anderson, G.S. Minimum and maximum development rates of some forensically important Calliphoridae (Diptera). *J. Forensic Sci.* **2000**, *45*, 824–832. [CrossRef]

29. Niederegger, S.; Pastuschek, J.; Mall, G. Preliminary studies of the influence of fluctuating temperatures on the development of various forensically relevant flies. *Forensic Sci. Int.* **2010**, *199*, 72–78. [CrossRef]

30. Martín-Vega, D.; Simonsen, T.J.; Wicklein, M.; Hall, M.J. Age estimation during the blow fly intra-puparial period: A qualitative and quantitative approach using micro-computed tomography. *Int. J. Leg. Med.* **2017**, *131*, 1429–1448. [CrossRef]

31. Kamal, A.S. Comparative study of thirteen species of sarcosaprophagous Calliphoridae and Sarcophagidae (Diptera) I. Bionomics. *Ann. Entomol. Soc. Am.* **1958**, *51*, 261–271. [CrossRef]

32. Donovan, S.; Hall, M.; Turner, B.; Moncrieff, C. Larval growth rates of the blowfly, *Calliphora vicina*, over a range of temperatures. *Med. Vet. Entomol.* **2006**, *20*, 106–114. [CrossRef]

33. Davies, L.; Ratcliffe, G. Development rates of some pre-adult stages in blowflies with reference to low temperatures. *Med. Vet. Entomol.* **1994**, *8*, 245–254. [CrossRef]

34. Kaneshrajah, G.; Turner, B. *Calliphora vicina* larvae grow at different rates on different body tissues. *Int. J. Leg. Med.* **2004**, *118*, 242–244. [CrossRef]

35. Brown, K.; Thorne, A.; Harvey, M. *Calliphora vicina* (Diptera: Calliphoridae) pupae: A timeline of external morphological development and a new age and PMI estimation tool. *Int. J. Leg. Med.* **2015**, *129*, 835–850. [CrossRef] [PubMed]

36. Davies, K.; Harvey, M.L. Internal Morphological Analysis for Age Estimation of Blow Fly Pupae (Diptera: Calliphoridae) in Postmortem Interval Estimation. *J. Forensic Sci.* **2013**, *58*, 79–84. [CrossRef] [PubMed]

37. Niederegger, S.; Wartenberg, N.; Spiess, R.; Mall, G. Influence of food substrates on the development of the blowflies *Calliphora vicina* and *Calliphora vomitoria* (Diptera, Calliphoridae). *Parasitol. Res.* **2013**, *112*, 2847–2853. [CrossRef] [PubMed]

38. Baqué, M.; Filmann, N.; Verhoff, M.A.; Amendt, J. Establishment of developmental charts for the larvae of the blow fly *Calliphora vicina* using quantile regression. *Forensic Sci. Int.* **2015**, *248*, 1–9. [CrossRef]
39. Salimi, M.; Rassi, Y.; Oshaghi, M.; Chatrabgoun, O.; Limoee, M.; Rafizadeh, S. Temperature requirements for the growth of immature stages of blowflies species, Chrysomya albiceps and *Calliphora vicina*, (Diptera: Calliphoridae) under laboratory conditions. *Egypt. J. Forensic Sci.* **2018**, *8*, 1–6. [CrossRef]
40. Defilippo, F.; Bonilauri, P.; Dottori, M. Effect of Temperature on Six Different Developmental Landmarks within the Pupal Stage of the Forensically Important Blowfly *Calliphora vicina* (Robineau-Desvoidy) (Diptera: Calliphoridae). *J. Forensic Sci.* **2013**, *58*, 1554–1557. [CrossRef]
41. Reiter, C. Zum wachstumsverhalten der maden der blauen schmeißfliege *Calliphora vicina*. *Zeitschrift für Rechtsmedizin* **1984**, *91*, 295–308. [CrossRef]
42. Hwang, C.; Turner, B. Small-scaled geographical variation in life-history traits of the blowfly *Calliphora vicina* between rural and urban populations. *Entomol. Exp. Appl.* **2009**, *132*, 218–224. [CrossRef]
43. Limsopatham, K.; Hall, M.J.; Zehner, R.; Zajac, B.K.; Verhoff, M.A.; Sontigun, N.; Sukontason, K.; Sukontason, K.L.; Amendt, J. A molecular, morphological, and physiological comparison of English and German populations of *Calliphora vicina* (Diptera: Calliphoridae). *PLoS ONE* **2018**, *13*, e0207188. [CrossRef]
44. Greenberg, B.; Tantawi, T.I. Different developmental strategies in two boreal blow flies (Diptera: Calliphoridae). *J. Med. Entomol.* **1993**, *30*, 481–484. [CrossRef]
45. Ireland, S.; Turner, B. The effects of larval crowding and food type on the size and development of the blowfly, *Calliphora vomitoria*. *Forensic Sci. Int.* **2006**, *159*, 175–181. [CrossRef]
46. Grassberger, M.; Friedrich, E.; Reiter, C. The blowfly Chrysomya albiceps (Wiedemann)(Diptera: Calliphoridae) as a new forensic indicator in Central Europe. *Int. J. Leg. Med.* **2003**, *117*, 75–81. [CrossRef]
47. Vélez, M.C.; Wolff, M. Rearing five species of Diptera (Calliphoridae) of forensic importance in Colombia in semicontrolled field conditions. *Papéis Avulsos de Zoologia* **2008**, *48*, 41–47. [CrossRef]
48. Shiravi, A.; Mostafavi, R.; Akbarzadeh, K.; Oshaghi, M.A. Temperature requirements of some common forensically important blow and flesh flies (Diptera) under laboratory conditions. *Iran. J. Arthropod-Borne Dis.* **2011**, *5*, 54.
49. Beuter, L.; Mendes, J. Development of *Chrysomya albiceps* (Wiedemann)(Diptera: Calliphoridae) in different pig tissues. *Neotrop. Entomol.* **2013**, *42*, 426–430. [CrossRef]
50. Rashed, S.S.; Yamany, A.S.; El-Basheir, Z.M.; Zaher, E.E. Influence of fluctuated room conditions on the development of the forensically important *Chrysomya albiceps* (Wiedemann)(Diptera: Calliphoridae). *J. Adv. Med. Med. Res.* **2015**, *5*, 1413–1421. [CrossRef]
51. Queiroz, M.M.d.C. Temperature requirements of Chrysomya albiceps (Wiedemann, 1819)(Diptera, Calliphoridae) under laboratory conditions. *Memórias do Instituto Oswaldo Cruz* **1996**, *91*, 785–788. [CrossRef]
52. Grassberger, M.; Reiter, C. Effect of temperature on *Lucilia sericata* (Diptera: Calliphoridae) development with special reference to the isomegalen-and isomorphen-diagram. *Forensic Sci. Int.* **2001**, *120*, 32–36. [CrossRef]
53. Tarone, A.; Picard, C.; Spiegelman, C.; Foran, D. Population and temperature effects on *Lucilia sericata* (Diptera: Calliphoridae) body size and minimum development time. *J. Med. Entomol.* **2011**, *48*, 1062–1068. [CrossRef]
54. Tarone, A.M.; Foran, D.R. Components of developmental plasticity in a Michigan population of *Lucilia sericata* (Diptera: Calliphoridae). *J. Med. Entomol.* **2006**, *43*, 1023–1033. [CrossRef] [PubMed]
55. Picard, C.J.; Deblois, K.; Tovar, F.; Bradley, J.L.; Johnston, J.S.; Tarone, A.M. Increasing precision in development-based postmortem interval estimates: What's sex got to do with it? *J. Med. Entomol.* **2013**, *50*, 425–431. [CrossRef]
56. Introna, F.; Altamura, B.M.; Dell'Erba, A.; Dattoli, V. Time since death definition by experimental reproduction of *Lucilia sericata* cycles in growth cabinet. *J. Forensic Sci.* **1989**, *34*, 478–480. [CrossRef]
57. Gallagher, M.B.; Sandhu, S.; Kimsey, R. Variation in Developmental Time for Geographically Distinct Populations of the Common Green Bottle Fly, *Lucilia sericata* (Meigen). *J. Forensic Sci.* **2010**, *55*, 438–442. [CrossRef] [PubMed]
58. Nuorteva, P. Sarcosaprophagous insects as forensic indicators. In *Forensic Medicine: A Study in Trauma Environmental Hazards*; Tedeschi, C.G., Eckert, W.G., Tedeschi, L.G., Eds.; W.B. Saunders Co.: Philadelphia, PA, USA, 1977; pp. 1072–1095.
59. Ash, N.; Greenberg, B. Developmental temperature responses of the sibling species Phaenicia sericata and Phaenicia pallescens. *Ann. Entomol. Soc. Am.* **1975**, *68*, 197–200. [CrossRef]
60. Wall, R.; French, N.; Morgan, K. Effects of temperature on the development and abundance of the sheep blowfly *Lucilia sericata*(Diptera: Calliphoridae). *Bull. Entomol. Res.* **1992**, *82*, 125–131. [CrossRef]
61. Clark, K.; Evans, L.; Wall, R. Growth rates of the blowfly, *Lucilia sericata*, on different body tissues. *Forensic Sci. Int.* **2006**, *156*, 145–149. [CrossRef]
62. Rueda, L.C.; Ortega, L.G.; Segura, N.A.; Acero, V.M.; Bello, F. *Lucilia sericata* strain from Colombia: Experimental colonization, life tables and evaluation of two artifcial diets of the blowfy *Lucilia sericata* (Meigen)(Diptera: Calliphoridae), Bogotá, Colombia Strain. *Biol. Res.* **2010**, *43*, 197–203. [CrossRef]
63. Roe, A.; Higley, L.G. Development modeling of *Lucilia sericata* (Diptera: Calliphoridae). *PeerJ* **2015**, *3*, e803. [CrossRef]
64. El-Moaty, Z.A.; Abd Elmoneim, M.K. Developmental variation of the blow fly *Lucilia sericata* (Meigen, 1826)(Diptera: Calliphoridae) by different substrate tissue types. *J. Asia-Pac. Entomol.* **2013**, *16*, 297–300. [CrossRef]

65. Karabey, T.; Sert, O. The analysis of pupal development period in *Lucilia sericata* (Diptera: Calliphoridae) forensically important insect. *Int. J. Leg. Med.* **2018**, *132*, 1185–1196. [CrossRef]

66. Aubernon, C.; Charabidzé, D.; Devigne, C.; Delannoy, Y.; Gosset, D. Experimental study of *Lucilia sericata* (Diptera Calliphoridae) larval development on rat cadavers: Effects of climate and chemical contamination. *Forensic Sci. Int.* **2015**, *253*, 125–130. [CrossRef] [PubMed]

67. Cervantès, L.; Dourel, L.; Gaudry, E.; Pasquerault, T.; Vincent, B. Effect of low temperature in the development cycle of *Lucilia sericata* (Meigen)(Diptera, Calliphoridae): Implications for the minimum postmortem interval estimation. *Forensic Sci. Res.* **2018**, *3*, 52–59. [CrossRef] [PubMed]

68. Pruna, W.; Guarderas, P.; Donoso, D.A.; Barragán, Á. Life cycle of *Lucilia sericata* (Meigen 1826) collected from Andean mountains. *Neotrop. Biodivers.* **2019**, *5*, 3–9. [CrossRef]

69. Kim, H.-C.; Kim, S.-J.; Yun, J.-E.; Jo, T.-H.; Choi, B.-R.; Park, C.-G. Development of the greenbottle blowfly, *Lucilia sericata*, under different temperatures. *Korean J. Appl. Entomol.* **2007**, *46*, 141–145. [CrossRef]

70. Wang, M.; Wang, Y.; Hu, G.; Wang, Y.; Xu, W.; Wu, M.; Wang, J. Development of *Lucilia sericata* (Diptera: Calliphoridae) Under Constant Temperatures and its Significance for the Estimation of Time of Death. *J. Med. Entomol.* **2020**, *57*, 1373–1381. [CrossRef] [PubMed]

71. Byrd, J.H.; Allen, J.C. The development of the black blow fly, Phormia regina (Meigen). *Forensic Sci. Int.* **2001**, *120*, 79–88. [CrossRef]

72. Nabity, P.; Higley, L.G.; Heng-Moss, T.M. Effects of temperature on development of *Phormia regina* (Diptera: Calliphoridae) and use of developmental data in determining time intervals in forensic entomology. *J. Med. Entomol.* **2006**, *43*, 1276–1286. [CrossRef]

73. Nunez-Vazquez, C.; Tomberlin, J.K.; Cantu-Sifuentes, M.; Garcia-Martinez, O. Laboratory development and field validation of *Phormia regina* (Diptera: Calliphoridae). *J. Med. Entomol.* **2013**, *50*, 252–260. [CrossRef]

74. Grassberger, M.; Reiter, C. Effect of temperature on development of the forensically important holarctic blow fly *Protophormia terraenovae* (Robineau-Desvoidy)(Diptera: Calliphoridae). *Forensic Sci. Int.* **2002**, *128*, 177–182. [CrossRef]

75. Warren, J.A.; Anderson, G.S. The development of *Protophormia terraenovae* (Robineau-Desvoidy) at constant temperatures and its minimum temperature threshold. *Forensic Sci. Int.* **2013**, *233*, 374–379. [CrossRef]

76. Clarkson, C.; Hobischak, N.; Anderson, G. A comparison of the development rate of *Protophormia terraenovae* (Robineau-Desvoidy) raised under constant and fluctuating temperature regimes. *Can. Soc. Forensic Sci. J.* **2004**, *37*, 95–101. [CrossRef]

77. Grassberger, M.; Reiter, C. Effect of temperature on development of *Liopygia* (= *Sarcophaga*) *argyrostoma* (Robineau-Desvoidy) (Diptera: Sarcophagidae) and its forensic implications. *J. Forensic Sci.* **2002**, *47*, 1332–1336. [CrossRef] [PubMed]

78. Sert, O.; Örsel, G.M.; Şabanoğlu, B.; Özdemir, S. A Study of the pupal developments of Sarcophaga argyrostoma (Robineau-Desvoidy, 1830). *Forensic Sci. Med. Pathol.* **2020**, *16*, 12–19. [CrossRef] [PubMed]

79. Grzywacz, A. Thermal requirements for the development of immature stages of *Fannia canicularis* (Linnaeus)(Diptera: Fanniidae). *Forensic Sci. Int.* **2019**, *297*, 16–26. [CrossRef] [PubMed]

80. Meyer, J.A.; Mullens, B.A. Development of immature *Fannia* spp. (Diptera: Muscidae) at constant laboratory temperatures. *J. Med. Entomol.* **1988**, *25*, 165–171. [CrossRef] [PubMed]

81. Gruszka, J.; Matuszewski, S. Estimation of physiological age at emergence based on traits of the forensically useful adult carrion beetle *Necrodes littoralis* L.(Silphidae). *Forensic Sci. Int.* **2020**, *314*, 110407. [CrossRef] [PubMed]

82. Novák, M.; Frątczak-Łagiewska, K.; Mądra-Bielewicz, A.; Matuszewski, S. Eye-background contrast as a quantitative marker for pupal age in a forensically important carrion beetle *Necrodes littoralis* L. (Silphidae). *Sci. Rep.* **2020**, *10*, 1–12. [CrossRef]

83. Guerroudj, F.Z.; Berchi, S. Effect of temperature on the development of carrion beetle *Silpha rugosa* (Linnaeus, 1758) (Coleoptera: Silphidae) in Algeria. *J. Entomol. Zool. Stud.* **2016**, *4*, 920–922.

84. Montoya-Molina, S.; Jakubec, P.; Qubaiová, J.; Novák, M.; Šuláková, H.; Růžička, J. Developmental Models of the Forensically Important Carrion Beetle, *Thanatophilus sinuatus* (Coleoptera: Silphidae). *J. Med. Entomol.* **2020**, in press. [CrossRef]

85. Wang, Y.; Yang, J.B.; Wang, J.F.; Li, L.L.; Wang, M.; Yang, L.J.; Tao, L.Y.; Chu, J.; Hou, Y.D. Development of the forensically important beetle *Creophilus maxillosus* (Coleoptera: Staphylinidae) at constant temperatures. *J. Med. Entomol.* **2017**, *54*, 281–289. [CrossRef]

86. Watson-Horzelski, E.J. Survival and time of development for *Creophilus maxillosus* (L.) (Coleoptera: Staphylinidae) at three constant temperatures. *Coleopt. Bull.* **2012**, *66*, 365–370. [CrossRef]

87. Frątczak-Łagiewska, K.; Grzywacz, A.; Matuszewski, S. Development and validation of forensically useful growth models for Central European population of *Creophilus maxillosus* L. (Coleoptera: Staphylinidae). *Int. J. Leg. Med.* **2020**, *134*, 1–15.

88. Matuszewski, S.; Frątczak-Łagiewska, K. Size at emergence improves accuracy of age estimates in forensically-useful beetle *Creophilus maxillosus* L. (Staphylinidae). *Sci. Rep.* **2018**, *8*, 2390. [CrossRef]

89. Frątczak-Łagiewska, K.; Matuszewski, S. Sex-specific developmental models for *Creophilus maxillosus* (L.) (Coleoptera: Staphylinidae): Searching for larger accuracy of insect age estimates. *Int. J. Leg. Med.* **2018**, *132*, 887–895. [CrossRef]

90. Martín-Vega, D.; Díaz-Aranda, L.M.; Baz, A.; Cifrián, B. Effect of temperature on the survival and development of three forensically relevant *Dermestes* species (Coleoptera: Dermestidae). *J. Med. Entomol.* **2017**, *54*, 1140–1150. [CrossRef] [PubMed]

91. Amos, T. Some laboratory observations on the rates of development, mortality and oviposition of *Dermestes frischii* (Kug.)(Col., Dermestidae). *J. Stored Prod. Res.* **1968**, *4*, 103–117. [CrossRef]

92. Lambiase, S.; Murgia, G.; Sacchi, R.; Ghitti, M.; Di Lucia, V. Effects of different temperatures on the development of *Dermestes Frischii* and *Dermestes undulatus* (Coleoptera, Dermestidae): Comparison between species. *J. Forensic Sci.* **2018**, *63*, 469–473. [CrossRef] [PubMed]

93. Coombs, C. The effect of temperature and relative humidity upon the development and fecundity of *Dermestes lardarius* L.(Coleoptera, Dermestidae). *J. Stored Prod. Res.* **1978**, *14*, 111–119. [CrossRef]

94. Fleming, D.; Jacob, T. The influence of temperature and relative humidity upon the number and duration of larval instars in *Dermestes lardarius* L.(Col., Dermestidae). *Entomol. Mon. Mag.* **1986**, *122*, 43–50.

95. Wang, Y.; Wang, M.; Hu, G.; Xu, W.; Wang, Y.; Wang, J. Temperature-dependent development of *Omosita colon* at constant temperature and its implication for PMImin estimation. *J. Forensic Leg. Med.* **2020**, *72*, 101946. [CrossRef]

96. Hu, G.; Wang, M.; Wang, Y.; Tang, H.; Chen, R.; Zhang, Y.; Zhao, Y.; Jin, J.; Wang, Y.; Wu, M. Development of *Necrobia rufipes* (De Geer, 1775)(Coleoptera: Cleridae) under Constant Temperature and Its Implication in Forensic Entomology. *Forensic Sci. Int.* **2020**, *311*, 110275. [CrossRef] [PubMed]

97. Grassberger, M.; Frank, C. Temperature-related development of the parasitoid wasp *Nasonia vitripennis* as forensic indicator. *Med. Vet. Entomol.* **2003**, *17*, 257–262. [CrossRef] [PubMed]

98. Rivers, D.B.; Losinger, M. Development of the gregarious ectoparasitoid N asonia vitripennis using five species of necrophagous flies as hosts and at various developmental temperatures. *Entomol. Exp. Appl.* **2014**, *151*, 160–169. [CrossRef]

99. Voss, S.C.; Spafford, H.; Dadour, I.R. Temperature-dependant development of *Nasonia vitripennis* on five forensically important carrion fly species. *Entomol. Exp. Appl.* **2010**, *135*, 37–47. [CrossRef]

100. Zhang, Y.; Wang, Y.; Liu, C.; Wang, J.; Hu, G.; Wang, M.; Yang, L.; Chu, J. Development of *Nasonia vitripennis* (Hymenoptera: Pteromalidae) at constant temperatures in China. *J. Med. Entomol.* **2019**, *56*, 368–377. [CrossRef]

101. da Silva Mello, R.; Aguiar-Coelho, V.M. Durations of immature stage development period of *Nasonia vitripennis* (Walker)(Hymenoptera: Pteromalidae) under laboratory conditions: Implications for forensic entomology. *Parasitol. Res.* **2009**, *104*, 411–418. [CrossRef]

102. Matuszewski, S.; Hall, M.J.; Moreau, G.; Schoenly, K.G.; Tarone, A.M.; Villet, M.H. Pigs vs people: The use of pigs as analogues for humans in forensic entomology and taphonomy research. *Int. J. Leg. Med.* **2020**, *134*, 793–810. [CrossRef]

103. Matuszewski, S. Estimating the pre-appearance interval from temperature in *Necrodes littoralis* L. (Coleoptera: Silphidae). *Forensic Sci. Int.* **2011**, *212*, 180–188. [CrossRef] [PubMed]

104. Wells, J.D. A forensic entomological analysis can yield an estimate of postmortem interval, and not just a minimum postmortem interval: An explanation and illustration using a case. *J. Forensic Sci.* **2019**, *64*, 634–637. [CrossRef]

105. Matuszewski, S. A general approach for postmortem interval based on uniformly distributed and interconnected qualitative indicators. *Int. J. Leg. Med.* **2017**, *131*, 877–884. [CrossRef]

106. Matuszewski, S.; Mądra-Bielewicz, A. Validation of temperature methods for the estimation of pre-appearance interval in carrion insects. *Forensic Sci. Med. Pathol.* **2016**, *12*, 50–57. [CrossRef] [PubMed]

107. Matuszewski, S.; Szafałowicz, M.; Grzywacz, A. Temperature-dependent appearance of forensically useful flies on carcasses. *Int. J. Leg. Med.* **2014**, *128*, 1013–1020. [CrossRef] [PubMed]

108. Archer, M. Comparative Analysis of Insect Succession Data from Victoria (Australia) Using Summary Statistics versus Preceding Mean Ambient Temperature Models. *J. Forensic Sci.* **2014**, *59*, 404–412. [CrossRef] [PubMed]

109. VanLaerhoven, S. Blind validation of postmortem interval estimates using developmental rates of blow flies. *Forensic Sci. Int.* **2008**, *180*, 76–80. [CrossRef]

110. Peters, F.T.; Drummer, O.H.; Musshoff, F. Validation of new methods. *Forensic Sci. Int.* **2007**, *165*, 216–224. [CrossRef]

111. Goff, M.L. Early postmortem changes and stages of decomposition. In *Current Concepts in Forensic Entomology*; Amendt, J., Goff, M.L., Campobasso, C.P., Grassberger, M., Eds.; Springer: Dordrecht, Germany, 2010.

112. Schoenly, K.G.; Michaud, J.P.; Moreau, G. Design and analysis of field studies in carrion ecology. In *Carrion Ecology, Evolution, and Their Applications*; Benbow, M.E., Tomberlin, J.K., Tarone, A.M., Eds.; CRC Press: Boca Raton, FL, USA, 2016; pp. 129–148.

113. Schoenly, K.G.; Haskell, N.H.; Hall, R.D.; Gbur, J.R. Comparative performance and complementarity of four sampling methods and arthropod preference tests from human and porcine remains at the Forensic Anthropology Center in Knoxville, Tennessee. *J. Med. Entomol.* **2007**, *44*, 881–894. [CrossRef]

114. Schoenly, K.G.; Haskell, N.H.; Mills, D.K.; Bieme-Ndi, C.; Larsen, K.; Lee, Y. Recreating death's acre in the school yard: using pig carcasses as model corpses to teach concepts of forensic entomology & ecological succession. *Am. Biol. Teach.* **2006**, *68*, 402–410. [CrossRef]

115. Schoenly, K.; Griest, K.; Rhine, S. An experimental field protocol for investigating the postmortem interval using multidisciplinary indicators. *J. Forensic Sci.* **1991**, *36*, 1395–1415. [CrossRef]

116. Tomberlin, J.; Byrd, J.; Wallace, J.; Benbow, M. Assessment of decomposition studies indicates need for standardized and repeatable research methods in forensic entomology. *J. Forensic Res.* **2012**, *3*, 147. [CrossRef]

117. Schoenly, K.; Goff, M.L.; Early, M. A BASIC algorithm for calculating the postmortem interval from arthropod successional data. *J. Forensic Sci.* **1992**, *37*, 808–823. [CrossRef]

118. Schoenly, K.; Goff, M.L.; Wells, J.D.; Lord, W.D. Quantifying statistical uncertainty in succession-based entomological estimates of the postmortem interval in death scene investigations: A simulation study. *Am. Entomol.* **1996**, *42*, 106–112. [CrossRef]

119. Matuszewski, S.; Bajerlein, D.; Konwerski, S.; Szpila, K. Insect succession and carrion decomposition in selected forests of Central Europe. Part 3: Succession of carrion fauna. *Forensic Sci. Int.* **2011**, *207*, 150–163. [CrossRef]

120. Bonacci, T.; Brandmayr, P.; Greco, S.; Tersaruolo, C.; Vercillo, V.; Brandmayr, T.Z.B. A preliminary investigation of insect succession on carrion in Calabria (southern Italy). *Terr. Arthropod Rev.* **2010**, *3*, 97–110.

121. Prado e Castro, C.; Serrano, A.; Martins Da Silva, P.; García, M.D. Carrion flies of forensic interest: A study of seasonal community composition and succession in Lisbon, Portugal. *Med. Vet. Entomol.* **2012**, *26*, 417–431. [CrossRef]

122. Martín-Vega, D.; Nieto, C.M.; Cifrián, B.; Baz, A.; Díaz-Aranda, L.M. Early colonisation of urban indoor carcasses by blow flies (Diptera: Calliphoridae): An experimental study from central Spain. *Forensic Sci. Int.* **2017**, *278*, 87–94. [CrossRef]

123. Matuszewski, S.; Bajerlein, D.; Konwerski, S.; Szpila, K. An initial study of insect succession and carrion decomposition in various forest habitats of Central Europe. *Forensic Sci. Int.* **2008**, *180*, 61–69. [CrossRef] [PubMed]

124. Matuszewski, S.; Szafałowicz, M. Temperature-dependent appearance of forensically useful beetles on carcasses. *Forensic Sci. Int.* **2013**, *229*, 92–99. [CrossRef] [PubMed]

125. Frątczak-Łagiewska, K.; Matuszewski, S. Resource partitioning between closely related carrion beetles: *Thanatophilus sinuatus* (F.) and *Thanatophilus rugosus* (L.)(Coleoptera: Silphidae). *Entomol. Gen.* **2018**, *37*, 143–156. [CrossRef]

126. Prado e Castro, C.; García, M.D.; Martins da Silva, P.; Faria e Silva, I.; Serrano, A. Coleoptera of forensic interest: A study of seasonal community composition and succession in Lisbon, Portugal. *Forensic Sci. Int.* **2013**, *232*, 73–83. [CrossRef] [PubMed]

127. Matuszewski, S. Estimating the Preappearance Interval from Temperature in *Creophilus maxillosus* L. (Coleoptera: Staphylinidae). *J. Forensic Sci.* **2012**, *57*, 136–145. [CrossRef] [PubMed]

128. Michaud, J.P.; Moreau, G. Predicting the visitation of carcasses by carrion-related insects under different rates of degree-day accumulation. *Forensic Sci. Int.* **2009**, *185*, 78–83. [CrossRef] [PubMed]

129. Higley, L.G.; Haskell, N.H. Insect development and forensic entomology. In *Forensic Entomology: Utility of Arthropods in Legal Investigations*; Byrd, J.H., Castner, J.L., Eds.; CRC Press: Boca Raton, FL, USA, 2001; pp. 287–302.

130. Archer, M.S. The effect of time after body discovery on the accuracy of retrospective weather station ambient temperature corrections in forensic entomology. *J. Forensic Sci.* **2004**, *49*, 553–559. [CrossRef] [PubMed]

131. Hofer, I.M.; Hart, A.J.; Martín-Vega, D.; Hall, M.J. Optimising crime scene temperature collection for forensic entomology casework. *Forensic Sci. Int.* **2017**, *270*, 129–138. [CrossRef] [PubMed]

132. Johnson, A.P.; Wallman, J.F.; Archer, M.S. Experimental and casework validation of ambient temperature corrections in forensic entomology. *J. Forensic Sci.* **2012**, *57*, 215–221. [CrossRef] [PubMed]

133. Hofer, I.M.; Hart, A.J.; Martín-Vega, D.; Hall, M.J. Estimating crime scene temperatures from nearby meteorological station data. *Forensic Sci. Int.* **2020**, *306*, 110028. [CrossRef]

134. Lutz, L.; Amendt, J. Stay cool or get hot? An applied primer for using temperature in forensic entomological case work. *Sci. Justice* **2020**, *60*, 415–422. [CrossRef] [PubMed]

135. Charabidze, D.; Hedouin, V. Temperature: The weak point of forensic entomology. *Int. J. Leg. Med.* **2019**, *133*, 633–639. [CrossRef]

136. Dourel, L.; Pasquerault, T.; Gaudry, E.; Benoît, V. Using Estimated On-Site Ambient Temperature Has Uncertain Benefit When Estimating Postmortem Interval. *Psyche* **2010**, *2010*, 1–7. [CrossRef]

137. Dadour, I.; Almanjahie, I.; Fowkes, N.; Keady, G.; Vijayan, K. Temperature variations in a parked vehicle. *Forensic Sci. Int.* **2011**, *207*, 205–211. [CrossRef]

138. Moreau, G.; Lutz, L.; Amendt, J. Honey, can you take out the garbage can? modeling weather data for cadavers found within containers. *Pure Appl. Geophys.* **2019**, 1–12, in press. [CrossRef]

139. Michalski, M.; Nadolski, J. Thermal conditions in selected urban and semi-natural habitats, important for the forensic entomology. *Forensic Sci. Int.* **2018**, *287*, 153–162. [CrossRef] [PubMed]

140. Calla, L.; Bohun, C.; LeBlanc, H. Advancing the Forensic Estimation of Time Since Death. *Pure Appl. Geophys.* **2021**, 1–11, in press.

141. Podhorna, J.; Aubernon, C.; Borkovcova, M.; Boulay, J.; Hedouin, V.; Charabidze, D. To eat or get heat: Behavioral trade-offs between thermoregulation and feeding in gregarious necrophagous larvae. *Insect Sci.* **2018**, *25*, 883–893. [CrossRef]

142. Aubernon, C.; Hedouin, V.; Charabidze, D. The maggot, the ethologist and the forensic entomologist: Sociality and thermoregulation in necrophagous larvae. *J. Adv. Res.* **2018**, *16*, 67–73. [CrossRef]

143. Aubernon, C.; Boulay, J.; Hédouin, V.; Charabidzé, D. Thermoregulation in gregarious dipteran larvae: Evidence of species-specific temperature selection. *Entomol. Exp. Appl.* **2016**, *160*, 101–108. [CrossRef]

144. Charabidze, D.; Bourel, B.; Gosset, D. Larval-mass effect: Characterisation of heat emission by necrophageous blowflies (Diptera: Calliphoridae) larval aggregates. *Forensic Sci. Int.* **2011**, *211*, 61–66. [CrossRef]

145. Gruner, S.V.; Slone, D.H.; Capinera, J.L.; Turco, M.P. Volume of larvae is the most important single predictor of mass temperatures in the forensically important calliphorid, *Chrysomya megacephala* (Diptera: Calliphoridae). *J. Med. Entomol.* **2016**, *54*, 30–34. [CrossRef] [PubMed]

146. Slone, D.H.; Gruner, S.V. Thermoregulation in larval aggregations of carrion-feeding blow flies (Diptera: Calliphoridae). *J. Med. Entomol.* **2007**, *44*, 516–523. [CrossRef] [PubMed]

147. Girard, M. *Etudes sur la Chaleur Libre Degagee par les Animaux Invertebres et Specialement les Insectes*; Victor Masson: Paris, France, 1869.

148. Turner, B.; Howard, T. Metabolic heat generation in dipteran larval aggregations: A consideration for forensic entomology. *Med. Vet. Entomol.* **1992**, *6*, 179–181. [CrossRef]

149. Rivers, D.B.; Thompson, C.; Brogan, R. Physiological trade-offs of forming maggot masses by necrophagous flies on vertebrate carrion. *Bull. Entomol. Res.* **2011**, *101*, 599–611. [CrossRef]

150. Heaton, V.; Moffatt, C.; Simmons, T. Quantifying the temperature of maggot masses and its relationship to decomposition. *J. Forensic Sci.* **2014**, *59*, 676–682. [CrossRef]

151. Johnson, A.P.; Wallman, J.F. Effect of massing on larval growth rate. *Forensic Sci. Int.* **2014**, *241*, 141–149. [CrossRef] [PubMed]

152. Kotzé, Z.; Villet, M.H.; Weldon, C.W. Heat accumulation and development rate of massed maggots of the sheep blowfly, Lucilia cuprina (Diptera: Calliphoridae). *J. Insect Physiol.* **2016**, *95*, 98–104. [CrossRef] [PubMed]

153. Matuszewski, S.; Mądra-Bielewicz, A. Heat production in a feeding matrix formed on carrion by communally breeding beetles. *Front. Zool.* **2021**, *18*, 5. [CrossRef]

154. Johnson, A.P.; Wighton, S.J.; Wallman, J.F. Tracking movement and temperature selection of larvae of two forensically important blow fly species within a "maggot mass". *J. Forensic Sci.* **2014**, *59*, 1586–1591. [CrossRef]

155. Heaton, V.; Moffatt, C.; Simmons, T. The movement of fly (Diptera) larvae within a feeding aggregation. *Can. Entomol.* **2018**, *150*, 326–333. [CrossRef]

156. Gruszka, J.; Krystkowiak-Kowalska, M.; Frątczak-Łagiewska, K.; Mądra-Bielewicz, A.; Charabidze, D.; Matuszewski, S. Patterns and mechanisms for larval aggregation in carrion beetle *Necrodes littoralis* (Coleoptera: Silphidae). *Anim. Behav.* **2020**, *162*, 1–10. [CrossRef]

157. Szpila, K. Key for identification of European and *Mediterranean blowflies* (Diptera, Calliphoridae) of medical and veterinary importance-adult flies. In *Forensic Entomology, an Introduction*; Gennard, D., Ed.; Willey-Blackwell: Chichester, UK, 2012; pp. 77–81.

158. Akbarzadeh, K.; Wallman, J.F.; Sulakova, H.; Szpila, K. Species identification of Middle Eastern blowflies (Diptera: Calliphoridae) of forensic importance. *Parasitol. Res.* **2015**, *114*, 1463–1472. [CrossRef] [PubMed]

159. Lutz, L.; Williams, K.A.; Villet, M.H.; Ekanem, M.; Szpila, K. Species identification of adult African blowflies (Diptera: Calliphoridae) of forensic importance. *Int. J. Leg. Med.* **2018**, *132*, 831–842. [CrossRef]

160. Prado e Castro, C.; Szpila, K.; Martínez-Sánchez, A.I.; Rego, C.; Silva, I.; Serrano, A.R.M.; Boieiro, M. The blowflies of the Madeira Archipelago: Species diversity, distribution and identification (Diptera, Calliphoridae sl). *ZooKeys* **2016**, *634*, 101–123. [CrossRef] [PubMed]

161. Wallman, J. A key to the adults of species of blowflies in southern Australia known or suspected to breed in carrion. *Med. Vet. Entomol.* **2001**, *15*, 433–437. [CrossRef]

162. Szpila, K. Key for the identification of third instars of European blowflies (Diptera: Calliphoridae) of forensic importance. In *Current Concepts in Forensic Entomology*; Amendt, J., Campobasso, C.P., Goff, M.L., Grassberger, M., Eds.; Springer: Dordrecht, Germany, 2010; pp. 43–56.

163. Szpila, K.; Richet, R.; Pape, T. Third instar larvae of flesh flies (Diptera: Sarcophagidae) of forensic importance—Critical review of characters and key for European species. *Parasitol. Res.* **2015**, *114*, 2279–2289. [CrossRef]

164. Grzywacz, A.; Hall, M.J.; Pape, T.; Szpila, K. Muscidae (Diptera) of forensic importance—An identification key to third instar larvae of the western Palaearctic region and a catalogue of the muscid carrion community. *Int. J. Leg. Med.* **2017**, *131*, 855–866. [CrossRef] [PubMed]

165. Giordani, G.; Grzywacz, A.; Vanin, S. Characterization and identification of puparia of *Hydrotaea* Robineau-Desvoidy, 1830 (Diptera: Muscidae) from forensic and archaeological contexts. *J. Med. Entomol.* **2019**, *56*, 45–54. [CrossRef] [PubMed]

166. Díaz-Aranda, L.M.; Martín-Vega, D.; Baz, A.; Cifrián, B. Larval identification key to necrophagous Coleoptera of medico-legal importance in the western Palaearctic. *Int. J. Leg. Med.* **2018**, *132*, 1795–1804. [CrossRef]

167. Park, S.-H.; Moon, T.-Y. Carrion Beetles (Coleoptera, Silphidae) of Potential Forensic Importance and Their Pictorial Identification Key by User-Friendly Characters in Korea. *Korean J. Leg. Med.* **2020**, *44*, 143–149. [CrossRef]

168. Novák, M.; Jakubec, P.; Qubaiová, J.; Šuláková, H.; Růžička, J. Revisited larval morphology of *Thanatophilus rugosus* (Coleoptera: Silphidae). *Int. J. Leg. Med.* **2018**, *132*, 939–954. [CrossRef] [PubMed]

169. Jakubec, P.; Novák, M.; Qubaiová, J.; Šuláková, H.; Růžička, J. Description of immature stages of *Thanatophilus sinuatus* (Coleoptera: Silphidae). *Int. J. Leg. Med.* **2019**, *133*, 1549–1565. [CrossRef] [PubMed]

170. Ma, T.; Huang, J.; Wang, J.F. Study on the pupal morphogenesis of *Chrysomya rufifacies* (Macquart) (Diptera: Calliphoridae) for postmortem interval estimation. *Forensic Sci. Int.* **2015**, *253*, 88–93. [CrossRef]

171. Flissak, J.; Moura, M. Intrapuparial development of *Sarconesia chlorogaster* (Diptera: Calliphoridae) for postmortem interval estimation (PMI). *J. Med. Entomol.* **2018**, *55*, 277–284. [CrossRef] [PubMed]

172. da Silva, S.M.; Moura, M.O. Intrapuparial development of *Hemilucilia semidiaphana* (Diptera: Calliphoridae) and its use in forensic entomology. *J. Med. Entomol.* **2019**, *56*, 1623–1635. [CrossRef] [PubMed]

173. Ramos-Pastrana, Y.; Londoño, C.A.; Wolff, M. Intra-puparial development of *Lucilia eximia* (Diptera, Calliphoridae). *Acta Amazon.* **2017**, *47*, 63–70. [CrossRef]

174. Wang, Y.; Gu, Z.-y.; Xia, S.-x.; Wang, J.-f.; Zhang, Y.-n.; Tao, L.-y. Estimating the age of *Lucilia illustris* during the intrapuparial period using two approaches: Morphological changes and differential gene expression. *Forensic Sci. Int.* **2018**, *287*, 1–11. [CrossRef] [PubMed]

175. Salazar-Souza, M.; Couri, M.S.; Aguiar, V.M. Chronology of the intrapuparial development of the blowfly *Chrysomya albiceps* (Diptera: Calliphoridae): Application in forensic entomology. *J. Med. Entomol.* **2018**, *55*, 825–832. [CrossRef] [PubMed]

176. Wydra, J.; Matuszewski, S. The optimal post-eclosion interval while estimating the post-mortem interval based on an empty puparium. *Forensic Sci. Med. Pathol.* **2020**, 1–7, in press. [CrossRef] [PubMed]

177. Frere, B.; Suchaud, F.; Bernier, G.; Cottin, F.; Vincent, B.; Dourel, L.; Lelong, A.; Arpino, P. GC-MS analysis of cuticular lipids in recent and older scavenger insect puparia. An approach to estimate the postmortem interval (PMI). *Anal. Bioanal. Chem.* **2014**, *406*, 1081–1088. [CrossRef] [PubMed]

178. Moore, H.E.; Pechal, J.L.; Benbow, M.E.; Drijfhout, F.P. The potential use of cuticular hydrocarbons and multivariate analysis to age empty puparial cases of Calliphora vicina and *Lucilia sericata*. *Sci. Rep.* **2017**, *7*, 1–11. [CrossRef]

179. Zhu, G.-H.; Jia, Z.-J.; Yu, X.-J.; Wu, K.-S.; Chen, L.-S.; Lv, J.-Y.; Benbow, M.E. Predictable weathering of puparial hydrocarbons of necrophagous flies for determining the postmortem interval: A field experiment using *Chrysomya rufifacies*. *Int. J. Leg. Med.* **2017**, *131*, 885–894. [CrossRef]

180. Wells, J.D.; LaMotte, N.L. Estimating the postmortem interval. In *Forensic Entomology: The Utility of Arthropods in Legal Investigations*, 2nd ed.; Byrd, J.H., Castner, J.L., Eds.; CRC Press: Boca Raton, FL, USA, 2010; pp. 367–388.

181. Wells, J.; LaMotte, L.R. Estimating maggot age from weight using inverse prediction. *J. Forensic Sci.* **1995**, *40*, 585–590. [CrossRef]

182. Reibe, S.; Doetinchem, P.v.; Madea, B. A new simulation-based model for calculating post-mortem intervals using developmental data for *Lucilia sericata* (Dipt.: Calliphoridae). *Parasitol. Res.* **2010**, *107*, 9–16. [CrossRef]

183. Mohr, R.M.; Tomberlin, J.K. Development and validation of a new technique for estimating a minimum postmortem interval using adult blow fly (Diptera: Calliphoridae) carcass attendance. *Int. J. Leg. Med.* **2015**, *129*, 851–859. [CrossRef] [PubMed]

184. Villet, M.H.; Richards, C.S.; Midgley, J.M. Contemporary precision, bias and accuracy of minimum post-mortem intervals estimated using development of carrion-feeding insects. In *Current Concepts in Forensic Entomology*; Amendt, J., Campobasso, C.P., Goff, M.L., Grassberger, M., Eds.; Springer: Dordrecht, Germany, 2010; pp. 109–138.

185. Kashyap, V.; Pillay, V. Efficacy of entomological method in estimation of postmortem interval: A comparative analysis. *Forensic Sci. Int.* **1989**, *40*, 245–250. [CrossRef]

186. Goff, M.L.; Odom, C.B. Forensic entomology in the Hawaiian Islands. Three case studies. *Am. J. Forensic Med. Pathol.* **1987**, *8*, 45–50. [CrossRef]

187. Lutz, L.; Amendt, J. Precocious egg development in wild *Calliphora vicina* (Diptera: Calliphoridae)—An issue of relevance in forensic entomology? *Forensic Sci. Int.* **2020**, *306*, 110075. [CrossRef]

188. Richards, C.S.; Simonsen, T.J.; Abel, R.L.; Hall, M.J.; Schwyn, D.A.; Wicklein, M. Virtual forensic entomology: Improving estimates of minimum post-mortem interval with 3D micro-computed tomography. *Forensic Sci. Int.* **2012**, *220*, 251–264. [CrossRef]

189. Richards, C.; Rowlinson, C.; Cuttiford, L.; Grimsley, R.; Hall, M.R. Decomposed liver has a significantly adverse affect on the development rate of the blowfly *Calliphora vicina*. *Int. J. Leg. Med.* **2013**, *127*, 259–262. [CrossRef]

190. Richards, C.; Rowlinson, C.; Hall, M.R. Effects of storage temperature on the change in size of *Calliphora vicina* larvae during preservation in 80% ethanol. *Int. J. Leg. Med.* **2013**, *127*, 231–241. [CrossRef]

191. Myskowiak, J.-B.; Doums, C. Effects of refrigeration on the biometry and development of *Protophormia terraenovae* (Robineau-Desvoidy)(Diptera: Calliphoridae) and its consequences in estimating post-mortem interval in forensic investigations. *Forensic Sci. Int.* **2002**, *125*, 254–261. [CrossRef]

192. Holmes, L.; Vanlaerhoven, S.; Tomberlin, J. Relative humidity effects on the life history of *Hermetia illucens* (Diptera: Stratiomyidae). *Environ. Entomol.* **2012**, *41*, 971–978. [CrossRef]

193. Bauer, A.; Bauer, A.M.; Tomberlin, J.K. Impact of diet moisture on the development of the forensically important blow fly *Cochliomyia macellaria* (Fabricius)(Diptera: Calliphoridae). *Forensic Sci. Int.* **2020**, *312*, 110333. [CrossRef] [PubMed]

194. Moffatt, C.; Heaton, V.; De Haan, D. The distribution of blow fly (Diptera: Calliphoridae) larval lengths and its implications for estimating post mortem intervals. *Int. J. Leg. Med.* **2016**, *130*, 287–297. [CrossRef]

195. Wells, J.D.; Lecheta, M.C.; Moura, M.O.; LaMotte, L.R. An evaluation of sampling methods used to produce insect growth models for postmortem interval estimation. *Int. J. Leg. Med.* **2015**, *129*, 405–410. [CrossRef]

196. Bernhardt, V.; Schomerus, C.; Verhoff, M.; Amendt, J. Of pigs and men—Comparing the development of *Calliphora vicina* (Diptera: Calliphoridae) on human and porcine tissue. *Int. J. Leg. Med.* **2017**, *131*, 847–853. [CrossRef]

197. Avila, F.W.; Goff, M.L. Arthropod succession patterns onto burnt carrion in two contrasting habitats in the Hawaiian Islands. *J. Forensic Sci.* **1998**, *43*, 581–586. [CrossRef] [PubMed]

198. Hewadikaram, K.A.; Goff, M.L. Effect of carcass size on rate of decomposition and arthropod succession patterns. *Am. J. Forensic Med. Pathol.* **1991**, *12*, 235–240. [CrossRef] [PubMed]

199. Shean, B.S.; Messinger, L.; Papworth, M. Observations of differential decomposition on sun exposed v. shaded pig carrion in coastal Washington State. *J. Forensic Sci.* **1993**, *38*, 938–949. [CrossRef] [PubMed]

200. Komar, D.; Beattie, O. Effects of Carcass Size on Decay Rates of Shade and Sun Exposed Carrion. *Can. Soc. Forensic Sci. J.* **1998**, *31*, 35–43. [CrossRef]

201. Shalaby, O.A.; deCarvalho, L.M.; Goff, M.L. Comparison of patterns of decomposition in a hanging carcass and a carcass in contact with soil in a xerophytic habitat on the Island of Oahu, Hawaii. *J. Forensic Sci.* **2000**, *45*, 1267–1273. [CrossRef]

202. VanLaerhoven, S.L.; Anderson, G.S. Insect succession on buried carrion in two biogeoclimatic zones of British Columbia. *J. Forensic Sci.* **1999**, *44*, 32–43. [CrossRef]

203. Schoenly, K.G.; Shahid, S.A.; Haskell, N.H.; Hall, R.D. Does carcass enrichment alter community structure of predaceous and parasitic arthropods? A second test of the arthropod saturation hypothesis at the Anthropology Research Facility in Knoxville, Tennessee. *J. Forensic Sci.* **2005**, *50*, 134–142. [CrossRef]

204. Shahid, S.A.; Schoenly, K.; Haskell, N.H.; Hall, R.D.; Zhang, W. Carcass enrichment does not alter decay rates or arthropod community structure: A test of the arthropod saturation hypothesis at the anthropology research facility in Knoxville, Tennessee. *J. Med. Entomol.* **2003**, *40*, 559–569. [CrossRef] [PubMed]

205. Archer, M.S. Annual variation in arrival and departure times of carrion insects at carcasses: Implications for succession studies in forensic entomology. *Aust. J. Zool.* **2004**, *51*, 569–576. [CrossRef]

206. Archer, M.S.; Elgar, M.A. Effects of decomposition on carcass attendance in a guild of carrion-breeding flies. *Med. Vet. Entomol.* **2003**, *17*, 263–271. [CrossRef]

207. Wang, J.; Li, Z.; Chen, Y.; Chen, Q.; Yin, X. The succession and development of insects on pig carcasses and their significances in estimating PMI in south China. *Forensic Sci. Int.* **2008**, *179*, 11–18. [CrossRef] [PubMed]

208. Sharanowski, B.J.; Walker, E.G.; Anderson, G.S. Insect succession and decomposition patterns on shaded and sunlit carrion in Saskatchewan in three different seasons. *Forensic Sci. Int.* **2008**, *179*, 219–240. [CrossRef]

209. McIntosh, C.S.; Dadour, I.R.; Voss, S.C. A comparison of carcass decomposition and associated insect succession onto burnt and unburnt pig carcasses. *Int. J. Leg. Med.* **2016**. [CrossRef]

210. Voss, S.C.; Cook, D.F.; Dadour, I.R. Decomposition and insect succession of clothed and unclothed carcasses in Western Australia. *Forensic Sci. Int.* **2011**, *211*, 67–75. [CrossRef] [PubMed]

211. Voss, S.C.; Spafford, H.; Dadour, I.R. Annual and seasonal patterns of insect succession on decomposing remains at two locations in Western Australia. *Forensic Sci. Int.* **2009**, *193*, 26–36. [CrossRef] [PubMed]

212. Voss, S.C.; Forbes, S.L.; Dadour, I.R. Decomposition and insect succession on cadavers inside a vehicle environment. *Forensic Sci. Med. Pathol.* **2008**, *4*, 22–32. [CrossRef]

213. Michaud, J.P.; Moreau, G. Effect of variable rates of daily sampling of fly larvae on decomposition and carrion insect community assembly: Implications for forensic entomology field study protocols. *J. Med. Entomol.* **2013**, *50*, 890–897. [CrossRef] [PubMed]

214. Michaud, J.P.; Moreau, G. A statistical approach based on accumulated degree-days to predict decomposition-related processes in forensic studies. *J. Forensic Sci.* **2011**, *56*, 229–232. [CrossRef] [PubMed]

215. Michaud, J.P.; Majka, C.G.; Prive, J.P.; Moreau, G. Natural and anthropogenic changes in the insect fauna associated with carcasses in the North American Maritime lowlands. *Forensic Sci. Int.* **2010**, *202*, 64–70. [CrossRef] [PubMed]

216. Matuszewski, S.; Bajerlein, D.; Konwerski, S.; Szpila, K. Insect succession and carrion decomposition in selected forests of Central Europe. Part 1: Pattern and rate of decomposition. *Forensic Sci. Int.* **2010**, *194*, 85–93. [CrossRef] [PubMed]

217. Reibe, S.; Madea, B. How promptly do blowflies colonise fresh carcasses? A study comparing indoor with outdoor locations. *Forensic Sci. Int.* **2010**, *195*, 52–57. [CrossRef]

218. Bugajski, K.N.; Seddon, C.C.; Williams, R.E. A comparison of blow fly (Diptera: Calliphoridae) and beetle (Coleoptera) activity on refrigerated only versus frozen-thawed pig carcasses in Indiana. *J. Med. Entomol.* **2011**, *48*, 1231–1235. [CrossRef]

219. Ahmad, A.; Ahmad, A.H.; Dieng, H.; Satho, T.; Ahmad, H.; Aziz, A.T.; Boots, M. Cadaver wrapping and arrival performance of adult flies in an oil palm plantation in northern peninsular Malaysia. *J. Med. Entomol.* **2011**, *48*, 1236–1246. [CrossRef]

220. Anderson, G.S. Comparison of decomposition rates and faunal colonization of carrion in indoor and outdoor environments. *J. Forensic Sci.* **2011**, *56*, 136–142. [CrossRef]

221. Kelly, J.A.; van der Linde, T.C.; Anderson, G.S. The influence of wounds, severe trauma, and clothing, on carcass decomposition and arthropod succession in South Africa. *Can. Soc. Forensic Sci. J.* **2011**, *44*, 144–157. [CrossRef]

222. Kelly, J.A.; van der Linde, T.C.; Anderson, G.S. The influence of clothing and wrapping on carcass decomposition and arthropod succession during the warmer seasons in central South Africa. *J. Forensic Sci.* **2009**, *54*, 1105–1112. [CrossRef]

223. Gunn, A.; Bird, J. The ability of the blowflies *Calliphora vomitoria* (Linnaeus), *Calliphora vicina* (Rob-Desvoidy) and *Lucilia sericata* (Meigen) (Diptera: Calliphoridae) and the muscid flies *Muscina stabulans* (Fallen) and *Muscina prolapsa* (Harris) (Diptera: Muscidae) to colonise buried remains. *Forensic Sci. Int.* **2011**, *207*, 198–204. [CrossRef]

224. Sutherland, A.; Myburgh, J.; Steyn, M.; Becker, P.J. The effect of body size on the rate of decomposition in a temperate region of South Africa. *Forensic Sci. Int.* **2013**, *231*, 257–262. [CrossRef] [PubMed]

225. Perez, A.E.; Haskell, N.H.; Wells, J.D. Commonly Used Intercarcass Distances Appear to Be Sufficient to Ensure Independence of Carrion Insect Succession Pattern. *Ann. Entomol. Soc. Am.* **2016**, *109*, 72–80. [CrossRef]

226. Perez, A.E.; Haskell, N.H.; Wells, J.D. Evaluating the utility of hexapod species for calculating a confidence interval about a succession based postmortem interval estimate. *Forensic Sci. Int.* **2014**, *241*, 91–95. [CrossRef] [PubMed]

227. Matuszewski, S.; Konwerski, S.; Fratczak, K.; Szafalowicz, M. Effect of body mass and clothing on decomposition of pig carcasses. *Int. J. Leg. Med.* **2014**, *128*, 1039–1048. [CrossRef] [PubMed]

228. Cammack, J.A.; Cohen, A.C.; Kreitlow, K.L.; Roe, R.M.; Watson, D.W. Decomposition of Concealed and Exposed Porcine Remains in the North Carolina Piedmont. *J. Med. Entomol.* **2016**, *53*, 67–75. [CrossRef]

229. Charabidze, D.; Hedouin, V.; Gosset, D. An experimental investigation into the colonization of concealed cadavers by necrophagous blowflies. *J. Insect Sci.* **2015**, *15*, 149. [CrossRef]

230. Matuszewski, S.; Mądra, A. Factors affecting quality of temperature models for the pre-appearance interval of forensically useful insects. *Forensic Sci. Int.* **2015**, *247*, 28–35. [CrossRef]

231. Wang, Y.; Wang, J.; Wang, Z.; Tao, L. Insect succession on pig carcasses using different exposure time-A preliminary study in Guangzhou, China. *J. Forensic Leg. Med.* **2017**, *52*, 24–29. [CrossRef]

232. Wang, Y.; Ma, M.Y.; Jiang, X.Y.; Wang, J.F.; Li, L.L.; Yin, X.J.; Wang, M.; Lai, Y.; Tao, L.Y. Insect succession on remains of human and animals in Shenzhen, China. *Forensic Sci. Int.* **2017**, *271*, 75–86. [CrossRef]

233. Mądra-Bielewicz, A.; Frątczak-Łagiewska, K.; Matuszewski, S. Sex-and Size-Related Patterns of Carrion Visitation in *Necrodes littoralis* (Coleoptera: Silphidae) and *Creophilus maxillosus* (Coleoptera: Staphylinidae). *J. Forensic Sci.* **2017**, *62*, 1229–1233. [CrossRef]

234. Cruise, A.; Hatano, E.; Watson, D.W.; Schal, C. Comparison of techniques for sampling adult necrophilous insects from pig carcasses. *J. Med. Entomol.* **2018**, *55*, 947–954. [CrossRef] [PubMed]

235. Knobel, Z.; Ueland, M.; Nizio, K.D.; Patel, D.; Forbes, S.L. A comparison of human and pig decomposition rates and odour profiles in an Australian environment. *Aust. J. Forensic Sci.* **2019**, *51*, 557–572. [CrossRef]

236. Steadman, D.W.; Dautartas, A.; Kenyhercz, M.W.; Jantz, L.M.; Mundorff, A.; Vidoli, G.M. Differential Scavenging among Pig, Rabbit, and Human Subjects. *J. Forensic Sci.* **2018**. [CrossRef]

237. Dautartas, A.; Kenyhercz, M.W.; Vidoli, G.M.; Meadows Jantz, L.; Mundorff, A.; Steadman, D.W. Differential Decomposition Among Pig, Rabbit, and Human Remains. *J. Forensic Sci.* **2018**. [CrossRef]

238. Connor, M.; Baigent, C.; Hansen, E.S. Testing the Use of Pigs as Human Proxies in Decomposition Studies. *J. Forensic Sci.* **2017**. [CrossRef]

239. Dawson, B.M.; Barton, P.S.; Wallman, J.F. Contrasting insect activity and decomposition of pigs and humans in an Australian environment: A preliminary study. *Forensic Sci. Int.* **2020**, *316*, 110515. [CrossRef] [PubMed]

240. Jarmusz, M.; Bajerlein, D. Decomposition of hanging pig carcasses in a forest habitat of Poland. *Forensic Sci. Int.* **2019**, *300*, 32–42. [CrossRef]

241. Bourel, B.; Callet, B.; Hedouin, V.; Gosset, D. Flies eggs: A new method for the estimation of short-term post-mortem interval? *Forensic Sci. Int.* **2003**, *135*, 27–34. [CrossRef]

242. Faris, A.; West, W.; Tomberlin, J.; Tarone, A. Field validation of a development data set for *Cochliomyia macellaria* (Diptera: Calliphoridae): Estimating insect age based on development stage. *J. Med. Entomol.* **2020**, *57*, 39–49. [CrossRef]

243. Faris, A.; Wang, H.-H.; Tarone, A.; Grant, W. Forensic entomology: Evaluating uncertainty associated with postmortem interval (PMI) estimates with ecological models. *J. Med. Entomol.* **2016**, *53*, 1117–1130. [CrossRef] [PubMed]

244. Harnden, L.M.; Tomberlin, J.K. Effects of temperature and diet on black soldier fly, *Hermetia illucens* (L.)(Diptera: Stratiomyidae), development. *Forensic Sci. Int.* **2016**, *266*, 109–116. [CrossRef] [PubMed]

245. Voss, S.C.; Magni, P.; Dadour, I.; Nansen, C. Reflectance-based determination of age and species of blowfly puparia. *Int. J. Leg. Med.* **2017**, *131*, 263–274. [CrossRef] [PubMed]

246. Tarone, A.M.; Foran, D.R. Generalized additive models and *Lucilia sericata* growth: Assessing confidence intervals and error rates in forensic entomology. *J. Forensic Sci.* **2008**, *53*, 942–948. [CrossRef] [PubMed]

247. Reibe-Pal, S.; Madea, B. Calculating time since death in a mock crime case comparing a new computational method (ExLAC) with the ADH method. *Forensic Sci. Int.* **2015**, *248*, 78–81. [CrossRef]

248. Weatherbee, C.R.; Pechal, J.L.; Stamper, T.; Benbow, M.E. Post-Colonization Interval Estimates Using Multi-Species Calliphoridae Larval Masses and Spatially Distinct Temperature Data Sets: A Case Study. *Insects* **2017**, *8*, 40. [CrossRef]

249. Pittner, S.; Bugelli, V.; Weitgasser, K.; Zissler, A.; Sanit, S.; Lutz, L.; Monticelli, F.; Campobasso, C.P.; Steinbacher, P.; Amendt, J. A field study to evaluate PMI estimation methods for advanced decomposition stages. *Int. J. Leg. Med.* **2020**, *134*, 1361–1373. [CrossRef]

250. Turner, B.; Wiltshire, P. Experimental validation of forensic evidence: A study of the decomposition of buried pigs in a heavy clay soil. *Forensic Sci. Int.* **1999**, *101*, 113–122. [CrossRef]

251. Bhadra, P.; Hart, A.; Hall, M. Factors affecting accessibility to blowflies of bodies disposed in suitcases. *Forensic Sci. Int.* **2014**, *239*, 62–72. [CrossRef]

252. Goff, M.L.; Omori, A.I.; Gunatilake, K. Estimation of postmortem interval by arthropod succession. Three case studies from the Hawaiian Islands. *Am. J. Forensic Med. Pathol.* **1988**, *9*, 220–225. [CrossRef]

253. Pohjoismäki, J.L.; Karhunen, P.J.; Goebeler, S.; Saukko, P.; Sääksjärvi, I.E. Indoors forensic entomology: Colonization of human remains in closed environments by specific species of sarcosaprophagous flies. *Forensic Sci. Int.* **2010**, *199*, 38–42. [CrossRef]

254. Bugelli, V.; Forni, D.; Bassi, L.A.; Di Paolo, M.; Marra, D.; Lenzi, S.; Toni, C.; Giusiani, M.; Domenici, R.; Gherardi, M. Forensic entomology and the estimation of the minimum time since death in indoor cases. *J. Forensic Sci.* **2015**, *60*, 525–531. [CrossRef] [PubMed]

255. Goff, M.L.; Flynn, M.M. Determination of postmortem interval by arthropod succession: A case study from the Hawaiian Islands. *J. Forensic Sci.* **1991**, *36*, 607–614. [CrossRef] [PubMed]

256. Goff, M.; Odom, C.; Early, M. Estimation of post-mortem interval by entomological techniques: A case study from Oahu, Hawaii. *Bull. Soc. Vector Ecol.* **1986**, *11*, 242–246.

Article

To Be There or Not to Be There, That Is the Question—On the Problem of Delayed Sampling of Entomological Evidence

Lena Lutz *, Marcel A. Verhoff and Jens Amendt

Institute of Legal Medicine, University Hospital Frankfurt, Goethe-University, Kennedyallee 104, D-60596 Frankfurt am Main, Germany; verhoff@med.uni-frankfurt.de (M.A.V.); amendt@em.uni-frankfurt.de (J.A.)
* Correspondence: lutz@med.uni-frankfurt.de; Tel.: +49-69-63017436

Simple Summary: Proper evidence sampling is at the heart of a sound forensic opinion and failure to follow the standards and guidelines can have serious consequences for the report and expert testimony in court. In casework, forensic entomologists often must base their expert opinion on information about the case and insect evidence provided by third parties, and this presents pitfalls. We analyzed two of those: delayed evidence sampling and the effect of low-temperature storage of the body prior to the autopsy. Our study shows that sampling at the scene is advisable to facilitate a sound entomological report and that the cooling sequence of a corpse must be completely tracked between its removal from the scene until the insect sampling.

Abstract: The aim of the current study was to analyze two major pitfalls in forensic entomological casework: delayed evidence sampling and the effect of low-temperature storage of the body. For this purpose, temperature profiles of heavily infested corpses during cooling and cases in which insect evidence was collected both at the scene and during autopsy were evaluated with regard to species composition and development stages found. The results show that the temperature in the body bags remained at higher average temperatures up to 10 °C relative to the mortuary cooler, therefore, sufficient for larval development, with significant differences in temperature between larval aggregations on one and the same body. In addition, we found large differences both in species number, species composition, and the developmental stages found at the scene and during the autopsy. These data and observations underscore the importance of sampling evidence at the scene and recording temperatures throughout the cooling period of a body.

Keywords: crime scene; autopsy; cooling period; entomological evidence; expertise

Citation: Lutz, L.; Verhoff, M.A.; Amendt, J. To Be There or Not to Be There, That Is the Question—On the Problem of Delayed Sampling of Entomological Evidence. *Insects* **2021**, *12*, 148. https://doi.org/10.3390/insects12020148

Academic Editors: Brian T. Forschler, Damien Charabidze and Daniel Martín-Vega

Received: 16 December 2020
Accepted: 6 February 2021
Published: 9 February 2021

Publisher's Note: MDPI stays neutral with regard to jurisdictional claims in published maps and institutional affiliations.

1. Introduction

Forensic entomology i.e., the use of insect evidence in legal investigations [1], has become one of the most accurate and precise methods to establish the minimum post-mortem interval (PMI_{min}), i.e., the time since the first insect colonization on a body, in the later stages of decomposition [2–6]. In addition to the worldwide scientific development in this field, with an average of approximate 100 publications per year since 2013 (Web of Science, 12 August 2020), forensic entomology is now more recognized in forensic casework by law enforcement: entomological reports are now an integral part of court proceedings in Europe [7–11], North [12] and South America [13], Asia [14–17], Africa [18], Australia and New Zealand [19], and the Middle East [20]. To present high quality entomological findings in the court, various standards and guidelines for sampling, analyzing, and reporting entomological evidence have been published in recent years [21–23]. These "rules" are usually easy to follow when a forensic entomologist is involved in a crime investigation from the beginning, but the reality in casework is different and it is sometimes far from "best practice" [24–26]. A forensic entomologist often must write a report based on the information provided to them regarding the case (photographs, videos, police reports) rather than first-hand experience, and after examination of insect evidence sampled by third parties, such as death investigators, medical examiners, or non-medical professionals. This could lead to a number of problems in the collection of insect evidence, e.g., missing the oldest developmental stage [24], contamination [27], or incorrect sample handling [28,29]. One of the biggest problems is that evidence is frequently not sampled at the scene but during the autopsy [24,28,30–34], often several days after the discovery of the body [31,32,35,36]. An analysis of 127 cases from the Harris County Institute of Forensic Sciences found that autopsy only sampling was performed in 42% of the cases while sampling at the scene and during the autopsy occurred only in 2% [24]. To our knowledge, no publications analyze how much the evidence collection differs between the scene and the autopsy on one and the same case. Sampling only at the autopsy may lead to a delay of several days before evidence collection, which could affect not only the composition of the entomological findings, but their stage of development. The impetus for this study was a 2018 death investigation that highlighted the importance to PMI_{min} estimates of who takes entomological samples and when.

Case Study

On 14 November 2018, the body of a 49-year-old man was found in an apartment after the janitor of the building complex reported a bad smell from his apartment. The body was in an advanced stage of decomposition with larval infestation, in supine position on the floor wearing just a T-shirt. In the apartment, paraphernalia (e.g., syringe and spoon) for drug use were found. The flat was in a poor state of cleanliness, the window was in a tilted, slightly open position and the heating was on. In addition to a medical examiner, a forensic entomologist was at the scene for sampling insect evidence. On the body itself, third instar larvae of blow flies (Diptera: Calliphoridae) were found and fresh, light colored pupae underneath a carpet close to the body. The samples were transferred on the same day to the insect laboratory. Half of the larval sample was killed with almost boiling water and then stored at 96% ethanol for length measurement and species identification, while the remaining larvae plus the sampled pupae were incubated at 25 °C until the emergence of the adult flies. Meanwhile, the body was stored in a plastic body bag in a cooler at 4 ± 2 °C.

On 19 November, i.e., five days after discovery of the body, an autopsy was performed and insect evidence (third instar larvae of blow flies), was collected. The insect samples from the autopsy were handled in the same way as those from the scene. The cause of death could not be determined in the autopsy due to the advanced decomposition of the body, but toxicological analyses showed an intoxication. The entomofauna of the body included larvae and pupae of *Calliphora vicina* and larvae of *Calliphora vomitoria*, *Lucilia ampullacea*, and *L. sericata*. *Calliphora vicina* was the numerically dominant species, accounting for 90% of the specimens. The species composition was almost identical at the scene and autopsy, except for piophilid larvae which were found on the body only during the autopsy. However, the crucial difference between the insect evidence from the scene and autopsy in relation to establishing a PMI_{min} was provided by the pupae from the scene. Due to the lack of sound temperature measurements at the scene, two different temperatures (20 °C and 25 °C) were used as the basis for the development of the most important species here, *C. vicina*, which were intended to reflect a range of possible room temperatures in the apartment. The PMI_{min} for the scene data was estimated to be 8–10 days, while using the data from the autopsy resulted in a PMI_{min} of 3–4 days. In addition to this large difference in the estimated PMI_{min} due to the different developmental stages, the sampled larvae also differed significantly in length (df = 64.98, $p < 0.001$), with the larvae from the scene being on average 3.3 mm longer than the larvae from the autopsy.

In addition to the incomplete and fragmentary insect evidence when obtained only during the autopsy, a second problem arises, namely the cooling time before the autopsy [34,37]. As already mentioned by Charabidze and Hedouin [36], temperature is still a weak point in forensic entomology due to numerous factors that influence the temperature the larvae are exposed to during their growth. Many studies have tried to establish the most accurate guidelines possible for temperature reconstruction and estimation [36,38–45], but all of them concentrate only on the temperature history prior to the discovery of the body. The time between the removal from the death scene and the autopsy is still a "gap of knowledge" when it comes to accurate temperature estimation and consideration of this cooling period on the development of forensically important species. Although there have been many studies on the influence of refrigeration on the development of necrophagous insects [28,37,46–49], data on the temperatures of heavily infested bodies inside body bags during the storage prior to an autopsy are still scarce, with only one study published [34].

The current study highlights two of the biggest problems in forensic entomological casework, i.e., evidence sampling and body storage temperature. Firstly, we analyzed the temperature profiles of heavily infested bodies stored in body bags in a walk-in cooler to describe the effect of cooling on the temperature inside the bags, i.e., on the temperature to which the larvae were exposed. Secondly, we analyzed the effect of who collects the insect evidence and when, on evidence composition (species and developmental stages), by examining cases in which insect evidence was collected both at the scene and the autopsy. Overall, common pitfalls are presented with data from everyday casework, which nourish and support the need for laboratory studies, along with guidelines for dealing with these pitfalls.

2. Materials and Methods

2.1. Body Cooling

We evaluated the temperature profiles of eight bodies during the summer months from May until August 2017. After the discovery of the bodies and a first examination at the scene, they were placed in white plastic body bags and transported to the Institute of Legal Medicine Frankfurt. All bodies were heavily infested with insects and stored in a walk-in cooler after their arrival at the institute. The walk-in cooler was set to a baseline temperature of 6 ± 2 °C. The day and time of delivery of the body to the institute and the collection of the body by the mortician, i.e., the beginning and end of cooling, were noted. As soon as possible after arrival, prior to the temperature measurements, the bodies were inspected, i.e., pictures were taken, the number of maggot masses was noted, and the insect evidence was sampled. Since the examination of the bodies had to be included in the routine activities of the institute, the evaluations and measurements for research purposes could not always be carried out immediately after the arrival of the bodies, but only on the next day or the Monday after the weekend. For this reason, the actual temperature measurement on the bodies may differ from the entire cooling period of the body (Table 1). For bodies that were cooled more than 6 days (n = 2), insect samples were also taken a second time during cooling to examine the effect of cold storage on larval development.

Table 1. Information of the bodies (n = 8) used for the temperature profiles in the walk-in cooler. "temperature measurement" refers to the actual duration of the measurement with the temperature logger on the bodies, while "cooling period" describes the time between the arrival of the body at the institute, i.e., beginning of cold storage, and the collection of the body by the mortician, i.e., end of cold storage. N° is the number of body and n the sample size.

N° Body	Maggot Masses [n]	Position of Temperature Measurement on the Body			Temperature Measurement [h]	Cooling Period [h]
		1	2	3		
1	1	mouth	chest	femoral	189	271
2	6	eye cavity	right flank	left flank	22	57
3	1	eye cavity	right neck	thighs	142	160
4	2	femoral	groin	mouth	267	280
5	4	chest	armpit	legs	192	213
6	2	left neck	mouth	thighs	25	48
7	2	chest	mouth	armpit	142	157
8	3	mouth	eye cavity	groin	76	187

The temperature was measured hourly with three iButtons (DS1922L-F5, Maxim Integrated, San Jose, CA, USA) placed in the body bags. If visually clearly defined maggot masses were present, the iButtons were placed directly into them, otherwise they were located at sites of moderate maggot aggregations. For this reason, the position of the iButtons differs between the examined bodies (Table 1). Before placing the data logger in the body bags, the initial temperature of the walk-in cooler close to the body was noted/confirmed. The temperature profiles of all eight bodies were visually analyzed and the temperature difference of the three body parts for each of the cases was tested for significance using a Kruskal–Wallis test ($p = 0.05$). The insect samples, i.e., third instar blow fly larvae, were killed with almost boiling water and then stored in 96% ethanol. Species identification was performed based on morphological characters with the current systematic literature [50]. In cases where samples of two events were available, i.e., after a short (14 and 40 h) and a long (160 and 280 h) cooling period, the length of 100 maggots of each event was determined using a geometric micrometer [51].

2.2. Scene vs. Autopsy

We evaluated insect-associated cases from the Institute of Legal Medicine Frankfurt in which entomological evidence was sampled at the scene and during the autopsy from November 2018 until October 2020 (n = 29). In most cases, the sampling at the scene was performed by a medical examiner as part of the postmortem inspection, and sampling during autopsy was performed by a forensic entomologist. Part of the fly larvae were killed with almost boiling water and afterwards stored in 96% ethanol, and the remaining specimens were transferred to minced meat and bred under controlled temperature in the laboratory until the adult stage. Adult specimens as well as beetle larvae were killed by freezing at $-20\,^{\circ}$C and then stored in 96% ethanol. Species identification was performed on the basis of morphological characters with the current systematic literature [50,52] and voucher specimens. Besides species identification, information on the developmental stages found at the scene and the autopsy was noted. For cases (n = 13) where a sufficient number of maggots was present, for both samples, scene and the autopsy, the larval length was measured with a geometric micrometer [51]. In addition, the date (day, month, year) of discovery and autopsy, the sex and age of the deceased, the place of discovery (indoor or outdoor), the presumed PMI, and the type of death (natural, unnatural, unclear) were noted for each case. We used a paired *t* test to examine the difference in larval length for samples from the scene and the autopsy. All data were analyzed and charted with R version 3.6.2 [53].

3. Results

3.1. Body Cooling—Temperature Profiles

All eight temperature profiles inside the body bags showed higher average temperatures (Table 2, Figure 1) of up to $\approx 10\,^{\circ}$C more, than the baseline temperature of $6\,^{\circ}$C of the walk-in cooler, independent of the duration of temperature measurement, the entire cooling period, or the position of the iButton on the body.

In most cases, there was a steep decline in temperature for the first 50 h and then the temperature fluctuated ($\pm 2\,^{\circ}$C) around a minimum (Figure 1). The temporal decrease to the minimum temperature inside the body bags took, depending on the case, from ≈ 80 h (Figure 1e,g), over ≈ 100 h (Figure 1c) up to ≈ 140 h (Figure 1d). However, in most cases the temperature did not reach a stable minimum, even after long cooling periods of 271 (Figure 1a) or 280 h (Figure 1d). Overall, the temperature profiles were highly variable over time (Figure 1). In the two cases of short cooling periods of up to 57 h the temperature decreased on average with a rate of $0.25\,^{\circ}$C per hour (Figure 1b,f).

Table 2. Temperature difference between the three positions used for temperature measurement on each body. Information on the exact position and measurement periods are given in Table 1. "Temperature" shows the mean values over the measurement period with standard deviation; "Initial temperature" is the temperature in the cooler directly before the start of the measurements in the body bags. Its variation is explained by the routine activity and different opening and closing times of the walk-in cooler in the context of the delivery and examination of other bodies.

N° Body	Initial Temperature [°C]	Temperature [°C]		
		iButton 1	iButton 2	iButton 3
1	8.5	15.4 ± 4.2	11.3 ± 1.9	11.1 ± 2.5
2	9.4	20.9 ± 2.8	13.9 ± 2.6	12.8 ± 1.6
3	11.3	9.7 ± 1.3	11.9 ± 1.3	11.4 ± 2
4	9.2	11.6 ± 1.9	9.0 ± 2.6	7.8 ± 2.3
5	7.0	8.9 ± 0.9	9.0 ± 0.9	10.4 ± 1.6
6	15.2	13.3 ± 1.8	15.4 ± 1.7	12.8 ± 1.2
7	-	11.9 ± 1.9	9.9 ± 2.2	11.58 ± 2.0
8	8.3	12.4 ± 0.9	9.9 ± 0.8	9.3 ± 0.9

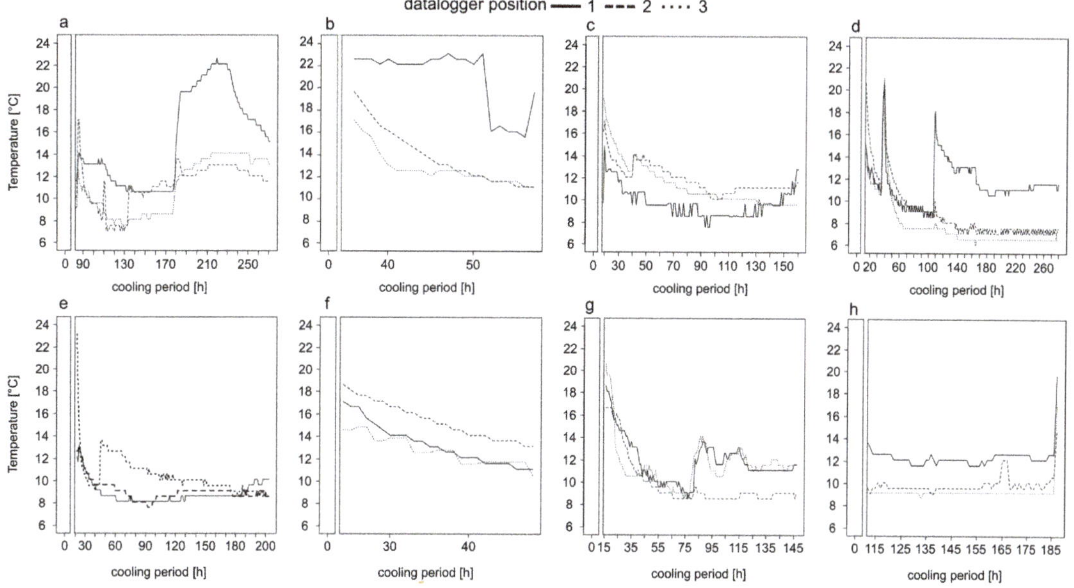

Figure 1. Temperature profiles of bodies heavily infested with insect larvae stored in a walk-in cooler in the Institute of Legal Medicine Frankfurt from May until August 2017. Information on the exact position of the iButton on the body, the period of temperature measurement and the entire cooling period are given in Table 1. The temperature profiles belong to corpse 1 (**a**), corpse 2 (**b**), corpse 3 (**c**), corpse 4 (**d**), corpse 5 (**e**), corpse 6 (**f**), corpse 7 (**g**) and corpse 8 (**h**).

3.2. Body Cooling—Temperature Difference of Body Positions

In addition to the high variability of the temperature profiles within every single profile, there were large differences between the temperature profiles of the recorded positions on one and the same body (Figures 1 and 2, Table 2). In all cases, at least two of the three recorded positions were significantly different to each other ($p < 0.001$), which in extreme cases resulted in the mean temperature at one recorded position being 8.1 °C or 7 °C warmer than at the other positions. In 50% of all cases (Figure 2c–f) the temperature differed significantly ($p < 0.001$) between all positions. There was no clear trend that a certain position of a body, i.e., eye or oral cavity, always has the highest temperature.

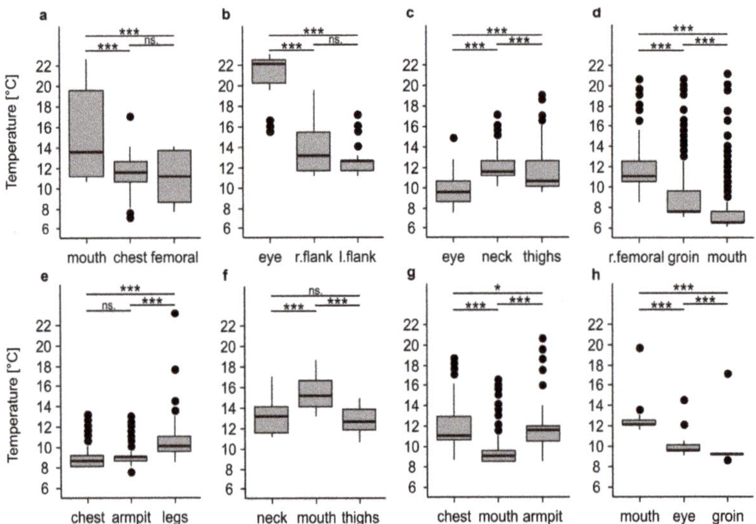

Figure 2. Temperature differences between the maggot masses on bodies stored in a walk-in cooler. Information on the period of temperature measurement and the entire cooling period are given in Table 1. The temperature profiles belong to corpse 1 (**a**), corpse 2 (**b**), corpse 3 (**c**), corpse 4 (**d**), corpse 5 (**e**), corpse 6 (**f**), corpse 7 (**g**), and corpse 8 (**h**). The asterisks representing *p*-values with *** <0.001, ** <0.01, * <0.05, *n.s* not significant.

3.3. Body Cooling—Temperature Difference of a Body with and without Larvae

For one of the corpses (corpse 4) we had information on the temperature in the body bag prior to the autopsy and even after the autopsy. The comparison of the temperature after 24 h of cooling with maggot masses on the body (left side Figure 3) showed that the decrease over time happens quite slowly with a rate of 0.12 °C per hour and results in a high minimum temperature of 11.07 °C, which is ≈5 °C above the cooler temperature.

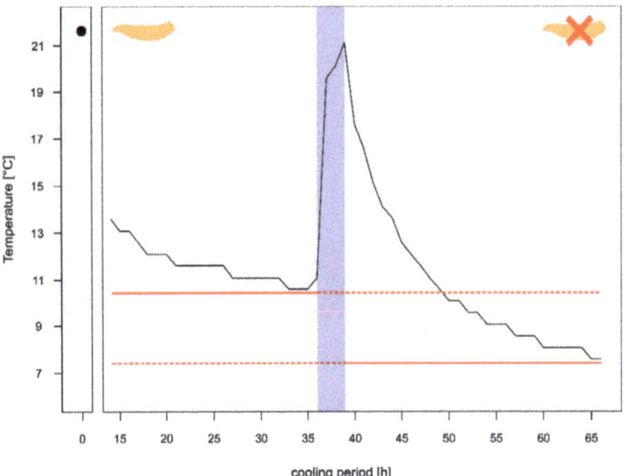

Figure 3. Temperature profile of a body prior to autopsy (with heavy larval infestation and several larval aggregations) and after the autopsy (almost no larvae left). The red solid line shows minimum temperature of the body prior autopsy and after the autopsy, respectively. The blue box represents the temperature measurements during the autopsy. The maggot represents the body with heavy larval infestation and the maggot with the red X represents the body after the removal of larvae.

Whereas after the autopsy, when the body was washed and most larvae were removed, the decline in temperature was faster with 0.41 °C per hour resulting in a minimum temperature of 7.56 °C after 24 h (right side Figure 3), being just 1 °C warmer than the overall minimum temperature during the entire cooling.

3.4. Body Cooling—Entomological Evidence and Larval Length

Overall, seven blow flies: *C. vicina*, *C. vomitoria*, *Ch. albiceps*, *L. ampullacea*, *L. sericata*, *P. regina*, *P. terraenovae*, and one flesh fly namely *S. argyrostoma* were identified (Table S1). The most common species was *L. sericata*, being present on each of the eight bodies. All measured temperatures (Table 2) were sufficient for larval development of *C. vicina*, *C. vomitoria*, *P. terraenovae* and *L. sericata* inside the body bags, assuming a lower threshold of 2 °C, 3 °C, 7.8 °C and 9 °C respectively [54].

In two cases, we were able to compare the larval lengths after both a short and a long cooling period. All larvae belonged to *L. sericata*. The comparison showed that in one case the specimens underwent significant growth during the cooling (Figure S1a). The larvae with a short cooling period of 40 h were on average 2.52 mm shorter than after 160 h of cooling (df = 191.77, $p < 0.005$). In the second case we found a significant ($p < 0.001$) decrease in larval length (Figure S1b). The larvae were on average 1.46 mm shorter after 280 h of cooling than after 14 h.

3.5. Scene vs. Autopsy

In total, we investigated 29 cases from November 2018 to October 2020. Of the 29 cases, 72.4% (n = 21) were male and 27.6% (n = 8) female with an average age of 66.4 years. All bodies were found indoor and the majority (75.9 %, n = 22) were found during the summer months from April until September whereas 24.1% (n = 7) were found from October until February. The suspected PMI averaged 17 days, with two cases showing a rather long PMI ranging from three months (case 26) to two years (case 13). An autopsy was performed on average 6.2 days after the bodies were found. In more than 50% (n = 17) of the cases the autopsy could not find a clear cause of death due to the advanced decomposition of the body. In 37.9% (n = 11) of the cases the person died of natural causes (e.g., heart failure, internal causes) and in one case the person died of unnatural causes by intoxication.

3.6. Scene vs. Autopsy—Entomological Evidence

Overall, we found a large difference both in species number, species composition, and the developmental stages found at the scene and during the autopsy (Table 3). Only in 6.9% (n = 2; case 8, 14) of all cases exactly the same species were found. In 93.1%, the species composition was very different from each other, in an extreme case (case 15) even with a difference of up to six species between the entomological evidence taken at the scene and during autopsy. There was a general trend that more species were found during the autopsy. The development stages found, were identical between the scene and autopsy in more than 50% of the cases, but in almost one third of these cases the PMI was around 7–10 days, so that probably only one developmental stage (larvae) was available. In 48.3% of the cases, we saw differences in the developmental stages found, with a slight trend that older developmental stages such as pupae or puparia were found at the scene but not during the autopsy. Considering both the species found and the stages of development together, there were differences in 96.5% (n = 28) of all cases resulting in just one case (case 14) where the entomological evidence (insect species, developmental stage) was exactly the same both at the scene and at the autopsy.

Table 3. Summary of cases where entomological samples were taken at the scene and during autopsy from November 2018 to October 2020. Information on the number of insect species is given. In brackets species unique to the scene or autopsy are listed. The stages of development found, where A= adult specimen, PR = puparia, P = pupae, L = larvae, are listed. In bold are all cases where the larval length was measured. The check mark indicates if a developmental stage was found and the dark gray filling if it was missing.

Scene						Autopsy				
A	PR	P	L	Species [n]	Case	Species [n]	L	P	PR	A
		✓	✓	4 (0)	**1**	5 (1)	✓			
	✓		✓	4 (1)	2	3 (0)	✓	✓	✓	
		✓	✓	2 (1)	**3**	3 (1)	✓	✓	✓	
		✓	✓	5 (0)	4	5 (0)	✓			
	✓	✓	✓	3 (1)	**5**	5 (3)	✓			
	✓		✓	1 (0)	6	2 (1)	✓	✓	✓	
			✓	1 (0)	7	3 (2)	✓	✓		
	✓	✓	✓	1 (0)	**8**	1 (0)	✓	✓		
✓	✓			2 (1)	9	1 (0)				✓
		✓	✓	3 (0)	10	4 (1)	✓	✓		
		✓	✓	3 (1)	11	2 (0)	✓			
	✓	✓	✓	3 (0)	12	6 (3)	✓	✓	✓	
		✓	✓	2 (0)	13	3 (1)	✓	✓		
			✓	1 (0)	**14**	1 (0)	✓			
		✓	✓	2 (0)	15	8 (6)	✓	✓	✓	
		✓	✓	5 (3)	**16**	2 (0)	✓			
		✓	✓	4 (0)	17	6 (2)	✓	✓		
		✓	✓	3 (1)	**18**	4 (2)	✓	✓	✓	
		✓	✓	2 (0)	**19**	4 (2)	✓	✓		
			✓	2 (1)	20	2 (1)	✓			
			✓	1 (0)	**21**	3 (2)	✓			
			✓	1 (0)	22	3 (2)	✓			
		✓	✓	2 (0)	**23**	3 (1)	✓	✓		
			✓	3 (0)	**24**	5 (2)	✓			
			✓	3 (1)	25	2 (2)	✓			
		✓	✓	4 (1)	26	5 (2)	✓	✓		
		✓	✓	4 (1)	**27**	4 (1)	✓	✓		
		✓	✓	5 (3)	**28**	3 (1)	✓			
			✓	2 (0)	29	3 (1)	✓			

3.7. Scene vs. Autopsy—Larval Length

For 13 cases we had a sufficient number of larvae both from the scene and the autopsy to compare the larval length. In eight cases larvae of *L. sericata* were measured, in four cases, larvae of *C. vicina*, and in one case, larvae of *Ch. albiceps*.

We found significant differences in larval length in 46.1 % (n = 6) of the cases (Figure 4). In three cases (Figure 4a,c,e) the larvae underwent significant development during the time between the discovery of the body and its autopsy. In these cases, the larvae from the autopsy were significantly larger ($p < 0.01$) with a mean difference of up to 1.8 mm (Figure 4e). In the other three cases (Figure 4b,d,f) we found a significant ($p < 0.001$) decrease in larval length. The larvae from the autopsy were significantly smaller with a mean difference of up to 3.2 mm (Figure 4f). In 53.9 % (n = 7) of the cases we found no significant differences in larval length (Figure S2).

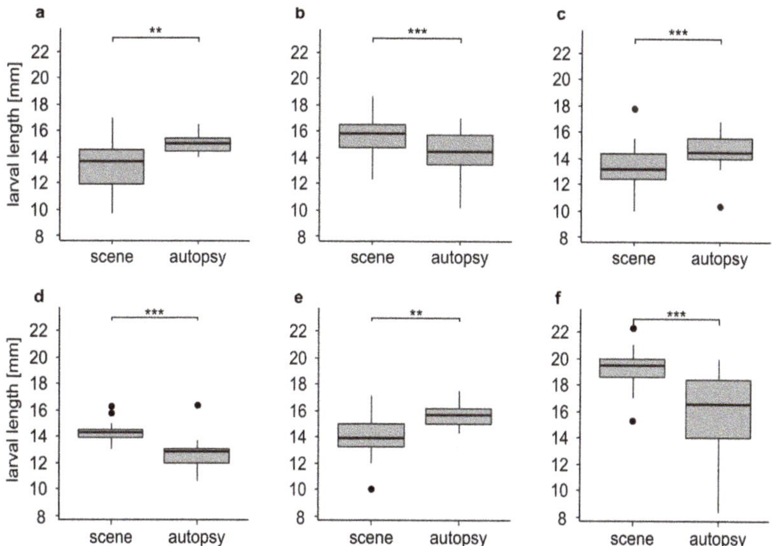

Figure 4. Larval length in mm for insect samples from the scene and the autopsy (n-values = identical for the scene and the autopsy; (**a**) larvae of *L. sericata* (n = 20) from case 3 after 0 and 8 days of cooling; (**b**) larvae of *L. sericata* (n = 60) from case 5 after 0 and 8 days of cooling; (**c**) larvae of *L. sericata* (n = 44) from case 23 after 0 and 5 days of cooling; (**d**) larvae of *L. sericata* (n = 13) from case 24 after 0 and 4 days of cooling; (**e**) larvae of *L. sericata* (n = 14) from case 27 after 0 and 8 days of cooling; (**f**) larvae of *C. vicina* (n = 50) from case 1 after 0 and 5 days of cooling. The asterisks representing *p*-values with *** <0.001, ** <0.01.

4. Discussion

This study presents real case data on two major pitfalls in forensic entomological casework: firstly, the thermal history of maggots on a body, i.e., the effects of storage in a cooler prior to the autopsy, and secondly the sampling of insect evidence with regard to the place and time of the sampling (at the scene right after the discovery of the body, or during the autopsy), as well as the training of the person performing the sampling.

4.1. Body Cooling

Knowing and understanding temperature is a key element in forensic entomological casework due to its effect on carcass decomposition, activity, oviposition, and succession of insects on a cadaver and especially on the growth of forensically important species. Therefore, the documentation and estimation of temperature insects experience during their growth is important, recommended in every standard book of forensic entomology, and crucial for an accurate PMI_{min} estimation. One temperature-relevant aspect that has received little attention so far is the storage of the insect-infested body in a cooled morgue prior to the autopsy, which is of major importance especially in cases where the entomological evidence is obtained just during the autopsy. Even in quite recent guidelines for the use of temperature in casework [36], the use of a margin error of $\pm 2\,°C$ on the refrigerator temperature is given as a solution. Our results showed that all temperature profiles inside the body bags showed average temperatures that were up to 10 °C higher than the baseline temperature of the walk-in cooler, no matter the duration of the temperature measurement, the entire cooling period or the position on the body. Even the lowest mean average temperature was 1.8 °C higher than the cooler. Therefore, adding just 2 °C to the temperature of the cooler will possibly not reflect the actual temperature the larvae experienced during growth and could lead to an erroneous PMI estimation depending on the duration of the impact. Our results are in line with Huntington et al. [34] and Thevan et al. [55] who reported a significant temperature difference of up to 8 °C between the inside of the body bag and the storage cooler's temperature.

These major temperature differences are caused by the maggot masses on the body, which have been reported to exceed the surrounding environment in the field much as 10–30 °C [56,57] and maintain temperature even under cooling [58]. This effect can be clearly seen when we compare the temperatures of one and the same body with and without maggots. The decline in temperature without maggots was much faster (0.41 °C/h) and reached a stable minimum close to the cooler temperature after 24 h, while the decline in temperature with maggots was very slow (0.12 °C/h) and reached a minimum temperature still 5 °C higher than that of the cooler. In our study, we focused not only on temperature profiles in the body bags, but also on recording the temperature inside the maggot masses during cooling, and especially on the variability on one and the same body. Our results show that in all cases, at least two of the three recorded maggot masses were significantly different to each other which, in extreme cases, resulted in the mean temperature at one recorded position being 8.1 °C or 7 °C warmer than at the other positions.

This problem is further complicated by the fact that specimens of different ages and sizes can withstand cooling quite differently [48,59], and flies may be able to handle the temperature decrease at least temporarily with their behavior or physiology [60]. Diapause, marked by the reduction of metabolic activity, is mainly regulated by the photoperiod acting on the maternal generation of blow flies and by the thermal conditions of development of their larvae. Vinogradova and Reznik [61] showed the occurrence of diapause in *C. vicina* field populations already from the middle of August, with temperatures relevant for the larvae in the low double-digit range. Hence, depending on the time of the year (day length) and the duration of the cold storage (temperature) larval diapause cannot be ruled out during storage in the morgue. This might be no serious problem in summer and especially during further breeding of living specimens after sampling at a temperature of normally >20 °C, as specimens might acclimatize quickly and terminate diapause. However, we admit that knowledge about species-specific induction and termination of diapause (and dormancy) in blow flies is still limited. Due to these kinds of impacts and variations and the fact that the actual dynamics are more complex [62], it will be difficult or even impossible to establish a serious correction factor for calculating the temperature experienced by the larvae during storage, not least as not all of them have experienced the same temperature [34]. Nevertheless, it is important to know about the temporal decline of temperature inside a body bag after the transfer to the cooler, i.e., the period larvae experience still high temperatures. The cooling rate of a body without insect infestation is, depending on the weight, 1–2 °C per hour [63]. Our results showed that the cooling rate of a heavily infested body is around 0.12 °C per hour and that it can take up to 100 h until the temperatures inside the bag reaches its minimum - even the lowest temperature inside the body bag is usually higher than the cooler temperature. Does that special kind of temperature regime have an effect on maggot development? Our results show that blow fly larvae can undergo significant development despite the cooling, resulting in a difference in length of up to 2 mm, which can lead to erroneous PMI estimations depending on species and temperature. These results are in line with other studies [34,55], according to which maggots clearly continue to develop during cooling and even complete their development. In our study, comparison of scene- and autopsy-based PMI_{min} estimations based only on larval length (results not shown) indicates that discrepancies are less than 24 h even for significantly different larval lengths and therefore likely to be negligible, as PMI_{min} is usually narrowed down to the day. However, further studies need to be conducted to quantify the effect of cooling on different species and their developmental stages.

The ongoing developmental and feeding activity of the insects along with the decay of the body is not just a problem for the forensic entomologist, but also for the medical examiner. Since a high larval activity and progressing decay can impact autopsy findings [64] due to substantial tissue loss [34]. Especially maggot masses around the head i.e., in the mouth, the eyes, and around the neck, can easily destroy signs of bruises or skin lesions that would be helpful to determine a cause of death. We have seen some cases where the corpses at the scene were just in an early stage of decomposition, with beginning larval aggregation on the head, but at the autopsy several days later all the tissue of the face was lost, and the head was almost skeletonized. We believe that these problems are more common than expected, especially in the summer and recommend that heavily infested bodies be autopsied as soon as possible.

Besides further growth, i.e., an incline in larval length during the cooling, we also saw the opposite i.e., a decline in larval length of up to 3.2 mm. This can probably be explained by the developmental stage that the specimens on the body had already reached before storage [48]. Archer et al. [28] found that late third instar larvae of *C. vicina* shrink on average of −1.2 mm during cooling, whereas second and mid third instar larvae grow overall during cooling. If the maggots on the body are already in post feeding stage then they could be preparing for pupation during cooling, which in general is associated with larval shrinkage of several mm depending on species and temperature [65]. An example of this phenomenon is provided by the case study presented in the introduction. The majority of larvae at the scene was probably just before post-feeding, resulting in the maximum length during their development. Normally, there is a decrease in larval length during post feeding until the onset of pupation. During the cold storage until the autopsy, the larvae of the cold adapted species *C. vicina* continued its development. Reiter [66] observed shrinkage of 2.9 mm at 6.5 for *C. vicina* mm during this phase in the development, which is in a similar range to our results. For casework, it is therefore necessary to know at least the prevailing development stages prior to cooling and a subsequent preservation.

4.2. Scene vs. Autopsy

Proper evidence sampling is at the heart of a sound forensic entomologist's opinion and failure to follow the protocol can have serious consequences for the report and expert testimony in court [28]. Hence, the gold standard is the sampling of entomological evidence at the scene and during the autopsy [21], and in cases of low population due to season or accessibility of the body, even only scene collection by a trained expert may be sufficient—but this will be the exception rather than the rule: many studies demonstrated that insect evidence is frequently sampled just during the autopsy [24,28,30–34] and sometimes even several days after the discovery of the body [31,32,35,36]. Our results show that such practices can lead to large differences in species diversity and the developmental stages found, leading to a biased or wrong entomological report. Missing the oldest developmental stage will lead to erroneous and especially compromised interpretation of the entomological evidence. As shown in the case study (Figure S3), the lack of pupae, i.e., the oldest developmental stage, at the autopsy, resulted in an underestimation of the PMI_{min} of up to 5 days, depending on the temperature used for the estimation. This difference will become larger the older, i.e., further developed, the missed developmental stage is. In our study, using entomological evidence collected only at autopsy would have led to an underestimation of the PMI_{min} in 24.4% (n = 7) of cases.

Additionally, our survey revealed that during the autopsy more species were found. This is probably because in our study a trained forensic entomologist took the samples during the autopsy, while a medical examiner was in charge at the scene. Reasons for poor evidence sampling can be the lack of experience of the persons in charge at the scene of death (police, crime scene technicians, medical examiner), the lack of time and, above all, the unawareness of the diversity of entomological evidence. However, to have just a limited view on the species composition on a body, due to, e.g., missing indicator species for certain stages of decomposition, seasons of the year, or for post-mortem transfer, can have a major effect on PMI estimation and case evaluation.

5. Conclusions

Forensic entomology works with live animals that interact constantly and closely with their environment and faces a diversity of species and morphological manifestations that is difficult for an entomological non-expert to handle. It therefore must deal with many pitfalls when it comes to casework and knowing them will decrease their negative impact.

Our study shows that sampling at the scene is advisable if not mandatory to facilitate a sound entomological report. Most sub-optimal sampling happens because of the lack of knowledge and experience. As a forensic entomologist will not attend every scene, educating and training the police, crime scene technicians and medical examiners is one important approach to cope this issue.

If on-site sampling has not occurred, the first samples should be taken immediately upon arrival of the body at the mortuary, even (or especially!) if the autopsy is not performed until days later. During autopsy, sampling should focus not only on the body itself but also on the body bag, the clothes and other belongings of the deceased as the fauna perhaps responded to the changed environment and temperature and left the body for shelter or preparing for pupation.

A better understanding of the influence of (rapidly and drastically) fluctuating temperatures on the growth of necrophagous insects is urgently required in forensic entomology. For the evaluation of the entomological evidence and writing the report, the temperature profile and cooling sequence of the body must be completely tracked from its removal of the scene until the insect sampling. If possible, temperature measurements in the body bags itself but more important in the larval aggregation should be carried out. In the absence of these data, the cooling and resulting slower growth should be discussed in the report, but always the fastest possible development (i.e., without cooling effects, for example) should be calculated to meet the concept of PMI_{min}.

Even though this study is not based on a designed experiment, but analyzes data from everyday casework, it is very well suited to illustrate current problems, particularly in temperature reconstruction and evidence collection. It presents detailed data that ultimately demonstrate the need to track the temperature profiles of stored bodies and sample insect evidence both at the scene and during autopsy. Most importantly, it is a clear call and provides a basis for further research focusing on the link between laboratory studies and casework.

Returning to the question of "to be there or not to be there"; it is always better for the forensic entomologist to be at the scene and sample the evidence by him or herself, but this will not be a regular option. We just need to know then what to do if we were not there, and what information is mandatory.

Supplementary Materials: The following are available online at https://www.mdpi.com/2075-4450/12/2/148/s1, Figure S1: Larval length in mm for insect samples after different cooling periods, Figure S2: Larval length in mm for insect samples from the scene and the autopsy, Figure S3: Comparison of scene-based and autopsy-based PMI_{min} with temperatures of 20 °C and 25 °C. Table S1: Insect species found on the seven bodies used for the temperature profiles.

Author Contributions: Conceptualization, J.A. and L.L.; methodology, J.A.; formal analysis, L.L.; investigation, L.L.; resources, J.A. and M.A.V.; writing—original draft preparation, L.L.; writing—review and editing, J.A., M.A.V., and L.L.; visualization, L.L.; supervision, J.A. All authors have read and agreed to the published version of the manuscript.

Funding: This research received no external funding.

Institutional Review Board Statement: Not applicable.

Informed Consent Statement: Not applicable.

Data Availability Statement: The data presented in this study are available on request from the corresponding author.

Acknowledgments: We thank L. Taube and A. Majkowski for their assistance with the temperature measurement and insect sampling during this study as well as our medical colleagues for their assistance with the evidence sampling at the scene.

Conflicts of Interest: The authors declare no conflict of interest.

References

1. Magni, P.; Guercini, S.; Leighton, A.; Dadour, I. Forensic Entomologists: An Evaluation of their Status. *J. Insect Sci.* **2013**, *13*, 78. [CrossRef]
2. Catts, E.P.; Goff, M.L. Forensic Entomology in Criminal Investigations. *Annu. Rev. Entomol.* **1992**, *37*, 253–272. [CrossRef]
3. Buchan, M.J.; Anderson, G.S. Time since death: A review of the current status of methods used in the later postmortem interval. *J. Can. Soc. Forensic Sci.* **2001**, *34*, 1–22. [CrossRef]
4. Amendt, J.; Krettek, R.; Zehner, R. Forensic entomology. *Naturwissenschaften* **2004**, *91*, 51–65. [CrossRef]
5. Greenberg, B. Flies as forensic indicators. *J. Med. Entomol.* **1991**, *28*, 565–577. [CrossRef]
6. Pittner, S.; Bugelli, V.; Weitgasser, K.; Zissler, A.; Sanit, S.; Lutz, L.; Monticelli, F.; Campobasso, C.P.; Steinbacher, P.; Amendt, J. A field study to evaluate PMI estimation methods for advanced decomposition stages. *Int. J. Leg. Med.* **2020**, *134*, 1361–1373. [CrossRef]
7. Gennard, D. *Forensic Entomology: An Introduction*; John Wiley & Sons, Ltd.: Chichester, UK, 2007; Volume 8, ISBN 9788578110796.
8. Gaudry, E.; Dourel, L. Forensic entomology: Implementing quality assurance for expertise work. *Int. J. Leg. Med.* **2013**, *127*, 1031–1037. [CrossRef] [PubMed]
9. Bourguinon, L.; Braet, Y.; Hubrecht, F.; Vanpoucke, S. Belgium, the Netherlands, and Luxembourg. In *Forensic Entomology International Dimensions and Frontiers*, 1st ed.; Tomberlin, J.K., Benbow, M.E., Eds.; CRC Press: Boca Raton, FL, USA, 2015; pp. 101–108.
10. Lindström, A. Sweden, Finland, Norway, and Denmark. In *Forensic Entomology International Dimensions and Frontiers*, 1st ed.; Tomberlin, J.K., Benbow, M.E., Eds.; CRC Press: Boca Raton, FL, USA, 2015; pp. 109–114.
11. Amendt, J.; Cherix, D.; Grassberger, M. Austria, Switzerland, and Germany. In *Forensic Entomology International Dimensions and Frontiers*, 1st ed.; Tomberlin, J.K., Benbow, M.E., Eds.; CRC Press: Boca Raton, FL, USA, 2015; pp. 127–131.
12. Faris, A.M.; Wang, H.H.; Tarone, A.M.; Grant, W.E. Forensic Entomology: Evaluating Uncertainty Associated with Postmortem Interval (PMI) Estimates with Ecological Models. *J. Med. Entomol.* **2016**, *53*, 1117–1130. [CrossRef] [PubMed]
13. Moretti, T.C.; Godoy, W.A.C. South America. In *Forensic Entomology: International Dimensions and Frontiers*, 1st ed.; Tomberlin, J.K., Benbow, M.E., Eds.; CRC Press: Boca Raton, FL, USA, 2015; pp. 175–186.
14. Wang, M.; Chu, J.; Wang, Y.; Li, F.; Liao, M.; Shi, H.; Zhang, Y.; Hu, G.; Wang, J. Forensic entomology application in China: Four case reports. *J. Forensic Leg. Med.* **2019**, *63*, 40–47. [CrossRef]
15. Wang, J. China. In *Forensic Entomology International Dimensions and Frontiers*; Tomberlin, J.K., Benbow, M.E., Eds.; CRC Press: Boca Raton, FL, USA, 2015; pp. 7–19.
16. Said, S.M.; Bashah, R.M.Z.R.K. Exploring the legal aspects and court process of forensic entomology from the Malaysia's perspective. *Serangga* **2018**, *23*, 268–278.
17. Sukontason, K.; Sukontason, K.L. Thailand. In *Forensic Entomology International Dimensions and Frontiers*, 1st ed.; Tomberlin, J.K., Benbow, M.E., Eds.; CRC Press: Boca Raton, FL, USA, 2015; pp. 37–43.
18. Williams, K.A.; Villet, M.H. A history of southern African research relevant to forensic entomology. *S. Afr. J. Sci.* **2006**, *102*, 59–65.
19. Wallman, J.F.; Archer, M.S. Australia and New Zealand. In *Forensic Entomology International Dimensions and Frontiers*, 1st ed.; Tomberlin, J.K., Benbow, M.E., Eds.; CRC Press: Boca Raton, FL, USA, 2015; pp. 53–63.
20. Hall, R.D.; Greenberg, B.; Kunich, J.C. Entomology and the Law—Flies as Forensic Indicators. *J. Med. Entomol.* **2006**. [CrossRef]
21. Amendt, J.; Campobasso, C.P.; Gaudry, E.; Reiter, C.; LeBlanc, H.N.; Hall, M.J.R. Best practice in forensic entomology—Standards and guidelines. *Int. J. Leg. Med.* **2007**, *121*, 90–104. [CrossRef] [PubMed]
22. Byrd, J.; Sutton, L. Forensic entomology for the investigator. *WIREs Forensic Sci.* **2020**, *2*, 1–14. [CrossRef]
23. Amendt, J.; Richards, C.S.; Campobasso, C.P.; Zehner, R.; Hall, M.J.R. Forensic entomology: Applications and limitations. *Forensic Sci. Med. Pathol.* **2011**, *7*, 379–392. [CrossRef] [PubMed]
24. Sanford, M.R. Forensic Entomology in the Medical Examiner's Office. *Acad. Forensic Pathol.* **2015**, *5*, 306–317. [CrossRef]
25. Campobasso, C.P.; Introna, F. The forensic entomologist in the context of the forensic pathologist's role. *Forensic Sci. Int.* **2001**, *120*, 132–139. [CrossRef]
26. Amendt, J., Anderson, G.; Campobasso, C.P.; Dadour, I.; Gaudry, E.; Hall, M.J.R.; Moretti, T.C.; Sukontason, K.L.; Villet, M.H. Chapter 29—Standard Practices. In *Forensic Entomology: International Dimensions and Frontiers*, 1st ed.; Tomberlin, J.K., Benbow, M.E., Eds.; CRC Press: Boca Raton, FL, USA, 2015; pp. 381–394.
27. Archer, M.S.; Elgar, M.A.; Briggs, C.A.; Ranson, D.L. Fly pupae and puparia as potential contaminants of forensic entomology samples from sites of body discovery. *Int. J. Leg. Med.* **2006**, *120*, 364–368. [CrossRef]
28. Archer, M.S.; Jones, S.D.; Wallman, J.F. Delayed reception of live blowfly (*Calliphora vicina* and *Chrysomya rufifacies*) larval samples: Implications for minimum postmortem interval estimates. *Forensic Sci. Res.* **2018**, *3*, 27–39. [CrossRef]
29. Bugelli, V.; Campobasso, C.P.; Verhoff, M.A.; Amendt, J. Effects of different storage and measuring methods on larval length values for the blow flies (Diptera: Calliphoridae) *Lucilia sericata* and *Calliphora vicina*. *Sci. Justice* **2017**, *57*, 159–164. [CrossRef]
30. Wells, J.; Lamotte, L. Estimating the Postmortem Interval. *Forensic Entomol.* **2009**, 367–388. [CrossRef]
31. Lindström, A.; Jaenson, T.G.T.; Lindquist, O. Forensic Entomology—First Swedish Case Studies. *J. Can. Soc. Forensic Sci.* **2003**, *36*, 207–210. [CrossRef]
32. Pohjoismäki, J.L.O.; Karhunen, P.J.; Goebeler, S.; Saukko, P.; Sääksjärvi, I.E. Indoors forensic entomology: Colonization of human remains in closed environments by specific species of sarcosaprophagous flies. *Forensic Sci. Int.* **2010**, *199*, 38–42. [CrossRef]

33. Vanin, S.; Tasinato, P.; Ducolin, G.; Terranova, C.; Zancaner, S.; Montisci, M.; Ferrara, S.D.; Turchetto, M. Use of Lucilia species for forensic investigations in Southern Europe. *Forensic Sci. Int.* **2008**, *177*, 37–41. [CrossRef]

34. Huntington, T.E.; Higley, L.G.; Baxendale, F.P. Maggot development during morgue storage and its effect on estimating the post-mortem interval. *J. Forensic Sci.* **2007**, *52*, 453–458. [CrossRef] [PubMed]

35. Baumjohann, K.; Benecke, M.; Rothschild, M.A. Schussverletzungen oder Käferfraß. *Rechtsmedizin* **2014**, *24*, 114–117. [CrossRef]

36. Charabidze, D.; Hedouin, V. Temperature: The weak point of forensic entomology. *Int. J. Leg. Med.* **2018**, 633–639. [CrossRef]

37. Hill, L.; Gilbert, A.E.; Coetzee, M. Modeling Temperature Variations Using Monte Carlo Simulation: Implications for Estimation of the Postmortem Interval Based on Insect Development Times. *J. Forensic Sci.* **2020**. [CrossRef] [PubMed]

38. Lutz, L.; Amendt, J. Stay cool or get hot? An applied primer for using temperature in forensic entomological case work. *Sci. Justice* **2020**, *60*, 415–422. [CrossRef]

39. Hofer, I.M.J.; Hart, A.J.; Martín-Vega, D.; Hall, M.J.R. Optimising crime scene temperature collection for forensic entomology casework. *Forensic Sci. Int.* **2017**, *270*, 129–138. [CrossRef] [PubMed]

40. Hofer, I.M.J.; Hart, A.J.; Martín-Vega, D.; Hall, M.J.R. Estimating crime scene temperatures from nearby meteorological station data. *Forensic Sci. Int.* **2020**, *306*. [CrossRef]

41. Archer, M.S. The effect of time after body discovery on the accuracy of retrospective weather station ambient temperature corrections in forensic entomology. *J. Forensic Sci.* **2004**, *49*, 553–559. [CrossRef] [PubMed]

42. Dabbs, G.R. How Should Forensic Anthropologists Correct National Weather Service Temperature Data for Use in Estimating the Postmortem Interval? *J. Forensic Sci.* **2015**, *60*, 581–587. [CrossRef]

43. Dourel, L.; Pasquerault, T.; Gaudry, E.; Vincent, B. Using estimated on-site ambient temperature has uncertain benefit when estimating postmortem interval. *Psyche (Stuttg.)* **2010**, *2010*. [CrossRef]

44. Johnson, A.P.; Wallman, J.F.; Archer, M.S. Experimental and Casework Validation of Ambient Temperature Corrections in Forensic Entomology. *J. Forensic Sci.* **2012**, *57*, 215–221. [CrossRef] [PubMed]

45. Moreau, G.; Lutz, L.; Amendt, J. Honey, Can You Take Out the Garbage Can? Modeling Weather Data for Cadavers Found Within Containers. *Pure Appl. Geophys.* **2019**. [CrossRef]

46. Vinogradova, E.B. Methods of short-term storage of cultures of the blowfly *Calliphora Vicina* R.-D. (Diptera, Calliphoridae). *Entomol. Rev.* **2010**, *89*, 1019–1024. [CrossRef]

47. Johl, H.K.; Anderson, G.S. Effects of refrigeration on development of the blow fly, *Calliphora vicina* (Diptera: Calliphoridae) and their relationship to time of death. *J. Entomol. Soc. Br. Columbia* **1996**, *93*, 93–98.

48. Myskowiak, J.B.; Doums, C. Effects of refrigeration on the biometry and development of *Protophormia terraenovae* (Robineau-Desvoidy) (Diptera: Calliphoridae) and its consequences in estimating post-mortem interval in forensic investigations. *Forensic Sci. Int.* **2002**, *125*, 254–261. [CrossRef]

49. Pitts, K.M.; Wall, R. Cold shock and cold tolerance in larvae and pupae of the blow fly, *Lucilia sericata. Physiol. Entomol.* **2006**, *31*, 57–62. [CrossRef]

50. Szpila, K. Key for the identification of third instars of european blowflies (Diptera: Calliphoridae) of forensic importance. In *Current Concepts in Forensic Entomology*; Springer: Dordrecht, The Netherlands, 2010; pp. 43–56. ISBN 9781402096846.

51. Villet, M.H. An inexpensive geometrical micrometer for measuring small, live insects quickly without harming them: Technical note. *Entomol. Exp. Appl.* **2007**, *122*, 279–280. [CrossRef]

52. Grzywacz, A.; Hall, M.J.R.; Pape, T.; Szpila, K. Muscidae (Diptera) of forensic importance-an identification key to third instar larvae of the western Palaearctic region and a catalogue of the muscid carrion community. *Int. J. Leg. Med.* **2017**, *131*, 855–866. [CrossRef] [PubMed]

53. Team, R.C. *R: A Language and Environment for Statistical Computing 2019*; R Foundation for Statistical Computing: Vienna, Austria, 2019; Available online: https://www.R-project.org/ (accessed on 10 August 2020).

54. Marchenko, M.I. Medicolegal relevance of cadaver entomofauna for the determination of the time of death. *Forensic Sci. Int.* **2001**, *120*, 89–109. [CrossRef]

55. Thevan, K.; Ahmad, A.H.; Che, C.S.; Singh, B. Growth of Chrysomya megacephala (Fabricius) Maggots in a Morgue Cooler. *J. Forensic Sci.* **2010**, *55*, 1656–1658. [CrossRef]

56. Turner, B.; Howard, T. Metabolic heat generation in dipteran larval aggregations: A consideration for forensic entomology. *Med. Vet. Entomol.* **1992**, *6*, 179–181. [CrossRef]

57. Rivers, D.B.; Thompson, C.; Brogan, R. Physiological trade-offs of forming maggot masses by necrophagous flies on vertebrate carrion. *Bull. Entomol. Res.* **2011**, *101*, 599–611. [CrossRef]

58. Bugelli, V.; Papi, L.; Fornaro, S.; Stefanelli, F.; Chericoni, S.; Giusiani, M.; Vanin, S.; Campobasso, C.P. Entomotoxicology in burnt bodies: A case of maternal filicide-suicide by fire. *Int. J. Leg. Med.* **2017**, *131*, 1299–1306. [CrossRef]

59. Bugelli, V.; Campobasso, C.P.; Zehner, R.; Amendt, J. How should living entomological samples be stored? *Int. J. Leg. Med.* **2019**, *133*, 1985–1994. [CrossRef]

60. Aubernon, C.; Boulay, J.; Hédouin, V.; Charabidzé, D. Thermoregulation in gregarious dipteran larvae: Evidence of species-specific temperature selection. *Entomol. Exp. Appl.* **2016**, *160*, 101–108. [CrossRef]

61. Vinogradova, E.B.; Reznik, S.Y. Induction of larval diapause in the blowfly, *Calliphora vicina* R.-D. (Diptera, Calliphoridae) under field and laboratory conditions. *Entomol. Rev.* **2013**, *93*, 935–941. [CrossRef]

62. Wu, T.H.; Shiao, S.F.; Okuyama, T. Development of insects under fluctuating temperature: A review and case study. *J. Appl. Entomol.* **2015**, *139*, 592–599. [CrossRef]
63. Wardak, K.S.; Cina, S.J. Algor Mortis: An Erroneous Measurement Following Postmortem Refrigeration. *J. Forensic Sci.* **2011**, *56*, 1219–1221. [CrossRef] [PubMed]
64. Viero, A.; Montisci, M.; Pelletti, G.; Vanin, S. Crime scene and body alterations caused by arthropods: Implications in death investigation. *Int. J. Leg. Med.* **2019**, *133*, 307–316. [CrossRef] [PubMed]
65. Grassberger, M.; Reiter, C. Effect of temperature on *Lucilia sericata* (Diptera: Calliphoridae) development with special reference to the isomegalen- and isomorphen-diagram. *Forensic Sci. Int.* **2001**, *120*, 32–36. [CrossRef]
66. Reiter, C. Zum Wachstumsverhalten der Maden der blauen Schmeißfliege *Calliphora vicina*. *Rechtsmedizin* **1984**, *91*, 295–308. [CrossRef]

Review

The Pitfalls in the Path of Probabilistic Inference in Forensic Entomology: A Review

Gaétan Moreau

Département de biologie, Université de Moncton, Moncton, NB E1A 3E9, Canada; gaetan.moreau@umoncton.ca; Tel.: +1-506-8584975

Simple Summary: Experimental studies in forensic entomology must follow a series of rules to generate accurate predictions about criminal cases. These rules are reviewed and some approaches that have been used to solve experimental problems in forensic entomology are presented. Finally, recommendations are provided to avoid the publication and possible use in court of forensic studies that fall short of experimental standards.

Abstract: To bridge the gap between experimentation and the court of law, studies in forensic entomology and other forensic sciences have to comply with a set of experimental rules to generate probabilistic inference of quality. These rules are illustrated with successional studies of insects on a decomposing substrate as the main example. The approaches that have been used in the scientific literature to solve the issues associated with successional data are then reviewed. Lastly, some advice to scientific editors, reviewers and academic supervisors is provided to prevent the publication and eventual use in court of forensic studies using poor research methods and abusing statistical procedures.

Keywords: animal carcass; cadaver; decaying substrate; insect succession; postmortem interval; successional studies; vertebrate decomposition

Citation: Moreau, G. The Pitfalls in the Path of Probabilistic Inference in Forensic Entomology: A Review. *Insects* **2021**, *12*, 240. https://doi.org/10.3390/insects12030240

Academic Editors: Damien Charabidze and Daniel Martín-Vega

Received: 10 February 2021
Accepted: 8 March 2021
Published: 12 March 2021

Publisher's Note: MDPI stays neutral with regard to jurisdictional claims in published maps and institutional affiliations.

1. Forensic Entomology, an Inferential Science

Inference is defined as an operation by which one moves from one assertion that is considered true to another assertion by means of rules that make that second assertion equally true. Considering this, much of the research carried out in forensic entomology, as in other forensic sciences, is fundamentally inferential because experiments carried out on samples are often used to draw conclusions about the general population. For example, it is common to carry out studies using cadavers [1,2], whole animal carcasses [3,4] or animal tissues (reviewed in [5]) with the aim of later making extrapolations from experimental data to estimate a postmortem interval (PMI), a period of insect activity (PIA) or a post-colonization interval (PCI) in a criminal case. Reasoning from data to an individual case presents considerable challenges [6]. In addition, as the conditions of the experiment do not correspond in all points to those of the case, compliance with a set of rules is required to ensure the quality of the inference being drawn. These rules have been thoroughly described for over 100 years in treatises on experimental design and statistics [7–10], and this is where the problem lies: most practitioners in forensic sciences have limited statistical literacy. While several recent publications have sought to provide tools to forensic researchers wishing to improve the quality of methods, experimental design and statistics [11–14], forensic publications and conference presentations with poor research design and data analysis are still common. It is important to recognize that studies that fall short of experimental standards encourage false-positive findings, which can lead to wrongful convictions or exonerations. Incidentally, recent miscarriages of justice have often been attributable to flawed expert testimony [15].

What are the main methodological issues observed in the forensic entomology literature that make inference impossible? The foremost issue is a lack of compliance with Fisher's four cornerstones of experimental design: randomization, replication, independence, and controls [8,9]. This has already been discussed in recent literature reviews [12–14] and although some progress has been made in the last nine years, the absence of treatment replication, also called simple pseudoreplication, is still rampant in forensic entomology. A common form of simple pseudoreplication is to use one replicate per treatment (e.g., one carcass per condition, a single growth chamber per temperature for a study on maggot development, a single field and forest to test for differences between habitats, collecting data for a single year, i.e., one spring, summer, fall and winter for a study on seasonal effects) and consider subsamples to be true replicates when interpreting the data. The second issue is an absence of inferential statistical tests permitting the generation of probabilistic inference. It has been argued elsewhere that probabilistic inference procedures are an appropriate approach for measuring uncertainty in forensic science [16,17]. However, forensic entomology remains a descriptive science where inferential statistical tests are the exception rather than the rule and where the uncertainty associated with indicators of PMI, PIA or PCI is seldom disclosed. A good example of a widespread practice which impedes probabilistic inference is the presentation of summary tables or histograms pooling the information of several cadavers/carcasses [12].

There are however cases where even with a well-planned study, the very nature of the experiment is binding and imposes some adjustments to allow for inference. This is particularly true for studies of insect succession on cadavers or vertebrate carcasses, probably the most complex type of experiment from which one would extract a probabilistic inference in forensic entomology. Below, I substantiate this assertion by laying out the methodological issues associated with successional data recovered during studies of succession on necromass. While forensic entomology is used as an example, most of the elements discussed here apply to other forensic sciences as well.

2. Issues Associated with Successional Data

Successional data are samples, measurements, statistics or any factual information recovered sequentially from a decaying substrate. In forensic entomology, a good example of successional data is the maggot sample recovered at different time intervals from a cadaver/carcass to describe fly ontogenic and morphological changes [18,19]. Another example of successional data is the insect sample [1,2] or the visual count of insects [20] recovered through time from a cadaver/carcass. Then again, samples of microorganisms/bacteria [21], fungus [22], volatile organic compounds [23], synovial fluid and vitreous humour [24], or taphonomical changes (i.e., body scores, decomposition stages [25–27]) recovered from decomposing necromass are also successional data. To be clear, most of the time, information used to estimate the postmortem interval is successional data. What many overlook is that generating inferences from these datasets represents a challenge because of their inherent properties, which are listed in Table 1 and described below.

2.1. Data Measured Repetitively from a Small Number of Sampling Units and Field Sites

Most of the time, forensic studies use few cadavers/carcasses as sampling units because of the challenge associated with obtaining them and the time required to sample them. In addition, studies are usually conducted in the same environment year after year due to limitations associated with body farms or the difficulty in obtaining and accessing different experimental sites [28]. This has several consequences other than considerations linked with carcass enrichment [29,30], the first one being that records of successional data will be interdependent. Statistically, interdependence means that successive samples are correlated. This interdependence can sometimes be noticed visually during an experiment; if a rare insect is observed on a given carcass on day 2 of the study, it is more likely that the same insect will be observed the next day on the same carcass than on another carcass. A second consequence is the development of an autoregressive covariance structure, which means that interdependence is higher in two adjacent time periods and systematically decreases as the distance between the time periods increases. Again, this can be detected during an experiment as species recovered from a carcass/cadaver sampled on two successive days will be more similar than those collected over a longer time interval, such as day 1 and day 6 of the study. Both conditions indicate that the variance is not randomly distributed in the data, contravening a fundamental assumption of several statistical tests. A third consequence of having a small number of sampling units is that analyses will have limited statistical power. Statistical power is the likelihood that a hypothesis test will detect an effect if there is one. It depends on sample size and effect size, the latter being a quantitative measure of the magnitude of the experimental effect. This means that if a small difference exists between two sets of conditions (i.e., a small effect size), a large number of samples is necessary for this difference to be significant (for an example of insufficient sample size as revealed by power calculation in forensic entomology, see [13]). Conversely, if the effect size is large, few samples are needed to detect a difference between two sets of conditions (for an example, see [5]). A fourth consequence of having a small number of sampling units is that the study has a low internal validity because there is a greater chance that bias or random effects will have consequences on a significant portion of the sample. Internal validity describes the extent to which a study establishes a trustworthy cause-and-effect relationship and is not influenced by other factors. A low internal validity means that our confidence that manipulated variables caused a change in measured variables is compromised. In experiments with few carcasses/cadavers, it is much more difficult to detect bias and ascertain whether the data collected are valid and only affected by the variables under the control of the researcher than in larger studies.

Table 1. Issues and consequences of successional samples or measurements recovered sequentially from a decaying substrate.

Issues	Consequences
1. Data measured repetitively from a small number of sampling units and field sites	Data interdependence Autoregressive covariance structure Low statistical power Low internal validity
2. Data presenting non-linear trends	Non-linear effects Overfitting Non-Gaussian distribution
3. Datasets with a relatively large proportion of unexplained variance	High proportion of systematic variance Low external validity
4. Data affected by temporal and spatial effects	Data interrelated in time Data interrelated in space
5. Datasets including many independent and dependent variables	Autocorrelation Multicollinearity Overfitting Alpha inflation

2.2. Non-Linearity

Generally, successional data exhibit non-linear trends. Instead of increasing or decreasing linearly (first-order regression) or quadratically (second-order regression), variables often exhibit complex trends that can only be fitted with high-order polynomial regressions. For example, insect occurrence or abundance on cadavers/carcasses is bell-shaped or multimodal, with few insects documented at the beginning and end of succession, and many insects documented at an intermediate time interval [20,31]. High-order polynomial models are difficult to translate into biological terms, and their adjustment can lead to statistical model overfitting. Overfitting means that an overly complex model has been produced which fits the underlying relationship between the study variables as well as the noise unique to each sample. As a result, overfitted models produce misleading coefficients and although they generally perform perfectly with the experimental data, they are likely to perform poorly when other data (such as from a criminal case) are used [32]. In addition, when successional records of insects or residuals of analyses on successional records are plotted using frequency distribution histograms, a normal or Gaussian distribution is seldom obtained. Instead, distributions of forensic data tend to conform to the binomial distribution, the Poisson distribution or the quasi-Poisson distribution. The Poisson distribution is one of the best statistical distributions to model the number of times an event occurs in an interval of time or space, such as the number of times that *Phormia regina* (Diptera: Calliphoridae) is observed each day during succession on a given carcass in my study area. The first consequence of the two conditions identified above is that statistical models developed without considering non-linearity and non-Gaussian data distribution tend to have a poor fit. A second consequence is that approximations provided by the central limit theorem are likely to be inadequate, which signifies that issues linked to the lack of normality cannot be disregarded. A third consequence is that, usually, transformations applied to push the data closer to a Gaussian distribution perform poorly. A final consequence is that classical linear models (e.g., analysis of variance, regression) are often ineffective with such data.

2.3. Datasets with a Relatively Large Proportion of Unexplained Variance

Statistical tests can generally separate variance into two components: the explained and unexplained variance. The explained variance is the variance associated with the study variables whose influence is being studied such as the PMI and postmortem accumulation of degree-days in successional studies of insects. The unexplained variance is any residual variance associated with random variance that occurs because experimental units always exhibit some differences, even in homogeneous conditions, as well as variance associated with unknown variables, called systematic variance. Systematic variance is non-random variance due to factors not manipulated or measured during the study. In forensic entomology, systematic variance can overly dominate the explained variance because insect occurrence is influenced by meteorology, animal behavior, life histories, microhabitat, etc. This means that study variables often poorly explain the factors responsible for effects, and that the study has little external validity. External validity refers to the extent to which research findings can be generalized. A low external validity limits the ability to draw inferences from experiments and generalize the findings. This implies that it is incorrect to use the results of the study to explain a given situation occurring elsewhere, including a case study. For example, if only 15% of the variance is accounted for by a model developed in a laboratory study to predict the PIA from the size of maggots, this model is unusable in a forensic case because it has no predictive power.

2.4. Data Affected by Temporal and Spatial Effects

As a rule, all insect successional studies carried out in the field are inevitably affected by temporal and spatial effects. Temporal effects occur because the same experimental units (i.e., carcasses/cadavers) are sampled over and over during most forensic studies. In addition to sometimes causing oversampling problems [33], this leads to an interrelation of the samples over time. Moreover, repeated samples cannot be considered random because a sample taken at time 1 is necessarily collected before a sample at time 2. Such a dataset is unlikely to be stationary in the sense that the mean, variance, autocorrelation and other statistical properties are all constant over time, contravening another assumption of several statistical tests. Spatial effects occur because field conditions are often heterogeneous and because the placement of sampling units can create interdependence between units, thereby causing systematic spatial variation that results in observable data clusters. Cadavers/carcasses that are closely spaced can become linked by dispersing organisms that synchronize their dynamics [34]. For example, in a recent study [20], we noticed that less competitive carrion beetles such as *Necrophila americana* (Coleoptera: Silphidae) were kept at bay from carcasses by more competitive species such as *Necrodes surinamensis* (Coleoptera: Silphidae). The former species thus moved to less interesting habitat patches, producing a spatial structure explaining the results of the study. Overlooking or ignoring these spatial patterns, when they are present, could lead to false conclusions about cause-and-effect relationships [35]. This issue is particularly problematic for studies conducted in body farms or small field sites where cadavers/carcasses cannot be placed far apart.

2.5. Datasets Including Many Independent and Dependent Variables

Often, in forensic studies involving cadavers/carcasses, a large amount of data is collected from each experimental unit (i.e., cadavers/carcasses). For example, several studies record daily the presence of different taxa, the occurrence of egg masses, the size of maggots, the temperature of the cadaver, the stage of decomposition of the cadaver, the postmortem accumulation of degree days, etc. However, the scientific literature recommends that the number of variables remain low compared to the number of observations, because the chance of finding significant but biologically irrelevant relationships between the dependent and independent variables increases with the number of variables [36]. Ignoring this recommendation is likely to result in autocorrelation between processes as well as autocorrelation between response variables. When autocorrelation exists between two variables, this indicates that one of the study variables is a duplicate of the other variable more or less shifted in time or in space. For example, during the summer in my study area, the postmortem accumulation of degree days and the PMI are always autocorrelated because the temperature is relatively constant. Without proper statistical adjustment to deal with autocorrelation, multicollinearity, overfitting (discussed in Section 2.2) and Alpha inflation are inevitable. Multicollinearity is a process whereby dependence among the explanatory variables is strong enough that one explanatory variable can be accurately predicted from the others. Thus, the collinear explanatory variables contain the same information about the dependent variables. A model developed using collinear explanatory variables yields highly volatile parameter estimates with large standard errors [37]. With such a model, a small change in the data can result in a large change or even a change of sign in parameter estimates. Consequently, little confidence can be placed in a model affected by multicollinearity. Alpha inflation is also known as familywise error, experimentwise error or cumulative Type I error. Alpha (or α) refers to the significance level. In most experimental sciences, $\alpha = 0.05$, which implies that you have a 5% chance of incorrectly rejecting the true null hypothesis when you perform a statistical test (i.e., you have a 5% chance of detecting an effect when there is none). As more tests are conducted on the same dataset, the likelihood of obtaining at least one erroneous significant result increases. For example, if the response to a given treatment (e.g., shading) of 20 blowfly species sampled in an experiment is analyzed individually with $\alpha = 0.05$, there is 64.1% chance that one or several analyses of species responses will be significant due to chance. This number is obtained by the following calculation:

$$\alpha' = 1 - (1 - 0.05)^{20} = 0.641 \tag{1}$$

In short, multicollinearity, overfitting and Alpha inflation imply that the interpretation of the results will be too liberal, that statistical inferences will be less reliable and that the whole study will potentially be misleading.

3. Possible Remedies to the Issues Associated with Successional Data

To address the problems discussed above, I list in Table 2 and review below some approaches that have been used to deal with these issues in forensic entomology and in studies of insects in other degradative systems (i.e., dung, dead wood).

Table 2. Consequences and possible remedies to the issues associated with successional samples or measurements recovered sequentially from a decaying substrate.

Consequences of Successional Data	Remedies
1. Low statistical power Low internal validity Low external validity	Increase the sample size Increase the number of locations, times, and conditions
2. Data interdependence Autoregressive covariance structure Non-linear effects Non-Gaussian distribution	Generalized linear models (GLMs), generalized linear mixed models (GLMMs), generalized additive models (GAMs), generalized additive mixed models (GAMMs)
3. Autocorrelation Multicollinearity Overfitting Alpha inflation	Multivariate statistics
4. Data interrelated in time Data interrelated in space	Time series analysis Spatial statistics Repeated measures and/or spatially explicit GLMs, GLMMs, GAMs, GAMMs
5. High proportion of systematic variance	Ensure that all influential variables have been accounted for Use a model that is better suited to data Acknowledge this variability

3.1. How to Solve Problems Related to Low Statistical Power as well as to Low Internal and External Validity

These problems have a single cure: increase the sample size. If this is impossible because sampling is time-consuming, review the protocol to find out ways to speed up its implementation. Another method to increase external validity is to repeat the study in a variety of locations, times, and conditions. This may appear paradoxical because generally, one seeks to decrease rather than increase natural variability in experimental studies. However, the purpose of forensic science differs from process-driven science as the aim is generally not to conclude about the effects of a treatment, but rather to use the study results for the interpretation of data from a criminal case that occurred under different conditions than the study. A good example of the two precepts discussed above is given by Horgan [38] who studied dung beetles attracted to decomposing cow dung in 16 widely separated locations in the contrasting pasture landscapes of El Salvador and Atlantic Nicaragua. Because of all the variability that was consciously integrated into the design of this study, the observed trends are strong and easily transposable to other environments. For a similar example using saproxylic beetles, see [39].

3.2. How to Solve Problems Related to Interdependence between Records, Auto-Regressive Covariance Structure, Non-Linear Effects and Non-Gaussian Distributions

The solution to these problems is straightforward: use generalized linear models (GLMs), generalized linear mixed models (GLMMs), generalized additive models (GAMs) or generalized additive mixed models (GAMMs). Basically, these models work as analyses of variance, analyses of covariance or linear regressions but use non-linear link functions to allow for responses with nonlinear distributions such as Poisson, binomial, gamma, etc. These models can also account for data interdependence and for the autoregressive structure of the data. Moreover, additive models use non-parametric smoother functions to fit models with fewer assumptions. As the description of statistical models is not one of the objectives of this text, I take this opportunity instead to encourage readers to consult examples of the application of these models in recent forensic literature. For examples of GLMs/GLMMs, see [26,31,40,41]. For examples of GAMs/GAMMs, see [5,39,42,43].

3.3. How to Solve Problems Related to Autocorrelation, Multicollinearity, Overfitting and Alpha Inflation

The solution to these problems is forthright: use multivariate statistics. Multivariate statistics encompass all approaches that allow for the simultaneous analysis of several response variables. They can be grouped into different categories such as the descriptive or correlative methods (e.g., ordinations, canonical correlations and clustering methods), or the explanatory and predictive methods (e.g., regression trees, multivariate analysis of variance (MANOVA), discriminant analysis, random forests). These methods are useful to deal with autocorrelation, the choice of a method depending on the objectives of the study, the experimental design and the analyst's preferences. When multicollinearity and overfitting are an issue, multivariate statistics can help with variable selection to reduce dimensionality and allow for further exploration and analysis of the data [44]. As mentioned above, the description of statistical models is not one of the objectives of this text. Thus, I encourage readers to consult examples of the application of these models in recent studies in forensics or other degradative systems. For examples of ordination and canonical correlations in a forensic context, see [40,45–47]. For examples of multivariate regression trees, see studies on saproxylic beetles such as [48,49]. For examples of discriminant analysis, see studies on saproxylic beetles such as [50] or on dung beetles such as [51]. For examples of MANOVAs in a forensic context, see [33,40,52,53].

3.4. How to Solve Problems Related to Having Data Interrelated in Time or Space

If the objective of the study is to analyze only the temporal trends of a dataset, time series analysis such as used by Andow and Kiritani [54] in a study of saproxylic beetles is appropriate. If variables other than time need to be included in the model, then the GLM, GLMM, GAM and GAMM approaches suggested above are adequate (for an example, see [5,20]). On the other hand, potential spatial relationships require the use of spatial statistics that specifically describe and model localized or geo-referenced data. To date, forensic entomologists have not ventured into spatial statistics, but this is bound to change as studies encompassing several locations are taking place. In contrast, spatial analyses are frequently used in studies on other decomposing substrates. For redundancy analysis, see [55] for an example with saproxylic beetles and [56] for an example with dung beetles. For an example of a polynomial generalized linear model analysis of the position with saproxylic beetles, see [57]. For an example of a spatially explicit GAMM with saproxylic beetles, see [58].

3.5. How to Solve Problems Related to Having a Large Amount of Systematic Variance

First, ask yourself whether all the influential variables were measured, and if the best model is being used. A research protocol excluding some influential variables as well as an inappropriate model can limit the ability of statistical tests to account for quantifiable effects. If the best model and design has been applied, my second recommendation is to accept this unexplained variance. An even better recommendation is to embrace it! It would be preposterous to hope to account for all the variance knowing the numerous factors influencing decomposition and insect colonization [59]. Most insects aggregating on and around cadavers/carcasses are hardly predictable [20,31], which contribute to this unexplained variance. Thus, this situation is not alarming and not unique to forensic entomology; the same large unexplained variance also prevails in studies of saproxylic beetles [48,49]. In the applied context of forensic entomology, it then appears more important to distinguish between what can and what cannot be explained. I firmly believe that too much time and energy has been spent researching insect species that do not have real potential for the estimation of the PMI, PIA or PCI.

4. Advice to Scientific Editors, Reviewers and Academic Supervisors

Only 27 out of the 160 field studies published between 1985 and 2013 in forensic anthropology, forensic entomology and forensic taphonomy had an adequate design and analysis [14] and a consultation of the recent literature shows that the situation has changed little. Therefore, it is no exaggeration to assume that most of the experimental studies in these fields that are used in court should not be admissible because they cannot generate probabilistic inference. This embarrassing situation can be explained by the fact that there is often no negative consequence related to the publication of studies using poor research methods and abusing statistical procedures [60]. While this situation is not unique to forensic entomology [61], it has far greater consequences than simply misleading colleagues and the scientific community because false-positive findings are being used to solve criminal cases. We all have a responsibility to prevent the publication of studies that fall short of experimental standards, including false preliminary studies. The prefix "a preliminary study on" has been overused in the forensic entomology literature. According to the Merriam-Webster Dictionary, "preliminary" indicates that something is a prelude to something else. However, most preliminary studies have no follow-up, this term rather serving as a loophole to allow for the publication of pseudoreplicated studies.

To avoid the publication and possible use in court of forensic studies that are poorly designed or poorly analyzed, scientific editors and reviewers should make sure that the following conditions are met by publications:

1. *The study is devoid of experimental errors.* Scientific editors and reviewers should not be afraid to require from authors a detailed description and a map of the layout of the study. Regardless of the nature of the study, the experimental unit should always be clearly identified. To learn how to recognize the experimental unit and main experimental errors, read [12-14,62]. Pseudoreplicated studies should never be published, even as "preliminary studies".

2. If the nature of the study allows for it, an inferential statistical test that permits extrapolation of the results to case scenarios is presented. The statistical procedures should be described in detail and an estimate of the experimental error should be evident in the tables and figures of the manuscript. If successional data are involved, the statistical test should comply with elements presented in Table 2.

3. *If the nature of the study does not allow for it, no inferential statistical test is presented.* In a widely cited article, Hurlbert [62] suggested that good articles that refrain from using inferential statistics when these cannot be applied are worth publishing. However, the authors should explicitly recognize that the study is descriptive, thus not allowing for transposition of the results to other situations or use in court.

Inevitably, to ensure that forensic entomology studies generate probabilistic inference of quality, scientific editors and reviewers will have to step up and act as the watchdogs of the scientific method. Nobody should be reluctant to question why money, time and publication space is spent on a study that has an inappropriate design, analysis or a lack of analysis.

My last advice is for academic supervisors and concerns the training of students. To effectively understand and implement the analyzes discussed herein, forensic science students need statistical knowledge that goes far beyond classical frequentist statistics and linear models. Therefore, I urge academic supervisors to encourage their students to pursue advanced training in statistics. More than ever, as access to powerful statistical tools has become more democratized, knowledge of experimental design and statistics is proving to be an essential skill to bridge the gap between laboratory/field studies and court evidence.

Funding: This research was funded by The Natural Sciences and Engineering Research Council of Canada, grant number DDG-2020-00021.

Institutional Review Board Statement: Not applicable.

Data Availability Statement: Not applicable.

Insects 2021, 12, 240

Acknowledgments: The author is thankful to Jeffery Tomberlin as this manuscript was developed following his suggestion made in the context of the Digital Congress of Forensic Entomology in September 2020. The author also wishes to thank D. Boudreau, N. Hammami, K. LeBlanc, L. Tousignant and three anonymous reviewers for comments on a previous version of this manuscript.

Conflicts of Interest: The author declares no conflict of interest. The funders had no role in the writing of the manuscript, or in the decision to publish the manuscript.

References

1. Schoenly, K.G.; Haskell, N.H.; Hall, R.D.; Gbur, J.R. Comparative performance and complementarity of four sampling methods and arthropod preference tests from human and porcine remains at the Forensic Anthropology Center in Knoxville, Tennessee. *J. Med. Entomol.* **2007**, *44*, 881–894. [CrossRef] [PubMed]
2. Dawson, B.M.; Barton, P.S.; Wallman, J.F. Contrasting insect activity and decomposition of pigs and humans in an Australian environment: A preliminary study. *Forensic Sci. Int.* **2020**, *316*, 110515. [CrossRef] [PubMed]
3. Payne, J.A. A summer carrion study of the baby pig *Sus scrofa* Linnaeus. *Ecology* **1965**, *46*, 592–602. [CrossRef]
4. VanLaerhoven, S.L.; Anderson, G.S. Insect succession on buried carrion in two biogeoclimatic zones of British Columbia. *J. Forensic Sci.* **1999**, *44*, 32–43. [CrossRef]
5. LeBlanc, K.; Boudreau, D.; Moreau, G. Small bait traps may not accurately reflect the composition of necrophagous Diptera associated to remains. *Insects.*
6. Faigman, D.L.; Monahan, J.; Slobogin, C. Group to individual (G2i) inference in scientific expert testimony. *Univ. Chic. Law Rev.* **2014**, *81*, 417–480. [CrossRef]
7. Peirce, C.S. A theory of probable inference. In *Studies in Logic by Members of the Johns Hopkins University*; Peirce, C.S., Ed.; Little, Brown, and Company: Boston, MA, USA, 1883; pp. 126–181.
8. Fisher, R.A. The arrangement of field experiments. *J. Min. Agric.* **1926**, *33*, 503–513.
9. Fisher, R.A. *The Design of Experiments*; Oliver and Boyd: Edinburgh, UK.; London, UK, 1935.
10. Cochran, W.G.; Cox, G.M. *Experimental Designs*; Wiley: New York, NY, USA, 1950.
11. Amendt, J.; Campobasso, C.P.; Gaudry, E.; Reiter, C.; LeBlanc, H.N.; Hall, M.J. Best practice in forensic entomology—Standards and guidelines. *Int. J. Legal. Med.* **2007**, *121*, 90–104. [CrossRef]
12. Michaud, J.-P.; Schoenly, K.G.; Moreau, G. Sampling flies or sampling flaws? Experimental design and inference strength in forensic entomology. *J. Med. Entomol.* **2012**, *49*, 1–10. [CrossRef] [PubMed]
13. Moreau, G.; Michaud, J.-P.; Schoenly, K.G. Experimental design, inferential statistics, and computer modeling. In *Forensic Entomology: International Dimensions and Frontiers*; Tomberlin, J.K., Benbow, M.E., Eds.; CRC Press, Taylor & Francis Group: Boca Raton, FL, USA, 2015; pp. 205–230.
14. Schoenly, K.G.; Michaud, J.-P.; Moreau, G. Design and analysis of field studies in carrion ecology. In *Carrion Ecology, Evolution and Their Applications*; Benbow, M.E., Tomberlin, J.K., Tarone, A.M., Eds.; CRC Press: Boca Raton, FL, USA, 2015; pp. 129–148.
15. Imwinkelried, E.J. The best insurance against miscarriages of justice caused by junk science: An admissibility test that is scientifically and legally sound. *Albany Law Rev.* **2017**, *81*, 851–875. [CrossRef]
16. Robertson, B.; Vignaux, G.A. Probability—The logic of the law. *Oxf. J. Leg. Stud.* **1993**, *13*, 457–478. [CrossRef]
17. Aitken, C.G.G.; Taroni, F. *Statistics and the Evaluation of Evidence for Forensic Scientists, 2nd ed.*; John Wiley & Sons Ltd: Chichester, UK, 2004. [CrossRef]
18. O'Flynn, M.A. The succession and rate of development of blowflies in carrion in southern Queensland and the application of these data to forensic entomology *Aust. J. Entomol.* **1983**, *22*, 137–148. [CrossRef]
19. Tabor, K.L.; Fell, R.D.; Brewster, C.C.; Pelzer, K.; Behonick, G.S. Effects of antemortem ingestion of ethanol on insect successional patterns and development of *Phormia regina* (Diptera: Calliphoridae). *J. Med. Entomol.* **2005**, *42*, 481–489. [CrossRef] [PubMed]
20. Lutz, L.; Amendt, J.; Moreau, G. Carcass concealment alters assemblages and reproduction of forensically important beetles. *Forensic Sci. Int.* **2018**, *291*, 124–132. [CrossRef]
21. Pechal, J.L.; Crippen, T.L.; Benbow, M.E.; Tarone, A.M.; Dowd, S.; Tomberlin, J.K. The potential use of bacterial community succession in forensics as described by high throughput metagenomic sequencing. *Int. J. Legal Med.* **2014**, *128*, 193–205. [CrossRef]
22. Fu, X.; Guo, J.; Finkelbergs, D.; He, J.; Zha, L.; Guo, Y.; Cai, J. Fungal succession during mammalian cadaver decomposition and potential forensic implications. *Sci. Rep.* **2019**, *9*, 1–9. [CrossRef] [PubMed]
23. Vass, A.; Smith, R.; Thompson, C.; Burnett, M.; Wolf, D.; Synstelien, J.; Dulgerian, N.; Eckenrode, B. Decomposition odor analysis database. *J. Forensic Sci.* **2004**, *49*, 1–10. [CrossRef]
24. Tumram, N.K.; Bardale, R.V.; Dongre, A.P. Postmortem analysis of synovial fluid and vitreous humour for determination of death interval: A comparative study. *Forensic Sci. Int.* **2011**, *204*, 186–190. [CrossRef]
25. Megyesi, M.S.; Nawrocki, S.P.; Haskell, N.H. Using accumulated degree-days to estimate the postmortem interval from decomposed human remains. *J. Forensic Sci.* **2005**, *50*, 1–9. [CrossRef]
26. Michaud, J.-P.; Moreau, G. A statistical approach based on accumulated degree-days to predict decomposition-related processes. *J. Forensic Sci.* **2011**, *56*, 229–232. [CrossRef]
27. Moffatt, C.; Simmons, T.; Lynch-Aird, J. An improved equation for TBS and ADD: Establishing a reliable postmortem interval framework for casework and experimental studies. *J. Forensic Sci.* **2016**, *61*, S201–S207. [CrossRef] [PubMed]

28. Matuszewski, S.; Hall, M.J.; Moreau, G.; Schoenly, K.G.; Tarone, A.M.; Villet, M.H. Pigs vs. people: The use of pigs as analogues for humans in forensic entomology and taphonomy research. *Int. J. Legal Med.* **2020**, *134*, 793–810. [CrossRef]

29. Schoenly, K.G.; Shahid, S.A.; Haskell, N.H.; Hall, R.D. Does carcass enrichment alter community structure of predaceous and parasitic arthropods? A second test of the arthropod saturation hypothesis at the anthropology research facility in Knoxville, Tennessee. *J. Forensic Sci.* **2005**, *50*, 134–142. [CrossRef]

30. Damann, F.E.; Tanittaisong, A.; Carter, D.O. Potential carcass enrichment of the University of Tennessee Anthropology Research Facility: A baseline survey of edaphic features. *Forensic Sci. Int.* **2012**, *222*, 4–10. [CrossRef] [PubMed]

31. Michaud, J.-P.; Moreau, G. Predicting the visitation of carcasses by carrion-related insects under different rates of degree-day accumulation. *Forensic Sci. Int.* **2009**, *185*, 78–83. [CrossRef] [PubMed]

32. Lever, J.; Krywinski, M.; Altman, N. Model selection and overfitting. *Nat. Methods* **2016**, *13*, 703–704. [CrossRef]

33. Michaud, J.-P.; Moreau, G. Effect of variable rates of daily sampling of fly larvae on decomposition and carrion-insect community assembly: Implications for forensic entomology field study protocols. *J. Med. Entomol.* **2013**, *50*, 890–897. [CrossRef]

34. Leibold, M.A.; Holyoak, M.; Mouquet, N.; Amarasekare, P.; Chase, J.M.; Hoopes, M.F.; Holt, R.D.; Shurin, J.B.; Law, R.; Tilman, D.; et al. The metacommunity concept: A framework for multi-scale community ecology. *Ecol. Lett.* **2004**, *7*, 601–613. [CrossRef]

35. Lichstein, J.W.; Simons, T.R.; Shriner, S.A.; Franzreb, K.E. Spatial autocorrelation and autoregressive models in ecology. *Ecol. Monogr.* **2002**, *72*, 445–463. [CrossRef]

36. Sachs, L. *Angewandte Statistik*; Springer: Berlin/Heidelberg, Germany, 1978.

37. Farrar, D.E.; Glauber, R.R. Multicollinearity in regression analysis: The problem revisited. *Rev. Econ. Stat.* **1967**, *49*, 92–107. [CrossRef]

38. Horgan, F.G. Dung beetles in pasture landscapes of Central America: Proliferation of synanthropogenic species and decline of forest specialists. *Biodivers. Conserv.* **2007**, *16*, 2149–2165. [CrossRef]

39. Jonsell, M.; Abrahamsson, M.; Widenfalk, L.; Lindbladh, M. Increasing influence of the surrounding landscape on saproxylic beetle communities over 10 years succession in dead wood. *For. Ecol. Manag.* **2019**, *440*, 267–284. [CrossRef]

40. Michaud, J.-P.; Moreau, G. Facilitation may not be an adequate mechanism of community succession on carrion. *Oecologia* **2017**, *183*, 1143–1153. [CrossRef]

41. Aubernon, C.; Hedouin, V.; Charabidze, D. The maggot, the ethologist and the forensic entomologist: Sociality and thermoregulation in necrophagous larvae. *J. Adv. Res.* **2019**, *16*, 67–73. [CrossRef]

42. Tarone, A.M.; Foran, D.R. Generalized additive models and Lucilia sericata growth: Assessing confidence intervals and error rates in forensic entomology. *J. Forensic Sci.* **2008**, *53*, 942–948. [CrossRef]

43. Moreau, G.; Lutz, L.; Amendt, J. Honey, can you take out the garbage can? Modeling weather data for cadavers found within containers. *Pure Appl. Geophys.* **2019**. [CrossRef]

44. Cumming, J.A.; Wooff, D.A. Dimension reduction via principal variables. *Comput. Stat. Data Anal.* **2007**, *52*, 550–565. [CrossRef]

45. De Souza, M.S.; Pepinelli, M.; de Almeida, E.C.; Ochoa-Quintero, J.M.; Roque, F.O. Blow flies from forest fragments embedded in different land uses: Implications for selecting indicators in forensic entomology. *J. Forensic Sci.* **2016**, *61*, 93–98. [CrossRef] [PubMed]

46. Moore, H.E.; Pechal, J.L.; Benbow, M.E.; Drijfhout, F.P. The potential use of cuticular hydrocarbons and multivariate analysis to age empty puparial cases of Calliphora vicina and Lucilia sericata. *Sci. Rep.* **2017**, *7*, 1–11. [CrossRef] [PubMed]

47. Boudreau, D.R.; Hammami, N.; Moreau, G. Environmental and evolutionary factors favoring the coexistence of sarcos-aprophagous Calliphoridae species competing for animal necromass. *Ecol. Entomol..*

48. Gandiaga, F.; Moreau, G. How long are thinning-induced resource pulses maintained in plantation forests? *For. Ecol. Manag.* **2019**, *440*, 113–121. [CrossRef]

49. Chiasson, B.; Moreau, G. Assessing the lifeboat effect of retention forestry using flying beetle assemblages. *For. Ecol. Manag.* **2021**, *483*, 118784. [CrossRef]

50. Vogel, S.; Bussler, H.; Finnberg, S.; Müller, J.; Stengel, E.; Thorn, S. Diversity and conservation of saproxylic beetles in 42 European tree species: An experimental approach using early successional stages of branches. *Insect Conserv. Divers.* **2021**, *14*, 132–143. [CrossRef]

51. Jugovic, J.; Koprivnikar, N. Rolling in the deep: Morphological variation as an adaptation to different nesting behaviours of coprophagous Scarabaeoidea. *Biologia* **2020**, 1–13. [CrossRef]

52. Pechal, J.L.; Moore, H.; Drijfhout, F.; Benbow, M.E. Hydrocarbon profiles throughout adult Calliphoridae aging: A promising tool for forensic entomology. *Forensic Sci. Int.* **2014**, *245*, 65–71. [CrossRef] [PubMed]

53. McIntosh, C.S.; Dadour, I.R.; Voss, S.C. A comparison of carcass decomposition and associated insect succession onto burnt and unburnt pig carcasses. *Int. J. Legal Med.* **2017**, *131*, 835–845. [CrossRef] [PubMed]

54. Andow, D.A.; Kiritani, K. Density-dependent population regulation detected in short time series of saproxylic beetles. *Popul. Ecol.* **2016**, *58*, 493–505. [CrossRef]

55. Kaila, L.; Martikainen, P.; Punttila, P. Dead trees left in clear-cuts benefit saproxylic Coleoptera adapted to natural disturbances in boreal forest. *Biodivers. Conserv.* **1997**, *6*, 1–18. [CrossRef]

56. Louzada, J.; Lima, A.P.; Matavelli, R.; Zambaldi, L.; Barlow, J. Community structure of dung beetles in Amazonian savannas: Role of fire disturbance, vegetation and landscape structure. *Landsc. Ecol.* **2010**, *25*, 631–641. [CrossRef]

57. McGeoch, M.A.; Schroeder, M.; Ekbom, B.; Larsson, S. Saproxylic beetle diversity in a managed boreal forest: Importance of stand characteristics and forestry conservation measures. *Divers. Distrib.* **2007**, *13*, 418–429. [CrossRef]
58. Mourant, A.; Lecomte, N.; Moreau, G. Indirect effects of an ecosystem engineer: How the Canadian beaver can drive the reproduction of saproxylic beetles. *J. Zool.* **2018**, *304*, 90–97. [CrossRef]
59. Campobasso, C.P.; Di Vella, G.; Introna, F. Factors affecting decomposition and Diptera colonization. *Forensic Sci. Int.* **2001**, *120*, 18–27. [CrossRef]
60. Smaldino, P.E.; McElreath, R. The natural selection of bad science. *R Soc. Open Sci.* **2016**, *3*, 160384. [CrossRef] [PubMed]
61. Ioannidis, J.P. Why most published research findings are false. *PLoS Med.* **2005**, *2*, e124. [CrossRef] [PubMed]
62. Hurlbert, S.H. Pseudoreplication and the design of ecological field experiments. *Ecol. Monogr.* **1984**, *54*, 187–211. [CrossRef]

Case Report

Entomological Contributions to the Legal System in Southeastern Spain

María-Isabel Arnaldos [1,2,*] **and María-Dolores García** [1,2]

1 Area of Zoology, Faculty of Biology, University of Murcia, 30100 Murcia, Spain
2 Unit of Forensic Entomology and Evidence Microscopic Analysis, External Service of Forensic Sciences and Techniques, University of Murcia, 30100 Murcia, Spain; mdgarcia@um.es
* Correspondence: miarnald@um.es

Simple Summary: Insects and other arthropods found at a forensic scene are considered to represent relevant evidence regarding the time and place of death, a possible antemortem or postmortem treatment of the victim, the geographical origin of narcotic substances, drugs, etc. However, in order to derive firm conclusions from entomological evidence, it is critical to understand the different aspects of the insect biology of the whole area under consideration. Moreover, most forensic investigators are not specially trained in basic entomology procedures, which can result in limitations when trying to achieve an accurate expert report. The present work illustrates the utility and necessity of such entomological studies and the expert training that is required using actual forensic cases.

Abstract: The aim of this work is to present a number of forensic cases that took place in Southeastern Spain (Murcia province) in which the entomological evidence aided to fully solve the issues raised during the legal enquiry, enhancing the close interrelationships between experimental studies performed and actual forensic cases assessed. In all cases, the expert report was requested by the police agents or the medical examiners, the latter attempting to make stronger their own conclusions. The assessment of all cases was possible by comparing the evidence and circumstances of each one with the experimental data previously obtained in our laboratory concerning aspects such as faunistic, ecological, morphological, etc., and by considering data from other researchers. In all cases, the evidence could be addressed, although in some cases, it had not been properly collected or processed. Thus, the utility of the experimental studies in forensic practice, even when being considered merely biological, and without immediate practical application, can be demonstrated as well as the need for providing specialized instruction on Forensic Entomology procedures to the different agents involved in forensic investigation.

Keywords: Forensic Entomology; Spain; experimental studies; cases

Citation: Arnaldos, M.-I.; García, M.-D. Entomological Contributions to the Legal System in Southeastern Spain. *Insects* **2021**, *12*, 429. https://doi.org/10.3390/insects12050429

Academic Editors: Jeffery K. Tomberlin, Damien Charabidze and Daniel Martín-Vega

Received: 14 January 2021
Accepted: 29 March 2021
Published: 10 May 2021

Publisher's Note: MDPI stays neutral with regard to jurisdictional claims in published maps and institutional affiliations.

1. Introduction

Although the application of entomology to forensic practice has been largely referred to in the scientific literature in many countries since the 1970s [1], the legal system in Spain considers the study of entomological evidence as a supplementary technique. As a matter of fact, entomological evidence is rarely collected at the crime scene or during the autopsy procedure. For this reason, an important source of data may be lost during the enquiry, making it more difficult, or even impossible, to get an accurate diagnosis of the actual case under investigation. Until now, the scope of recognition and the inclusion of Forensic Entomology practices into the routine of a forensic investigation has been limited in Spain. Nevertheless, there are medical examiners and members of the State security forces who are willing to implement a standardized recovery of entomological evidence to aid in those forensic cases in which they participate. Sometimes the Forensic Entomology teams are required to study some entomological evidence from actual forensic cases; as an example, the medical examiner may be interested in complementary entomological evidence to reinforce their own conclusions and make their own results and conclusions of the entomologist's expertise [1]. In such cases, the problem can arise when the evidence has not been properly recovered or processed or does not represent the whole entomological fauna related to the corpse. This could occur as a consequence of the lack of specific training on the subject from those recovering the samples, as reported, e.g., in García et al. [1].

Regardless of the source of evidence, the role of forensic entomologists is to study it and provide an independent expert's opinion, most often related to the estimation of the postmortem interval (PMI). This estimation can be made on the basis of the age of an immature insect fed on the corpse, which provides a minimum PMI, and the carrion arthropod succession model, which could suggest a maximum PMI [2]. Succession data have been used to very accurately calculate a PMI as large as 52 days [3] and could be applied to a much larger interval [2] if the seasonal dynamics and succession of the entomosarcosaprophagous fauna in a given area are accurately known. In order to apply any of the two methods, previous information and knowledge are needed on larval growth rates, and how they are influenced by temperature, as well as on carrion succession. Then, previously existing experimental data are usually searched [2]. It needs to be considered that a PMI does not always correspond to the period of insect activity (PIA) [4] because the time when insects first colonized a body may not necessarily happen at the time of death. It could occur within minutes of being dead or could be delayed under certain circumstances (e.g., body buried or wrapped or concealed indoors) [4]. The PIA can eventually agree with the minimum PMI. As regards the developmental pattern of a certain species, there is abundant information for different taxa. Nevertheless, further efforts are needed in order to cover as many different species as possible as well as to properly understand if the developmental pattern is affected by regional or geographical variations. In fact, and according to Grassberger and Reiter [5], it seems that intrinsic factors, such as geographic adaptation, could explain a difference in temperature-dependent development.

With regard to the carrion arthropod succession, much is known on the importance of the geographical region or biogeoclimatic zone, and even the particular habitat concerning a concrete case, on insect colonization of such carrion (i.e., [6–9]). It has been proposed that different environments or habitats can influence the patterns of community assemblage [10,11], illustrating the importance of performing studies on carrion entomofauna in a variety of habitats and scenarios in all regions in which Forensic Entomology is used [6].

In the Iberian Peninsula, there are reports of the contributions to the knowledge of the structure and dynamics of the entomosarcosaprophagous fauna in several areas (i.e., [7,9,11–26]), and, in at least one case [14], a relationship between experimental study and forensic cases' resolution has been established. Moreover, studies on certain groups of insects have addressed the morphology of immature stages [27–30] and the development of certain species under controlled conditions (i.e., [31,32] among others). Finally, some case reports have also contributed to promoting the use of Forensic Entomology in Spain [1,33,34]. Despite all of the above, the use of entomological evidence for solving forensic cases in Spain is still almost a rarity due, at least in part, to the lack of proper taxonomic revisions of many groups of arthropods [35]. For this reason, disseminating cases resolved with the help of this entomological discipline is particularly relevant in the country.

The aim of this work is to present a number of forensic cases that occurred in Southeastern Spain (Murcia province) in which entomological evidence was involved to fully elucidate the issues raised during the legal enquiry. It is important to highlight the close interrelationship between the experimental studies on entomosarcosaprophagous fauna that have been previously carried out and the actual forensic cases to be solved. This relationship is demonstrated by the fact that the assessment of all cases was possible by comparing the evidence and circumstances of each one with the experimental data that had previously been obtained and reported.

2. Cases Presentation

The forensic cases presented below are grouped based on the type of previous studies that were found to have utility for their resolution.

2.1. Morphological and Faunistic Studies

Knowledge of the morphology of the different species is critical when dealing with actual forensic cases. In this respect, classical faunistic studies on sarcosaprophagous fauna provide excellent training for identifying such fauna. The following cases fit well in this section.

- Case 1. A partly burned woman corpse was found in the early morning hours on the side of a road in the province of Murcia at the end of March. The police agents did not recover any entomological evidence but made an excellent graphic report from which different fly images could be studied. From these images (Figure 1)., and once magnified, *Chrysomya albiceps* (Wiedemann, 1819) was identified on the basis of our own experience and the dark hind stripe of the abdominal terga [36].

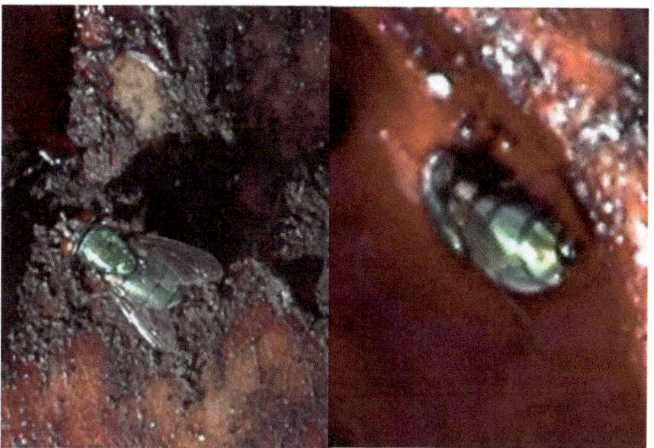

Figure 1. Photographs from graphic report by police agents.

Former faunistic studies [12,13] indicated that this species is the most common in the area, and efforts were directed to solving the case on the basis of this finding. None of the pictures revealed anything consistent with signs of activity of insects, such as eggs or larvae. According to what could be observed, the remains appeared to represent a fresh stage (despite having been superficially burned), suggesting a recent death. Because previous knowledge of fauna related to corpses in the area had pointed to a secondary character of *Ch. albiceps*, we consider the possibility that there could have also been some other, not graphically illustrated, activity of primary species. Nevertheless, and in agreement with Avila and Goff [37] and Pai et al. [38], who indicated that *Chrysomya* sp. can act as primary fly in burnt corpses, it could be concluded that the remains had been exposed for a short period. This conclusion was consistent with eyewitness testimony of a bonfire taking place in the location the same morning. The subsequent enquiry confirmed this conclusion.

- Case 2. García et al. [1] report the case of an unidentified mummified corpse found in a rural area of difficult access in Murcia province at the end of August. The police agents recovered some entomological evidence in situ from the corpse itself and below it, but this evidence was inadequate, and, as a result, two entomologists had to recover additional evidence from the corpse at the place where it was guarded. Among this evidence, a number of small fly pupae and puparia were studied. They could be identified as belonging to *Piophila megastigmata* McAlpine, 1978 on the basis of the already-existing morphological study of preimaginal stages of this species that allowed the identification of the found pupae thanks to the very wrinkled tegument (25–30 wrinkles, or even more, per abdominal segment) [29]. This finding represented the first report of such species on a corpse in Spain, and its presence was considered to estimate the PMI. For this purpose, a faunistic study of Diptera performed in the geographical area [39] was considered. This study stated that *P. megastigmata* could be present on corpses since the beginning of May. Given that the main question in the inquiry was whether the person had died on dates close to disappearance or at a later time, our findings were consistent with death likely occurring by the date of disappearance (at the end of May). Further details on the recovered fauna and the estimation of PMI can be found in García et al. [1].

2.2. Faunistic Succession and Developmental Data

Patterns of arthropod succession are considered a good method to estimate the PMI and can also provide valuable clues to an investigation [40] together with the essential accurate identification of taxa and with developmental data of species breeding on the corpse. Because some variation can occur in the colonization patterns due to the environmental and seasonal differences, PMI estimation on the basis of the succession data requires previous studies performed throughout the year on local carrion fauna [6]. According to Anderson [6] and Kreitlow [40], one of the most important challenges of Forensic Entomology is to provide the means for an estimation of the PMI or PIA as accurately as possible. In Murcia province, some projects of this kind have been addressed [13,41–43] and some forensic cases related to the aspects above mentioned are reported.

- Case 3. This old case (2001) was initially referred to by Arnaldos et al. [14], but some of its results were later revised in light of new scientific findings (see below) that allowed properly referring to some taxa initially misidentified or simply not identified. The corpse of a homeless man was found in the middle of November at the bottom of a dry pot with some still water, next to Murcia City. The corpse was partially clothed and showed an incised wound in the abdomen. A recovery of entomological evidence was performed because two samples (one fixed and one alive) were provided from the autopsy procedure, although the larvae had been fixed in boiling ethanol. The fixed sample consisted of larvae of *Calliphora vicina* Robineau-Desvoidy, 1830 (LIII), *Lucilia sericata* (Meigen, 1826) (LII, II-III, and III), *Chrysomya albiceps* (LI-II and II), *Muscina stabulans* (Fallén, 1826) (LII, II-III, and III), *Synthesiomya nudiseta* (Wulp, 1883) (LII, II-III, and III) and *Sarcophaga argyrostoma* (Robineau-Desvoidy, 1830), Phoridae (adults, pupae, and LI and II), and an Acari unidentified. The living sample was kept in an incubator in the laboratory, where adults of several species (*Ch. albiceps, L. sericata, S. nudiseta*, and Phoridae) were obtained. Most larvae of the fixed sample were initially identified considering Smith [44,45], Reiter [46], and Greenberg [47]. At the moment of the expertise, *S. nudiseta* specimens were not identified at the specific level, and, in the case of Sarcophagidae larvae, accurate identification was impossible due to the lack of appropriate keys at that moment. The identifications of larvae were recently reexamined in light of findings from Velásquez et al. [48] mainly for *S. nudiseta*, Grzywacz et al. [49] for *Muscina* and Szpila et al. [50], and Ubero-Pascal et al. [28] for Sarcophagidae and confirmed the identification provided in the previous paragraph. Adults Calliphoridae were identified according to González-Mora and Peris [36] and Peris and González-Mora [51]. In the case of *S. nudiseta*, a first approach was made using keys from Gregor et al. (2002) [52] (they do not include that species) and later checked with the identified larvae and confirmed with Pont [53]. In this forensic case, the recovered fauna fits with those characteristic of autumn in Murcia, formerly studied by Arnaldos et al. [13]. The corpse presented one open wound, which allowed the implication that the oviposition could have occurred immediately after death, in particular for *C. vicina* and *L. sericata*, primary species present on the corpses from the first day [12,13] in the region and that is known can coexist [54,55]. The species, as a whole, have been considered as belonging to the first waves of arthropods visiting a corpse [56]. Taking into account the oldest larvae of the primary species (*C. vicina* and *L. sericata*) and the mean temperatures registered in the area (at around 15 °C), and according to developmental data due to different authors, [57] among them, the PMI_m was estimated to be 11–17 days before the corpse was found. Despite the newly identified taxa, similar conclusions to previous ones were reached because these were based on *L. sericata* and *C. vicina* developmental data.

- Case 4. A male corpse was found at a home in Yecla (Murcia province) in the middle of September. The corpse was partly clothed and appeared to have been attacked by an animal, likely a dog. The man was last seen alive about 15 days earlier. Entomological evidence was recovered during the autopsy procedure and consisted of an adult female, three pupae, and several puparia of *Chrysomya albiceps*, one adult specimen of *Necrobia rufipes* (De Geer, 1875), and only five larvae, well developed, of *Dermestes* sp. (Figure 2).

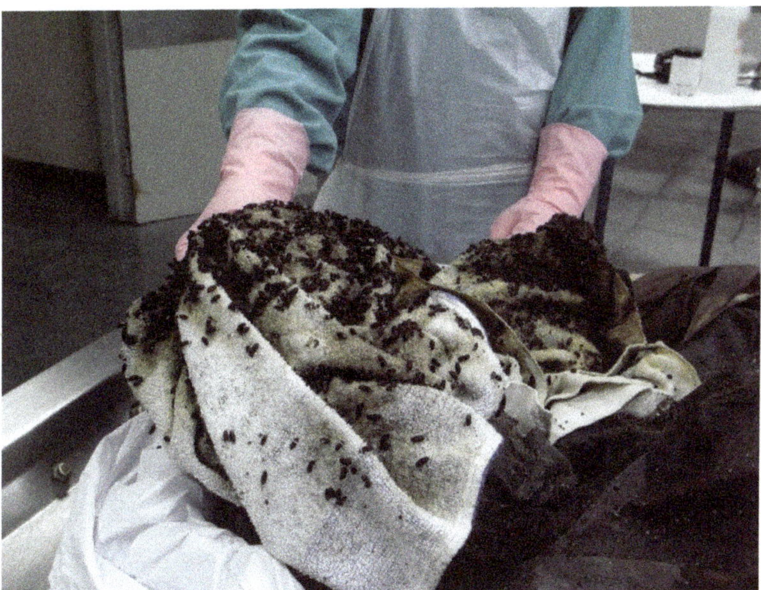

Figure 2. Autopsy procedure showing puparia from *Chrysomya albiceps*.

The best available evidence to estimate the PMI would be *Ch. albiceps* (for identification keys see above) that had clearly come to the end of its cycle. This species is the most common and abundant at the end of summer and during autumn [12]. Therefore, its presence in the corpse is congruent with the circumstances of the case. Taking into account the mean temperatures in Yecla during those days (20–25 °C) and developmental data of the species from several authors [58] and [59] among them, a PMI of around the time when the man was last seen alive was estimated. The fact of being inside a home was not considered relevant because the windows were open. Thus, it was reasonable to assume that the inside temperature was heavily influenced by the outside one.

- Case 5. A male corpse was found inside an abandoned old factory in Murcia province at the end of April. Entomological evidence was recovered during the autopsy procedure. As supplementary evidence, a complete graphical report was provided by the medical examiner in which abundant entomological specimens could be observed.

The decedent was last seen alive about one month and a half earlier. The evidence consisted of four different samples, two of them directly fixed in ethanol (one from the clothes and the other from the body itself) and two being alive (one from the clothes and the other from the body). The fixed samples included *Synthesiomya nudiseta* (pupae), *Chrysomya albiceps* (LI, II, and III and pupae), *Piophila megastigmata* (pupae and adults), *Dermestes frischi* Kugel, 1892 (adults and larvae), and adults of *Necrobia rufipes* and Histeridae (*Saprinus* sp.). (Figure 3) Diptera species were identified as indicated above. Coleoptera species were identified according to Bajerlein et al. [60] and Hackston [61].

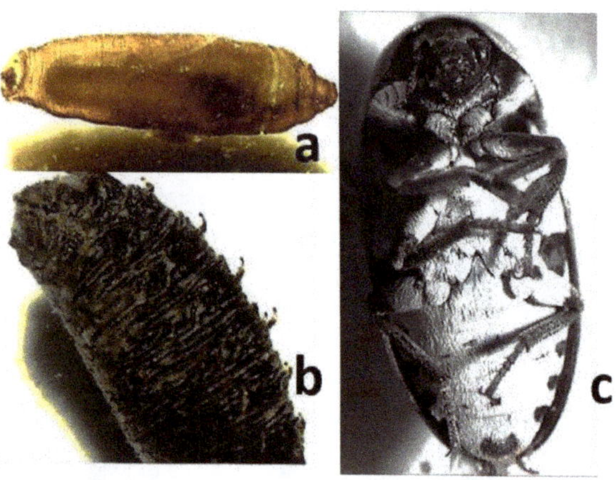

Figure 3. Entomological evidences (**a**) Piophilidae pupae (**b**) *Chrysomya albiceps* pupae (**c**) adult of *Dermestes frischi*.

From the living samples, kept at the lab, *Lucilia sericata*, *Chrysomya albiceps*, Phoridae, Chalcidoidea, and some unidentified Muscidae were obtained. The overall evidence was consistent with the sarcosaprophagous fauna characteristic of the season in the area [13] and suggested an advanced decomposition stage because Piophilidae, Dermestidae, and Histeridae, as well as aged preimaginal stages of Diptera, were present. Among all the taxa present, the most valuable for PMI estimation purposes were *Ch. albiceps*, of which some pupae were recovered. Adults emerged at the lab only two days after being kept in the incubator, suggesting that the pupae were very close to the end of the cycle. Taking into account the mean temperatures during the considered period (13.2 °C during March and 15 °C during April) and developmental available data (see former Case), a PMI_m consistent with the last time in which the decedent was seen alive was estimated.

- Case 6. A male corpse was found in an arid outdoor environment of Murcia province at the end of September. Entomological evidence recovered during the autopsy procedure was provided. Upon the request of the Forensic Entomology team, the medical examiner provided supplementary entomological evidence, proceeding of the moment of the corpse removal, and a graphical report, the latter providing an illustration of the dry decomposition stage of the corpse. The evidence from the autopsy consisted of two containers, one containing living specimens and the other individuals fixed in 80% ethanol. The evidence collected in situ consisted of one container with individuals that, although kept alive first, died at an unknown time and were preserved in 80% ethanol. The living sample was opened at the breeding lab, where it was found not to contain preimaginal stages of Diptera. This sample contained adults of *Necrobia rufipes* and Histeridae and adults and larvae of *Dermestes frischi*. The fixed sample from the autopsy contained adults of *Chrysomya albiceps*, *Cataglyphis* sp., and Histeridae and larvae of *Sarcophaga* sp. (LIII), *Lucilia sericata* (LIII), Piophilidae

(LIII), and Dermestidae. The overall evidence is consistent with the dry stage of the corpse because Dermestidae larvae are characteristics of such stage, as well as *N. rufipes* and LIII of Piophilidae [62]). Aged larvae of Calliphoridae and Sarcophagidae were seen as residual elements because it is more likely that they had completed their developmental cycle at the time that the evidence was recovered. In this case, the PMI estimation was performed on the basis of developmental data of *Dermestes* species as described by Martín-Vega et al. [63] and considering the environmental temperatures of the site during the summer (about 25–27 °C as a mean), concluding a period of about 23 days from the oviposition. According to Arnaldos et al. [13], during the summer season, adults of *Dermestes* are present on a corpse from Day 7 onwards, and larvae were recorded from Day 22, reaching a maximum around Day 37. Thus, a PMI_m of about 30 days was estimated.

- Case 7. A male corpse was found inside his home in Jumilla (Murcia province), lying on the ground and wrapped in a blanket, on November 19. According to the testimonies, the man was last seen alive on November 8. Entomological evidence was collected during the removal of the body and from the autopsy procedure. In addition, a graphical report was provided. The evidence consisted of four containers, two of them with living larvae, one with adult specimens dead and dry, and the other with immature specimens in 70% ethanol. The two living samples were introduced in an incubator chamber and kept at a temperature of 25 °C and a RH of 60%. Most specimens had pupated in about 10–13 days, and the adults emerged about one month later. The emerged adults belonged to *Calliphora vicina*, *Sarcophaga argyrostoma* (Robineau-Desvoidy, 1830) and *Chrysomya albiceps*. Meanwhile, the dead and fixed samples contained *C. vicina*, *L. sericata*, *Ch. albiceps*, *S. nudiseta*, and *Polistes gallicus* (Linnaeus 1767), as well as larvae II and III of Sarcophagidae (probably *S. argyrostoma* because the emerged adults belonged to this species) and abundant unidentified Diptera eggs and young larvae. From an in-depth study of the graphical report, it could be deduced the existence of abundant evidence on the blanket wrapping the corpse, in particular on the areas where fluid loss had occurred, as well as in contact with the body surface. There was no evidence on the head (eyes, mouth, etc.). The overall evidence was dominated by Diptera species that are common during the season [13]. A period of about seven days was estimated on the basis of the development of *C. vicina* and *S. argyrostoma*, to which the most aged larvae belonged, according to data reported in [57] and [64]. The mean outdoor temperatures at those days oscillated between 11 and 16 °C. A more accurate PMI_m could not be estimated because no data on environmental indoor conditions were provided. Moreover, the fact of the house being closed, with windows and shutters also closed, must have made it difficult for the insects to access the corpse. Thus, it was considered reasonable to add a minimum period of three days, following the experience of Goff [65], to estimate the PMI_m. This delay (colonization period) was considered on the basis of the circumstances of the scene (completely sealed building and the corpse wrapped) when compared with data from Goff [65], who stated a delay of two and a half days in the colonization of a corpse also wrapped but exposed at much higher mean environmental temperatures (20.5–23.8 °C) than in our case (11–16 °C). Therefore, an additional half-day was considered to estimate a minimum colonization period, as stated above. All considered, allowed us to conclude that death occurred around the day of the decedent's disappearance.

3. Discussion

Basic sciences are the first step in the translation of scientific knowledge to practical application. Without basic research, scientific advances would be limited and, overall, knowledge would stagnate. This situation applies to Forensic Sciences and, in particular, Forensic Entomology. Thus, it is quite difficult to reach a valid conclusion concerning entomological evidence if no previous understanding of entomofauna composition, dynamics, or any other matter exists. These issues are particularly relevant in geographical areas displaying biodiversity hot spots that may contain a high number of species and where some of which have not been studied in detail. This is the case of the Iberian Peninsula.

In addition, and as it concerns Forensic Entomology practice, Spain, like other countries, presents additional difficulties because this discipline is not included in the forensic routine and is only considered a supplementary technique. Moreover, the collection of entomological evidence can only be done by the police officers responsible for visual inspection and medical examiners, and, unfortunately, just a few professionals are willing to spend the time and efforts that are needed to recover entomological evidence for expert studies. Finally, and with very few exceptions, neither police officers nor medical examiners have been appropriately trained for entomological purposes and lack expertise on what is needed for a Forensic Entomology expert report, as proposed by Amendt et al. [4,66].

We have presented some actual interesting forensic cases that exemplify the applicability of studying entomological evidence related to corpses regardless of the appropriate protocols that might have been followed by the police officers and medical examiners. Few forensic cases reports exist in Spain, which may be the result of a low number of researchers interested in Forensic Entomology, probably due to the difficult research conditions and the lack of specific academic programs focused on this discipline. According to Wang et al. [67], case reports are quite valuable because Forensic Entomology is constantly evolving and broadening with its application. Thus, in order to increase interest in the discipline in countries, like Spain, it would be important to make more information available on actual cases related to Forensic Entomology. Examples of forensic case reports of interest in a number of countries are presented [67–69].

We think that collecting environmental data in situ or worrying about obtaining it both in situ and from the meteorological station should be done, even when dealing with an indoor location. This is a recurring issue in all the cases we studied. In Case 1, the absence of entomological evidence other than a reasonable graphical report made the conclusions, by themselves, weak. Despite that, such conclusions allowed the judicial police officers to open a new line of research that was finally successful and supported our statements. It is curious that we are usually responsible for obtaining data from the weather station. This situation can be problematic, especially in cases such as Case 7, where the indoor environmental conditions were unknown and only an approximation to the PIA could be achieved.

As indicated above, another limitation of the current forensic practice dealing with Forensic Entomology concerns the lack of training in the procedures to be followed in collecting evidence at the scene or from the corpse. This issue has been highlighted in all the presented cases, where a lack of appropriate collection of evidence made it difficult to achieve expert conclusions. Some examples of nonoptimal techniques concern the collection of a nonrepresentative sample of the existing fauna on the corpse (Cases 2, 4, and 7), bad preservation of living samples that can arrive dead to the lab (Cases 6 and 7), inadequate fixing of larvae (Cases 3, 5, and 7), or the lack of entomological evidence except for a graphical report (Case 1), which was of great quality. All the above highlight the need for providing specialized training in Forensic Entomology procedures to the different agents involved in a forensic investigation if the aim is to apply entomology to such forensic investigation.

Despite all the above, an expert report could be provided in the presented cases thanks to the previous knowledge obtained from faunistic, ecological, biological, etc., studies on entomosarcosaprophagous fauna, not from our lab in Southeastern Spain but also from other researchers. Thus, when considering the fauna collected as a whole, we were able to compare it with the results of former studies on seasonal cadaveric fauna in the area, allowing us to confirm both the environmental and seasonal origin of the studied samples.

As indicated above, knowledge of the morphology of the different species allows the accurate identification of the specimens, even when no complete individuals are available but only fragments of activity remain. Such is the situation of Case 2.

Despite all the available data on the life cycle and developmental patterns of some of the main species of forensic interest, much research is needed for more species and in more geographical areas. This is an especially important issue in areas such as the Iberian Peninsula, with high biodiversity and significant endemism in certain groups. Furthermore, the Peninsula, due to its particular geographical situation, has a high potential level of exchange between populations of different ranges.

An example of a case that could be solved thanks to the different previous studies here mentioned (faunistic, morphological, and developmental) is reported in Arnaldos et al. [34]. Moreover, this case allowed reporting for the first time a species related to a corpse (*Telomerina flavipes* (Meigen, 1830)), enlarging the global list of Diptera of forensic importance. This case, in addition, emphasizes the need for a proper collection of evidence and the identification of all entomological evidence to the deepest possible level in forensic practice because the recognition of new species involved in this process could provide additional information related to PMI [34].

All the above issues have been previously discussed by other authors in different countries. Thus, not much novelty regarding new taxa or new developmental data is now presented, although it reinforces the need for performing basic studies on taxonomy, ecology, succession process, etc., that can be applied to forensic practice. As stated by Saloña-Bordas et al. [35], although major advances in Forensic Entomology have occurred in Spain, further efforts are needed in the basic aspects of fauna related to corpses in different environments and regions. New trends in Forensic Sciences related to entomology point to microbiome [70], genetics [71], and cuticular hydrocarbons [72,73] studies among others; however, if one does not know what insect species are involved the features of the succession process in a given area, all attempts to deepen in the other, and newest, aspects would be predicted to be unsuccessful. It is important to remember that the first and critical attribute to be discerned concerns the fauna related to a corpse and its state and both aspects vary according to the environments, regions, and seasons.

4. Conclusions

The cases presented here demonstrate the utility of performing biological studies of all the entomological fauna of forensic interest. While the general opinion may be that most of the obtained results of such studies are not directly applicable to the forensic practice, the tremendous variability of forensic scenes and situations (corpse size and condition, wounds, wrapping, type of habitat, season, particular environment, etc.) validate the fact that almost every result is useful for providing an accurate expert report when needed.

Author Contributions: Conceptualization, Methodology, Formal Analysis, Investigation, Resources, Writing—Original Draft Preparation, Review and Editing; M.-I.A. and M.-D.G. Both authors have read and agreed to the published version of the manuscript.

Funding: This research received no external funding.

Institutional Review Board Statement: Not applicable.

Informed Consent Statement: Not applicable.

Data Availability Statement: Not applicable.

Acknowledgments: We are grateful to M.L. García, senior scientist level, from Kanalis Consulting, for reviewing the wording.

Conflicts of Interest: The authors declare no conflict of interest.

References

1. García, M.-D.; Arnaldos, M.-I.; Lago, V.; Ramírez, M.; Ubero, N.; Prieto, J.; Presa, J.-J.; Luna, A. The paradigm of interdisciplinarity in forensic investigation. A case in Southeastern Spain. *Leg. Med.* **2021**, *48*, 101817. [CrossRef] [PubMed]
2. Wells, J.D.; LaMotte, L.R. Estimating the Postmortem Interval. In *Forensic Entomology. The Utility of Arthropods in Leg. Investigations*, 2nd ed.; Byrd, J.H., Castner, J.L., Eds.; CRC Press: Boca Raton, FL, USA, 2010; pp. 367–388.

3. Schoenly, K.; Goff, M.L.; Wells, J.D.; Lord, W. Quantifying statistical uncertainty in succession-based entomological estimates of the postmortem interval in dead scene investigations: A simulation study. *Am. Entomol.* **1996**, *42*, 106–112. [CrossRef]

4. Amendt, J.; Anderson, G.; Campobasso, C.P.; Dadour, I.; Gaudry, E.; Hall, M.J.R.; Moretti, T.C.; Sukontason, K.L.; Villet, M.H. Standard practices. In *Forensic Entomology. International Dimensions and Frontiers*; Tomberlin, J.K., Benbow, M.E., Eds.; CRC Press: Boca Raton, FL, USA, 2015; pp. 381–398.

5. Grassberger, M.; Reiter, C. Effect of temperature on development of *Liopygia* (=*Sarcophaga*) *argyrostoma* (Robineau-Desvoidy) (Diptera: Sarcophagidae) and its forensic implications. *J. Forensic Sci.* **2002**, *47*, 1332–1336. [CrossRef]

6. Anderson, G.S. Factors that influence insect succession on carrion. In *Forensic Entomology. The Utility of Arthropods in Leg. Investigations*, 2nd ed.; Byrd, J.H., Castner, J.L., Eds.; CRC Press: Boca Raton, FL, USA, 2010; pp. 201–250.

7. Prado e Castro, C.; Sousa, J.P.; Arnaldos, M.I.; Gaspar, J.; García, M.D. Blowflies (Diptera: Calliphoridae) activity in sun exposed and shaded carrion in Portugal. *Ann. Soc. Entomol. Fr.* **2011**, *47*, 128–139. [CrossRef]

8. Pérez-Marcos, M.; Arnaldos, M.I.; López-Gallego, E.; Luna, A.; Khedre, A.; García, M.D. An approach for identifying the influence of carcass type and environmental features on sarcosaprophagous Diptera communities. *Ann. Soc. Entomol. Fr.* **2018**, *54*, 367–380. [CrossRef]

9. Pérez-Marcos, M.; López-Gallego, E.; Arnaldos, M.I.; Martínez-Ibáñez, D.; García, M.D. Formicidae (Hymenoptera) community in corpses at different altitudes in a semiarid wild environment in the southeast of the Iberian Peninsula. *Entomol. Sci.* **2020**, *23*, 297–310. [CrossRef]

10. VanLaerhoven, S.L. Ecological theory and its application in forensic entomology. In *Forensic Entomology. The Utility of Arthropods in Leg. Investigations*, 2nd ed.; Byrd, J.H., Castner, J.L., Eds.; CRC Press: Boca Raton, FL, USA, 2010; pp. 493–517.

11. Zabala, J.; Díaz, B.; Saloña-Bordas, M.I. Seasonal blowfly distribution and abundance in fragmented landscapes. Is it useful in forensic inference about where a corpse has been decaying? *PLoS ONE* **2014**, *9*, e99668. [CrossRef]

12. Arnaldos, I.; Romera, E.; García, M.D.; Luna, A. An initial study on the succession of sarcosaprophagous Diptera (Insecta) on carrion in the southeastern Iberian península. *Int. J. Leg. Med.* **2001**, *114*, 156–162. [CrossRef] [PubMed]

13. Arnaldos, M.I.; Romera, E.; Presa, J.J.; Luna, A.; García, M.D. Studies on seasonal arthropod succession on carrion in the southeastern Iberian Peninsula. *Int. J. Leg. Med.* **2004**, *118*, 197–205. [CrossRef]

14. Arnaldos, M.I.; García, M.D.; Romera, E.; Presa, J.J.; Luna, A. Estimation of postmortem interval in real cases based on experimentally obtained entomological evidence. *For. Sci. Int.* **2005**, *149*, 57–65. [CrossRef]

15. Prado e Castro, C.; Serrano, A.; Martins da Silva, P.; García, M.D. Carrion flies of forensic interest: A study of seasonal community composition and succession in Lisbon, Portugal. *Med. Vet. Entomol.* **2012**, *26*, 417–431. [CrossRef]

16. Prado e Castro, C.; García, M.D.; Martins da Silva, P.; Faria e Silva, I.; Serrano, A. Coleoptera of forensic interest: A study of seasonal community composition and succession in Lisbon, Portugal. *For. Sci. Int.* **2013**, *232*, 73–83. [CrossRef]

17. Prado e Castro, C.; García, M.-D.; Palma, C.; Martínez-Ibáñez, M.-D. First report on sarcosaprophagous Formicidae from Portugal (Insecta: Hymenoptera). *Ann. Soc. Entomol. Fr.* **2014**, *50*, 51–58. [CrossRef]

18. Pérez Marcos, M. Estudio de la Fauna Entomológica Asociada a un Cadáver en un Enclave Natural Montañoso en Murcia (SE España). Ph.D. Thesis, Universidad de Murcia, Murcia, Spain, 2016.

19. Martín-Vega, D.; Baz, A. Sarcosaprophagous Diptera assemblages in natural habitats in central Spain: Spatial and seasonal changes in composition. *Med. Vet. Entomol.* **2013**, *27*, 64–76. [CrossRef]

20. Martín-Vega, D.; Baz Ramos, A.; Cifrián Yagüe, B.; Díaz Aranda, L.M. Long-term insect successional patterns on pig carcasses in central Spain. *Int. J. Leg. Med.* **2019**, *133*, 1581–1592. [CrossRef] [PubMed]

21. Díaz-Aranda, L.M.; Martín-Vega, D.; Gómez-Gómez, A.; Cifrián, B.; Baz, A. Annual variation in decomposition and insect succession at a periurban area of central Iberian Peninsula. *J. Forensic Leg. Med.* **2018**, *56*, 21–31. [CrossRef] [PubMed]

22. Díaz Martín, B.; Saloña-Bordas, M.I. Arthropods of forensic interest associated to pig carcasses in Aiako Harria Natural Park (Basque Country, North of Spain). *Cienc. Forense* **2015**, *12*, 211–232.

23. Morales Rayo, J.; San Martín Peral, G.; Saloña Bordas, M.I. Primer estudio sobre la reducción cadavérica en condiciones sumergidas en la Península Ibérica, empleando un modelo de cerdo doméstico (*Sus scrofa* L., 1758) en el Río Manzanares (Comunidad Autónoma de Madrid). *Cienc. Forense* **2014**, *11*, 271–304.

24. Peralta Álvarez, B.; GilArriortua, M.; Saloña-Bordas, M.I. Variabilidad espacial y temporal de Califóridos (Diptera) necrófagos de interés forense capturados en la Comunidad Autónoma del Principado de Asturias (CAPA). *Boln. Asoc. Esp. Ent.* **2013**, *37*, 301–314.

25. Saloña, M.; Moneo, J.; Díaz, B. Estudio sobre la distribución de califóridos en la Comunidad Autónoma del País Vasco. *Boln. Asoc. Esp. Ent.* **2009**, *33*, 63–89.

26. Pérez-Bote, J.L.; Rivera, V.; Santos Almeida, A. La comunidad sarcosaprófaga en un entorno periurbano. *Boln. Asoc. Esp. Ent.* **2012**, *36*, 299–314.

27. Ubero-Pascal, N.; López-Esclapez, R.; García, M.-D.; Arnaldos, M.I. Morphology of preimaginal stages of *Calliphora vicina* Robineau-Desvoidy, 1830 (Diptera, Calliphoridae). A comparative study. *For. Sci. Int.* **2012**, *219*, 228–243. [CrossRef]

28. Ubero-Pascal, N.; Paños, A.; García, M.-D.; Presa, J.-J.; Torres, B.; Arnaldos, M.-I. Micromorphology of immature stages of *Sarcophaga (Liopygia) cultellata* Pandellé, 1896 (Diptera: Sarcophagidae), a forensically important fly. *Microsc. Res. Tech.* **2015**, *78*, 148–172. [CrossRef] [PubMed]

29. Paños, A.; Arnaldos, M.I.; García, M.D.; Ubero-Pascal, N. Ultrastructure of preimaginal stages of *Piophila megastigmata* McAlpine, 1978 (Diptera, Piophilidae): A fly of forensic importance. *Parasitol. Res.* **2013**, *112*, 3771–3788. [CrossRef] [PubMed]

30. Paños-Nicolás, A.; Arnaldos, M.I.; García, M.D.; Ubero-Pascal, N. *Sarcophaga* (*Liosarcophaga*) *tibialis* Macquart 1851 (Diptera; Sarcophagidae): Micropmorphology of preimaginal stages of a fly of medical and veterinary interest. *Parasitol. Res.* **2015**, *114*, 4031–4050. [CrossRef]

31. Martín-Vega, D.; Baz, A. Spatiotemporal Distribution of Necrophagous Beetles (Coleoptera: Dermestidae, Silphidae) Assemblages in Natural Habitats of Central Spain. *Ann. Entomol. Soc. Am.* **2012**, *105*, 44–53. [CrossRef]

32. Arnaldos Sanabria, M.I.; Torres Tomás, B.; García García, M.D. Primeros datos sobre el desarrollo del ciclo de vida del díptero de importancia forense *Sarcophaga cultellata* Pandellé, 1896 (Sarcophagidae). *Cuad. Med. Forense* **2013**, *19*, 6–12. [CrossRef]

33. Arnaldos, M.I.; Sánchez, F.; Álvarez, P.; García, M.D. A forensic entomology case from the Southeastern Iberian Peninsula. *Anil Aggrawal's Internet J. Forensic Med. Toxicol.* **2004**, *5*, 22–25.

34. Arnaldos, M.-I.; Ubero-Pascal, N.; García, R.; Carles-Tolrá, M.; Presa, J.J.; García, M.-D. The first report of *Telomerina flavipes* (Meigen, 1830) (Diptera, Sphaeroceridae) in a forensic case, with redescription of its pupa. *For. Sci. Int.* **2014**, *242*, e22–e30. [CrossRef] [PubMed]

35. Saloña-Bordas, M.I.; Magaña-Loarte, C.; García-Rojo, A.M. Spain. In *Forensic Entomology. International Dimensions and Frontiers*; Tomberlin, J.K., Benbow, M.E., Eds.; CRC Press: Boca Raton, FL, USA, 2015; pp. 145–157.

36. González-Mora, D.; Peris, S.V. Los Calliphoridae de España. 1: Rhiniinae y Chrysomyinae. *Eos* **1988**, *64*, 91–139.

37. Avila, F.W.; Goff, M.L. Arthropod succession patterns onto burnt carrion in two contrasting habitats in the Hawaiian Islands. *J. Forensic Sci.* **1998**, *43*, 581–586. [CrossRef]

38. Pai, C.-Y.; Jien, M.-C.; Li, L.-H.; Cheng, Y.-Y.; Yang, C.-H. Application of forensic entomology to postmortem interval determination of a burned human corpse: A homicide case report from Southern Taiwan. *J. Formos Med. Assoc.* **2007**, *106*, 792–798. [CrossRef]

39. Carles-Tolrá, M.; Arnaldos, M.I.; Begoña, I.; García, M.D. Novedades faunísticas y entomosarcosaprófagas de la Región de Murcia, SE de España (Insecta: Diptera). *Boln. Soc. Esp. Hist. Nat.* **2014**, *108*, 21–35.

40. Tabor Kreitlow, K.L. Insect succession in a natural environment. In *Forensic Entomology. The Utility of Arthropods in Leg. Investigations*, 2nd ed.; Byrd, J.H., Castner, J.L., Eds.; CRC Press: Boca Raton, FL, USA, 2010; pp. 251–269.

41. Begoña Gaminde, I. Sucesión de la entomofauna cadavérica en un medio montañoso del Sureste de la Península Ibérica. Ph.D. Thesis, Universidad de Murcia, Murcia, Spain, 2015.

42. López Gallego, E. Sucesión de la Entomofauna Sarcosaprófaga en un Hábitat Forestal Mediterráneo. Ph.D. Thesis, Universidad de Murcia, Murcia, Spain, 2016.

43. Pérez-Marcos, M.; Arnaldos-Sanabria, M.I.; García, M.D.; Presa, J.J. Examining the sarcosaprophagous fauna in a natural mountain environment (Sierra Espuña, Murcia, Spain). *Ann. Soc. Entomol. Fr.* **2016**, *52*, 264–280. [CrossRef]

44. Smith, K.G.V. *A manual of Forensic Entomology*; The Trustees of the British Museum (Natural History); London and Comstock Publishing Associates: London, UK, 1986.

45. Smith, K.G.V. An introduction to the immature stages of British flies. Diptera larvae, with notes on eggs, puparia and pupae. In *Handbooks for the Identification of British Insects*; CABI: London, UK, 1989; Volume 10.

46. Reiter, C. Zum wachstumsverhalten der Maden der blauen Schmeissfliege *Calliphora vicina*. *Z. Rechtsmed* **1984**, *91*, 295–308. [CrossRef]

47. Greenberg, B. *Flies and Disease. Volume I. Ecology, Classification and Biotic Associations*; Princeton University Press: Princeton, NJ, USA, 1971.

48. Velásquez, Y.; Magaña, C.; Martínez-Sánchez, A.; Rojo, S. Diptera of forensic importance in the Iberian Peninsula: Larval identification key. *Med. Vet. Entomol.* **2010**, *4*, 293–308. [CrossRef]

49. Grzywacz, A.; Hall, M.J.R.; Pape, T. Morphology successfully separates third instar larvae of *Muscina*. *Med. Vet. Entomol.* **2015**, *29*, 314–329. [CrossRef] [PubMed]

50. Szpila, K.; Richet, R.; Pape, T. Third instar larvae of flesh flies (Diptera: Sarcophagidae) of forensic importance—Critical review of characters and key for European species. *Parasitol. Res.* **2015**, *114*, 2279–2289. [CrossRef] [PubMed]

51. Peris, S.V.; González Mora, D. Los Calliphoridae de España, III: Luciliini (Diptera). *Boln. Soc. Esp. Hist. Nat. (Sección Biológica)* **1991**, *87*, 187–207.

52. Gregor, F.; Rozkosny, R.; Bartak, M.; Vanhara, J. *The Muscidae (Diptera) of Central Europe*; Folia Facultatis Scientiarum Naturalium Universitatis Masarykianae Brunensis: Brno, Czech Republic, 2002; Volume 107, pp. 1–280.

53. Pont, A.C. A review of Fannidae and Muscidae (Diptera) of Arabian Peninsula. *Fauna Saudi Arabia* **1991**, *12*, 312–365.

54. Leclercq, M. *Entomologie et Médecine Legale. Datation de la Morte*; Masson et cie Paris: Paris, France, 1978.

55. Leclercq, M.; Brahy, G. Entomologie et médicine legale: Datation de la mort. *J. Med. Leg.* **1985**, *28*, 271–278.

56. Leclercq, M.; Verstraeten, C. Entomologie et médecine légale. L éntomofaune des cadavres humains: Sa succession par son interprétation, ses résultatsses perspectives. *J. Med. Leg. Droit Med.* **1993**, *36*, 205–222.

57. Anderson, G.S. Minimum and maximum development rates of some forensically important Calliphoridae (Diptera). *J. Forensic Sci.* **2000**, *45*, 824–832. [CrossRef] [PubMed]

58. Grassberger, M.; Friedrich, E.; Reiter, C. The blowfly *Chrysomya albiceps* (Wiedemann) (Diptera: Calliphoridae) as a new forensic indicator in Central Europe. *Int. J. Leg. Med.* **2003**, *117*, 75–81. [CrossRef]

59. Richards, C.S.; Paterson, I.D.; Villet, M.H. Estimating the age of immature *Chrysomya albiceps* (Diptera: Calliphoridae), correcting for temperature and geographical latitude. *Int. J. Leg. Med.* **2008**, *122*, 271–279. [CrossRef]
60. Bajerlein, D.; Konwerski, S.; Madra, A. European beetles of forensic importance. Identification guide. In Proceedings of the 9th Meeting of the Europen Association for Forensic Entomology, Torun, Poland, 18–21 April 2012.
61. Hackston, M. Family Dermestidae, Creative Commons. 2014. Available online: https://quelestcetanimal-lagalerie.com/wp-content/uploads/2012/11/Family-Dermestidae-illustrated-key-to-British-species.pdf (accessed on 15 July 2017).
62. Byrd, J.H.; Castner, J.L. Insects of forensic importance. In *Forensic Entomology. The Utility of Arthropods in Leg. Investigations*, 2nd ed.; Byrd, J.H., Castner, J.L., Eds.; CRC Press: Boca Raton, FL, USA, 2010; pp. 39–126.
63. Martín-Vega, D.; Díaz-Aranda, L.M.; Baz, A.; Cifrian, B. Effect of temperature on the survival and development of three forensically relevant Dermestes species (Coleoptera: Dermestidae). *J. Med. Entomol.* **2017**, *54*, 1140–1150. [CrossRef]
64. Gómez Martín, N.I. *Desarrollo del Ciclo Vital de Sarcophaga Argyrostoma (Robineau-Desvoidy, 1830) Bajo Condiciones Controladas de Temperatura. Importancia en las Ciencias Forenses. Trabajo Fin de Máster en Ciencias Forenses*; Universidad de Murcia: Murcia, Spain, 2014.
65. Goff, M.L. Problems in Estimation of Postmortem Interval Resulting from Wrapping of the Corpse: A Case Study from Hawaii'. *J. Agric. Entomol.* **1992**, *9*, 237–243.
66. Amendt, J.; Campobasso, C.P.; Gaudry, E.; Reiter, C.; LeBlanc, H.N.; Hall, M.J.R. Best practice in forensic entomology—Standards and guidelines. *Int. J. Leg. Med.* **2007**, *121*, 90–104. [CrossRef] [PubMed]
67. Wang, M.; Chu, J.; Wang, Y.; Li, F.; Lia, M.; Shi, H.; Zhang, Y.; Hu, G.; Wang, J. Forensic entomology in China: Four case reports. *J. Forensic Leg. Med.* **2019**, *63*, 40–47. [CrossRef]
68. Lo Pinto, S.; Giordani, G.; Tuccia, F.; Ventura, F.; Vanin, S. First records of *Synthesiomyia nudiseta* (Diptera: Muscidae) from forensic cases in Italy. *For. Sci. Int.* **2017**, *276*, e1–e7. [CrossRef] [PubMed]
69. Magni, P.; Pérez-Bañón, C.; Borrini, M.; Dadour, I.R. *Syritta pipiens* (Diptera: Syrphidae), a new species associated with human cadavers. *For. Sci. Int.* **2013**, *231*, e19–e23. [CrossRef] [PubMed]
70. Benbow, M.E.; Pechal, J.L. Forensic Entomology and the Microbiome. In *Forensic Entomology. The Utility of Arthropods in Leg. Investigations*, 3rd ed.; Byrd, J.H., Tomberlin, J.K., Eds.; CRC Press: Boca Raton, FL, USA, 2019; pp. 449–517.
71. Stevens, J.R.; Picard, C.J.; Wells, J.D. Molecular genetic methods for Forensic Entomology. In *Forensic Entomology. The Utility of Arthropods in Leg. Investigations*, 3rd ed.; Byrd, J.H., Tomberlin, J.K., Eds.; CRC Press: Boca Raton, FL, USA, 2019; pp. 253–268.
72. Drijfhout, F.P. Cuticular hydrocarbons: A new tool in Forensic Entomology? In *Current Concepts in Forensic Entomology*; Amendt, J., Campobasso, C.P., Goff, M.L., Grassberger, M., Eds.; Springer: Dordrecht, The Netherlands, 2009; pp. 179–204.
73. Moore, H.E.; Drijfhout, F.P. Surface hydrocarbons as min-PMI indicators. Fit for purpose? In *Forensic Entomology. International Dimensions and Frontiers*; Tomberlin, J.K., Benbow, M.E., Eds.; CRC Press: Boca Raton, FL, USA, 2015; pp. 361–380.

Article

Small Bait Traps May Not Accurately Reflect the Composition of Necrophagous Diptera Associated to Remains

Kathleen LeBlanc, Denis R. Boudreau and Gaétan Moreau *

Département de Biologie, Université de Moncton, Moncton, NB E1A 3E9, Canada; ekl4765@umoncton.ca (K.L.); edb5864@umoncton.ca (D.R.B.)
* Correspondence: gaetan.moreau@umoncton.ca; Tel.: +1-506-8584975

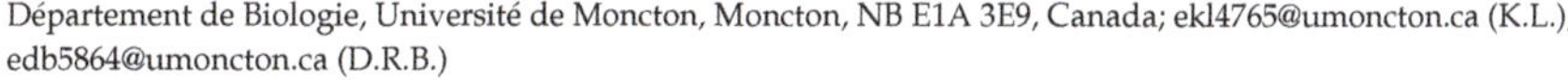

Simple Summary: Fly families such as Calliphoridae and Muscidae contribute to the decomposition of cadavers and play an important role in courtroom proceedings, in part because of the clues they provide to help determine the time of death. In forensic entomology studies, they are often sampled using small bait traps containing a small amount of decomposing animal tissue. To determine whether the fly assemblages recovered by small bait traps are similar to those found on whole remains, we simultaneously documented the flies found on domestic pig carcasses and within small traps baited with pork liver. Results indicated that the fly assemblages found in the small bait traps and on the carcasses were different and reinforced the fact that caution should be exercised when data obtained from small bait traps are used in court.

Abstract: Small bait traps are beginning to emerge in forensic entomology as a new approach to sample early-colonizing necrophagous Diptera species while reducing the investment in time and energy in obtaining information. To test the hypothesis conveyed by the literature that these traps can be a substitute for whole carcasses, we simultaneously documented the Diptera assemblages visiting and colonizing domestic pig carcasses and small traps baited with pork liver. Results indicated that Diptera species occurrence and assemblage composition in the small bait traps and on the carcasses differed, while they were similar when comparing only the pig carcasses. These results are in agreement with the literature that examined insect colonization of other decaying substrates. Although small bait traps can be useful tools to document the communities of necrophagous Diptera in a given area, we stress that caution must be exercised when extending the data obtained by these traps to courtroom proceedings.

Keywords: animal carcasses; bait attraction; decaying substrate; insect succession; forensic entomology; postmortem interval; vertebrate decomposition

Citation: LeBlanc, K.; Boudreau, D.R.; Moreau, G. Small Bait Traps May Not Accurately Reflect the Composition of Necrophagous Diptera Associated to Remains. *Insects* **2021**, *12*, 261. https://doi.org/10.3390/insects12030261

Academic Editors: Damien Charabidze and Daniel Martín-Vega

Received: 23 February 2021
Accepted: 17 March 2021
Published: 20 March 2021

Publisher's Note: MDPI stays neutral with regard to jurisdictional claims in published maps and institutional affiliations.

1. Introduction

Passive data retrieval methods are widely used in different fields of research to reduce the investment in time and energy in obtaining information [1–4]. This trend is for example present in the field of forensic entomology, which is the study of the association between insects and corpses for forensic and legal purposes, often to determine the post-mortem interval (PMI) of a corpse [5–7]. In this field of applied research, insect traps are increasingly used for the sampling of necrophagous Diptera [1,8], one of the key sarcosaprophagous faunal groups [5,9]. Different models of large and small bait traps have been developed, the more commonly used traps in forensic entomology being bottle traps, Schoenly traps, sticky target traps, wind-oriented traps, funnel traps, and bait bins [1,3,8–13]. Schoenly traps are baited with whole carcasses [3]. They have proven to be more representative of the colonization of carcasses by both Diptera and Coleoptera than traditional forensic methods. The latter involves direct manual capture using tools such as sweep nets, brushes, tweezers, and aspirators [9]. However, Schoenly traps represent a significative investment and their deployment in a large-scale entomological monitoring program or in a highly urbanized environment might be unrealistic.

Small bait traps are often advantageous compared to the human cadavers, vertebrate carcasses, or Schoenly traps because they emit a faint odor and there are no ethical, legal, or convenience implications associated to their use [1]. These traps mainly target Diptera, notably blowflies (Diptera: Calliphoridae), which are often the first arthropods to colonize carcasses [14]. They are generally baited with a small amount, being 50–200 g, of various animal organs such as pork muscle, pork liver, beef liver, chicken liver, or squid [1,12,15,16]. Synthetic compounds such as sodium sulfide, which act as constant chemical attractants to necrophagous Diptera, may also be used as bait, often with pork liver [8,11,13,17,18]. These small bait traps are emerging in forensic entomology as a method to record the local biodiversity of early-colonizing necrophagous Diptera species [1,15,19], which is known to vary according to the environmental conditions and the geographical region (e.g., [1,5,9]). The literature herein indeed shows that small bait traps are increasingly used to develop databases by forensic institutes and forensic medicine departments in various countries such as Australia, Austria, Brazil, Canada, Croatia, Germany, Portugal, Romania, Spain, Turkey, the United Kingdom, and the United States, to name a few. However, to our knowledge, no experiment has been done to statistically compare the composition of the insect assemblages found in these traps and on whole carcasses placed simultaneously in the same habitat. Nevertheless, some authors have claimed that small bait traps can be a replacement for human cadavers, based on the untested assumption that the bait must emit volatile organic compounds similar to those emitted by a cadaver/carcass in the early stages of decomposition [1,20]. This claim seems to be at odds with the literature indicating that the amount of animal carrion available has a significant impact on its visitation and colonization by Diptera in terms of their abundance and composition (e.g., [5,15,21,22]). As two very different animal substrates, sampling approaches and sampling intervals are compared, we have difficulty believing that previous studies have been able to obtain similar insect assemblages, even in the early stages of decomposition.

To test the claim that small bait traps can be a substitute for whole cadavers or carcasses, which affects the interpretation of data from these traps in case scenarios, we examined the visitation by necrophagous Diptera of domestic pig carcasses (*Sus scrofa domesticus*) and commercial traps baited with 60 g of pork liver. Domestic pig carcasses of ~22 kg are considered the best replacement for human cadavers, due to their anatomical characteristics, diet, replicability and practicality, and fur [10,23,24]. The hypothesis tested here is that the attractiveness of carrion to Diptera changes with species interactions, decomposition, and amount of available tissue. From this and the above cited literature, we predict that the Diptera assemblage collected on carcasses and in small bait traps will differ. If this prediction is rejected, we predict that the trapped insects will be representative of those which are collected on remains for a short period in early decomposition. Nevertheless, we not only documented early decomposition but prolonged the study until the dry stage to determine whether the small bait traps may be representative of a later decomposition stage, because they were baited with partially decomposed internal tissues.

2. Materials and Methods

2.1. Experimental Design

Domestic pig carcasses weighing 22.6 ± 3.3 kg were purchased from a certified hog farm located approximately 60 km from the study site. The animals were killed using a compressed air pistol with a retractable pin placed directly on the forehead. They were immediately placed in two airtight plastic bags to prevent colonization by insects prior to their arrival to the study site and transported there in less than 2 h. The study site was a large abandoned agricultural field located approximately 40 km northeast of Moncton in Cocagne, New Brunswick, Canada (46°20.364′ N, 64°39.724′ W). Abundant vegetation in the study field included Canada goldenrod (*Solidago canadensis* L.), dandelion (*Taraxacum officinale* Weber ex Wiggers), flat-top white aster (*Doellingeria umbellata* (Mill.) Nees), grasses from the Poaceae family (e.g., *Festuca trachyphylla* (Hack.) Krajina, *Phleum pratense* L.), and red clover (*Trifolium pratense* L.). A few spaced trees were present within the field, such as tamarack (*Larix laricina* (Du Roi) K.Koch), grey alder (*Alnus incana* (L.) Moench ssp. *rugosa* (Du Roi) R.T.Clausen), black spruce (*Picea mariana* (Mill.) Britton, Sterns & Poggenb.), dogwoods (*Cornus spp.*), and spruces (*Picea spp.*).

On 6 July 2020, two fresh carcasses and two small bait traps were set at the study site in direct sunlight. These four experimental units were placed in a way to maximize the distance which separated them (Figure S1). The carcasses were positioned 144 m away from one another, the traps were positioned 149 m from one another and at least 90 m from carcasses. A distance exceeding 50 m prevents cross-contamination by crawling fly larvae between carcasses [25,26]. This trial ended on 20 July 2020 and the carcasses, which had all reached the dry decay stage, were removed. A new set of four experimental units (i.e., two new carcasses and two traps) were set at the study site on 5 August 2020 (Figure S1). These carcasses were 142 m away from one another and the traps were 147 m from one another and at least 98 m from carcasses. This second trial ended on 19 August 2020, when the carcasses had all reached the dry decay stage.

The carcasses were placed on a 61 × 122 × 1 cm perforated plastic surface covered in soilless potting soil. The plastic surface allowed decomposition fluids and rain to leach away, delineated the sampling area, and facilitated the identification and recovery of insects that burrow in the soil or hide in the vegetation [27]. Each carcass was protected from vertebrate scavengers by a 61 × 61 × 122 cm cage made with chicken wire and anchored to the ground using pegs, bungee cords, and cinder blocks. Small bait traps consisted of "Flies Be Gone" commercial traps, modified to recover necrophagous Diptera. The closed bottom part of the trap was cut to create an opening and insert a 120 mL cylindrical plastic cup containing 60 g of minced pork liver as bait that was attached using duct tape. To hinder the direct colonization of the liver while allowing for the propagation of volatile organic compounds, the open end of the cup was wrapped in tulle fabric. Previously frozen baits were removed from the freezer and allowed to defrost and decompose at room temperature 24 h before the traps were set. When set, the traps were suspended from metal rods on-site, about 1.5 m from the ground and secured to the rod using zip ties.

2.2. Sampling Procedure

The sampling protocol was adapted from a previously published procedure [28]. During the decomposition process, the fauna visiting the carcasses was documented daily between 10:00 A.M. and 11:59 A.M. Carcasses were sampled in a random order. They were approached carefully to diminish disturbance of insects due to visitation. For up to 10 min, we observed the carcasses to tally the number of specimens per family. To complement this, a 10-min period was allocated for collection of flying insects using an entomological net. The few crawling insects that could not be identified in the field were collected using forceps for a maximum of 10 min. To minimize the impact of investigator disturbance on the decomposition process, less than 10% of any taxa was collected. The exact number of specimens collected was based on the number from each taxon that were visually tallied above. As can be seen in Section 3, only two families (i.e., Calliphoridae and Piophilidae) were abundant enough to be sampled from carcasses. Collected specimens were then placed in a jar containing 70% ethanol for preservation until further identification. Every three days, the small bait traps were replaced with new ones. This interval was used because in our geographic area, small bait traps are usually still empty after a single day of sampling.

All Calliphoridae and Piophilidae specimens were identified at the species level using Marshall et al. [29] and the Université de Moncton reference collection, with the exception of *Pollenia sp.* and *Protocalliphora sp.*, which were identified at the genus level. Other Diptera that were not abundant enough to be sampled from carcasses identified at the family level. Specimens were deposited in the Université de Moncton reference collection. Air temperature was obtained from a federal weather station located in Bouctouche, approximately 17 km from the study site. The average temperature and postmortem accumulation of degree-days over 5 °C (ADD$_5$) was calculated daily using the following formula: ADD$_5$ = [{T$_{min}$ + T$_{max}$}/2] − 5 °C, where T$_{min}$ and T$_{max}$ represent the daily minimum and maximum air temperatures, respectively, and 5 °C represents the minimum developmental threshold over which the accumulation was considered. The 5 °C threshold was selected because it led to the best models in terms of overall fit and percentage of variability explained by these models.

2.3. Statistical Analyses

To allow for the comparisons between traps and carcasses, the Diptera documented on each carcass as well as the postmortem ADD$_5$ during the trap sampling interval were individually summed up and the mean daily temperature was averaged for the same period. To examine the effects of postmortem ADD$_5$ on the abundance of Calliphoridae and other dipterans on carcasses and in small bait traps, we used generalized additive mixed models (GAMMs). The abundance of observed Diptera identified to the family level was used for this analysis. The GAMMs accounted for the autoregressive structure of the data, the repeated measures and the Poisson data distribution [30]. To examine the effects of the carcasses on small bait traps and vice-versa, a GAMM including the abundance on the other type of unit as a covariate was carried out. The Sørensen's and Percent similarity indices were calculated by comparing the species composition of adjacent pairs of carcasses and traps obtained from a combination of the abundance data from visual estimations and from the collection and subsequent identification of Diptera. A descriptive table presenting this data in terms of the percentage of the species encountered was also produced. The effects of postmortem ADD$_5$ on the index values were examined using the same GAMMs as above but with Gaussian (Sørensen) and quasipoisson (Percent similarity) data distribution. The Sørensen's and Percent similarity indices were also calculated by comparing species composition on carcasses exposed in the same month. Finally, we performed a canonical correlation analysis on data obtained from a combination of the abundance from visual estimations and from the collection and subsequent identification of Diptera using the *CCorA* function from the package *vegan*. Canonical correlation analysis allows an evaluation and an illustration of the relationships between two multivariate sets of variables, namely the study parameters (matrix of independent variables) and the abundance of each of the taxa (matrix of dependent variables). All statistical analyses were performed using R version 3.5.2 [31]. We ensured that all assumptions of each analysis were satisfied.

3. Results

3.1. Abundance Data

During the study, a total of 5001 Diptera specimens were captured using small bait traps and 4619 Diptera specimens were documented on the carcasses. Of the Diptera specimens collected in the traps, 98.0% were Calliphoridae, whereas they represented 40.3% of the specimens retrieved from the carcasses. Other Diptera represented 2.0% and 59.7% of the specimens in the baited traps and on the carcasses, respectively. For Calliphoridae, the abundance on the carcasses and in the baited traps reached a peak between 50 and 100 postmortem ADD$_5$ before plummeting (Figure 1a,b). The relationship between the Calliphoridae observed on the carcasses and postmortem ADD$_5$ was strong (i.e., r^2 = 0.63) but was comparatively weak for those captured by the baited traps (i.e., r^2 = 0.09) (Figure 1a,b). In fact, the abundance of Calliphoridae on the carcasses strongly influenced the abundance of Calliphoridae captured in the baited traps, with this being a positive relation (F = 21.602; $p < 0.01$). However, the presence of the traps did not have an effect on the abundance of Calliphoridae observed on the carcasses (F = 1.424; p = 0.22). Similarly, the abundance of other dipterans peaked between 50 and 200 postmortem ADD$_5$ on carcasses and between 50 and 100 postmortem ADD$_5$ in traps (Figure 1c,d). However, for both types of experimental units, the relationship between dipteran abundance and postmortem ADD$_5$ was weak (i.e., $r^2 \leq 0.10$).

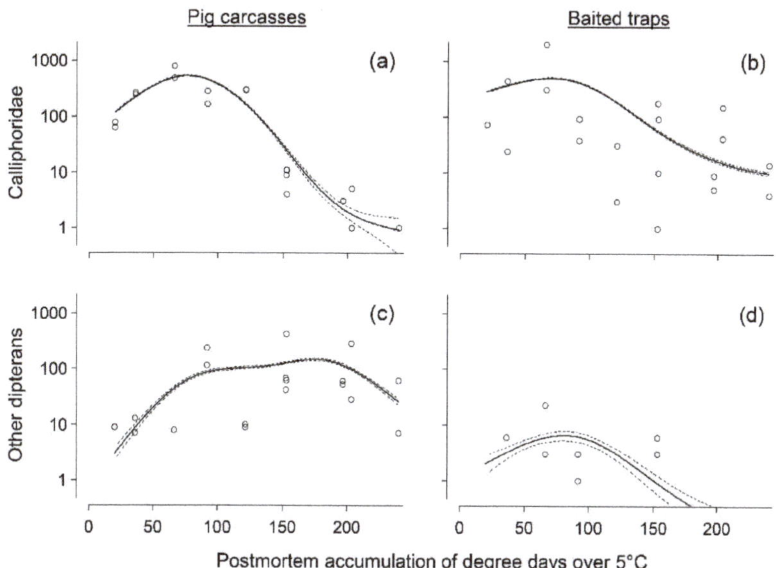

Figure 1. Dipterans documented as a function of postmortem accumulated degree-days over 5 °C, for: (**a**) Calliphoridae observed on carcasses ($r^2 = 0.63$; *edf* = 3.903; $F = 442$; $p < 0.01$), (**b**) Calliphoridae captured in baited traps ($r^2 = 0.09$; *edf* = 2.997; $F = 676.7$; $p < 0.01$), (**c**) other dipterans observed on carcasses ($r^2 = 0.05$; *edf* = 3.961; $F = 87.03$; $p < 0.01$) and (**d**) other dipterans captured in baited traps ($r^2 = 0.10$; *edf* = 2.555; $F = 8.87$; $p < 0.01$). The solid black line indicates the model prediction and the confidence interval of the prediction is illustrated with small dashes.

3.2. Comparison of Species Composition

The assemblages of Diptera documented on carcasses and in traps are presented in Table 1. The most abundant Calliphoridae associated with the carcasses were *Phormia regina* (25.6%), *Lucilia illustris* (9.0%), and *Pollenia sp.* (3.2%) (Table 1). Piophilidae (i.e., all *Stearibia nigriceps*: 48.6%), Sepsidae (4.1%), Muscidae (4.9%), Drosophilidae (0.9%), Sarcophagidae (0.8%), Anthomyiidae (0.4%), and Phoridae (0.1%) specimens were also sampled from the carcasses (Table 1). In the traps, the most commonly recovered Diptera were Calliphoridae, of which *L. illustris* (72.1%), *P. regina* (21.0%), and *Pollenia sp.* (4.1%) were most abundant (Table 1). Muscidae (0.8%), Sarcophagidae (0.5%), and Anthomyiidae (0.7%) were also recovered from the traps (Table 1). Five species captured by the traps were not sampled on the carcasses (Table 1), whereas one Calliphoridae genus was sampled only on the carcasses (*Protocalliphora sp.*). Drosophilidae, Sepsidae, Phoridae, and Piophilidae were never captured by the baited traps despite their presence on the carcasses (Table 1). When assemblages in traps and on carcasses were compared in terms of Diptera occurrence (Sørensen's index, Figure 2a) and abundance (Percent similarity index, Figure 2b), the highest similarity between carcasses and traps was observed below 50 postmortem ADD$_5$. Sørensen's index indicated that a maximum of 60% of species or families were shared between traps and carcasses early in the decomposition process (Figure 2a). In contrast, the percent similarity index indicated that the level of similarity in species composition documented within traps and on carcasses in the same period was 30% at best (Figure 2b). Afterward, the Diptera assemblages increasingly differed between carcasses and traps (Figure 2a,b). By comparison, 90% of the species that were present early in decomposition were shared between pairs of carcasses, after which the level of similarity decreased steadily (Figure 2a). The species composition of necrophagous Diptera assemblages on the carcasses were also highly similar throughout decomposition (~75% similarity) until no insects were observed (i.e., postmortem ADD$_5$ > 200; Figure 2b).

Table 1. Percentage of specimens of each species, genera and family of Diptera recovered during successive sampling intervals (PMI) from carcasses ($n = 4$) and small traps baited with pork liver ($n = 4$) in Cocagne, New Brunswick in July and August 2020.

Experimental Unit	Carcasses												Traps											
Sampling Month	July					August					Average		July					August					Average	
Postmortem Interval (PMI)	0–2	3–5	6–8	9–11	12–14	0–2	3–5	6–8	9–11	12–14	0–2	0–14	0–2	3–5	6–8	9–11	12–14	0–2	3–5	6–8	9–11	12–14	0–2	0–14
Lucilia illustris (Meigen)	35.7	13.1	-	5.0	-	36.0	-	-	-	-	35.9	9.0	87.7	42.4	74.3	75.7	57.1	89.8	52.9	57.8	94.3	88.9	88.8	72.1
Phormia regina (Meigen)	14.3	81.0	84.6	10.0	7.1	22.0	34.8	2.4	-	-	18.1	25.6	1.4	54.2	14.3	16.4	42.9	5.9	37.0	28.4	4.2	5.6	3.7	21.0
Pollenia sp. Robineau-Desvoidy	14.3	-	2.6	5.0	7.1	2.0	1.1	-	-	-	8.1	3.2	11.0	2.0	5.7	7.3	-	2.5	2.2	3.4	1.0	5.6	6.8	4.1
Protophormia terraenovae (Robineau-Desvoidy)	-	2.4	5.1	-	-	2.0	2.2	-	-	-	1.0	1.2	-	0.2	-	-	-	-	-	-	-	-	-	0.0
Calliphora terraenovae Macquart	7.1	-	-	-	-	-	-	-	-	-	3.6	0.7	-	-	-	-	-	-	1.4	-	-	-	-	0.1
Calliphora vicina Robineau-Desvoidy	3.6	-	-	-	-	-	-	-	-	-	1.8	0.4	-	0.0	-	-	-	-	-	-	-	-	-	0.0
Lucilia sericata (Meigen)	-	-	-	-	-	-	-	-	-	-	-	-	-	-	2.9	-	-	-	0.7	-	-	-	-	0.4
Calliphora livida Hall	-	-	-	-	-	-	-	-	-	-	-	-	-	0.0	-	-	-	-	1.4	-	0.5	-	-	0.2
Protocalliphora sp. Hough	-	-	-	-	-	2.0	-	-	-	-	1.0	0.2	-	-	-	-	-	-	-	-	-	-	-	-
Lucilia silvarum (Meigen)	-	-	-	-	-	-	-	-	-	-	-	-	-	-	-	0.6	-	-	-	-	-	-	-	0.1
Calliphora vomitoria (Linnaeus)	-	-	-	-	-	-	-	-	-	-	-	-	-	-	-	-	-	0.2	-	-	-	-	0.1	0.0
Lucilia magnicornis (Siebke)	-	-	-	-	-	-	-	-	-	-	-	-	-	0.0	-	-	-	-	-	-	-	-	-	0.0
Piophilidae—*Stearibia nigriceps* (Meigen)	3.6	3.6	-	60.0	78.6	14.0	33.7	92.9	100.0	100.0	8.8	48.6	-	-	-	-	-	-	-	-	-	-	-	-
Muscidae	10.7	-	2.6	5.0	7.1	16.0	7.9	-	-	-	13.4	4.9	-	1.1	-	-	-	0.8	2.2	4.3	-	-	0.4	0.8
Sepsidae	3.6	-	2.6	10.0	-	2.0	18.0	4.8	-	-	2.8	4.1	-	-	-	-	-	-	-	-	-	-	-	-
Sarcophagidae	3.6	-	-	-	-	4.0	-	-	-	-	3.8	0.8	-	0.0	-	-	-	0.4	0.7	3.4	-	-	0.2	0.5
Anthomyiidae	3.6	-	-	-	-	-	-	-	-	-	1.8	0.4	-	-	2.9	-	-	-	1.4	2.6	-	-	-	0.7
Drosophilidae	-	-	2.6	5.0	-	-	1.1	-	-	-	-	0.9	-	-	-	-	-	0.2	-	-	-	-	0.1	0.0
Phoridae	-	-	-	-	-	-	1.1	-	-	-	-	0.1	-	-	-	-	-	-	-	-	-	-	-	-

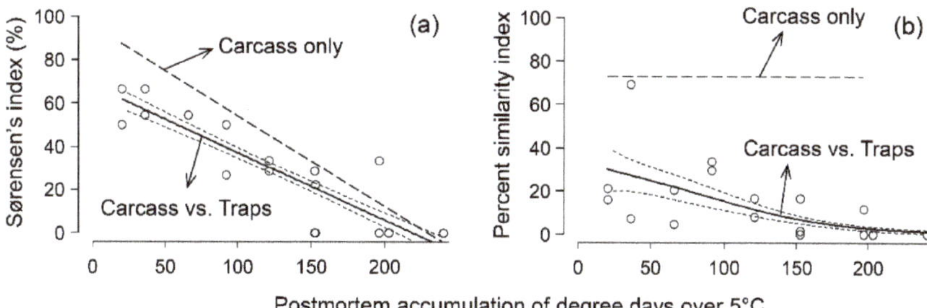

Postmortem accumulation of degree days over 5°C

Figure 2. Comparison of Diptera documented in traps and on carcasses using the (**a**) Sorensen's index ($r^2 = 0.78$; $edf = 1$; $F = 85.42$; $p < 0.01$) and the (**b**) percent similarity index ($r^2 = 0.29$; $edf = 1.584$; $F = 14.72$; $p < 0.01$) as a function of postmortem accumulated degree-days over 5 °C. The solid black line indicates the model prediction and the confidence interval of the prediction is illustrated with small dashes. For comparison purposes, the large dash line indicates the value of the index when used for carcass to carcass comparison.

To identify the taxa responsible for these differences, we produced a canonical correlation (Figure 3). The first canonical axis separated the Diptera assemblages from carcasses and traps (Figure 3). All other explanatory variables (i.e., postmortem ADD_5, mean temperature, sampling month) loaded onto the second canonical axis, indicating that the effect of the type of experimental unit is largely independent of them (Figure 3). *Lucilia illustris* was largely overrepresented by the traps and *Calliphora vomitoria*, *C. livida*, *L. sericata*, *L. silvarum*, *L. magnicornis*, *Pollenia sp.*, and Anthomyiidae were also recovered more frequently in traps than on carcasses (Table 1; Figure 3). Conversely, *Protocalliphora sp.*, *C. terraenovae*, *C. vicina*, *Protophormia terraenovae*, Muscidae, Phoridae, and Drosophilidae were more abundant on the carcasses than in the traps (Table 1; Figure 3). Muscidae, which on average represented 4.9% and 13.4% of the dipterans found on the carcasses throughout the decomposition and in early decomposition respectively, represented only 0.8% of trapped dipterans (Table 1). Similarly, *C. terraenovae*, *C. vicina*, Sarcophagidae, and Anthomyiidae were common on carcasses (composition > 1%) early in decomposition (i.e., PMI 0-2) but were rarely retrieved from traps (composition < 1%) (Table 1). *Phormia regina* and Sarcophagidae were equally abundant in the traps and on the carcasses (Table 1; Figure 3). However, Sarcophagidae represented a greater proportion of the specimens recovered from the carcasses in early decomposition than was represented by the traps (Table 1).

4. Discussion

Small bait traps are useful in forensic entomology to maintain a database of local and seasonal forensically important necrophagous insects [20,32–34], which may be crucial for analysis of entomological evidence in a given area [20]. That being said, the correlation between assemblages of Diptera captured by small bait traps and found on cadavers/carcasses has never been studied simultaneously, although some authors have used pre-existing data or literature to compare data [1,20,21]. However, these methods may be problematic because geography, climate, species-specific habitat associations, and seasons are known to have an important impact on Diptera communities [11,19,35–38]. Additionally, some authors have demonstrated the importance of biases associated with small bait traps, such as those caused by weather factors, height of traps, and position, as well as biased sex ratios and age-classes of captured Diptera [11,39,40]. Furthermore, the flight patterns of Calliphoridae are known to vary throughout the day [41,42]. As such, from an experimental standpoint, small bait traps which sample constantly are more likely to capture taxa which may not be active when the carcasses are being sampled. The aforementioned factors which influence Diptera assemblages and the many possible biases made us question the reliability of the previous comparisons between carcasses/cadavers and traps. These doubts have arisen because the experimental units were often geographically distant, located within different environments, and exposed to different climates and seasons.

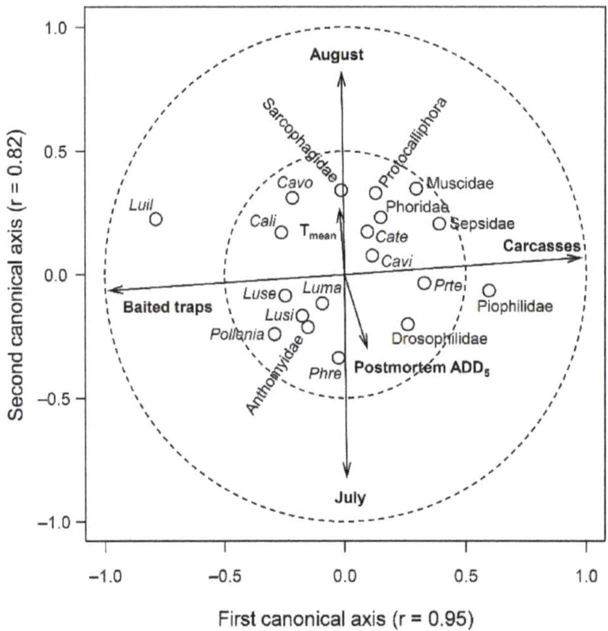

Figure 3. Canonical correlation analysis (Pillai's trace = 2.68; adjusted redundancy $r^2 = 0.37$; $p < 0.01$) between Diptera abundance (empty circles) and the study variables (i.e., type of experimental unit, sampling months, postmortem accumulated degree-days over 5 °C, and mean temperature—T_{mean}; arrows). Calliphoridae species are identified using the acronym of the first two letters of their genus and species.

The present study compared carcasses and small bait traps exposed to the same microclimate and community while avoiding cross-contamination by crawling insects. We documented a similar abundance of Diptera in early decomposition between traps and carcasses. However, the composition and richness of the Diptera assemblages found in the small bait traps and on the carcasses differed significantly even in early decomposition, while these were very similar when comparing only the carcasses. These results support our predictions and the hypothesis that the attractiveness of decomposing animal matter to Diptera changes with species interactions, decomposition, and amount of available tissue. Despite previous studies claiming the opposite [1,20], we found little evidence to support the idea of a resemblance between the Diptera assemblages during early decomposition on the carcasses and in small bait traps, at least in our study area. In addition, the data did not support the idea that the assemblages within small bait traps may be representative of a later stage of decomposition.

4.1. Discrepancies between Small Bait Traps and Remains

Our analysis indicated that the differences in species compositions were mainly caused by the overrepresentation of *L. illustris* in trap captures. This may be caused, in part, by the inability of this species to compete with the dominant species on the carcasses, such as *P. regina* [43]. Other Calliphoridae species such as *C. vomitoria* and *L. sericata* were also overrepresented in trap captures. This was expected because certain necrophagous insects are known to be more drawn to smaller baits [5], and because previous studies have suggested that the Diptera collected from baited traps may provide an exaggerated representation of those on cadavers [33]. Furthermore, studies have shown that certain Diptera species preferentially colonize carrion at greater heights from ground level than others [44]. Comparatively, other Diptera such as *Protocalliphora sp.* and *C. terraenovae* were sampled in greater abundance from the carcasses than by the traps, in agreement with field observations that many necrophagous Diptera species are probably more attracted to larger amounts of animal carrion [5,15,21,22]. Furthermore, a study in Brazil which compared the Diptera associated with 14 murder cases and traps set at the scene of the death found that many species were present in the traps while they were absent from the cadavers [33]. Similarly, recent work, which compared the initial species composition of piglet carcasses and baited beef liver traps placed in the same habitat but during subsequent years, demonstrated that the three dominant Calliphoridae species found on the carcasses did not all correspond to those captured by the baited traps [20]. However, unlike Weidner et al. [20], many of the early-colonizing Diptera retrieved in this study from the carcasses during early decomposition (i.e., PMI 0–2) were rarely recovered from the traps (e.g., *C. terraenovae*).

Diptera other than Calliphoridae were also underrepresented in trap captures. For example, the total proportion of Muscidae captured by the small bait traps was lower than that found on the carcasses by a ratio of 5:1. This may be related to the presence of excrement released by the carcasses [45,46] but further studies will be needed to test this hypothesis. We also suspect that the absence of smaller Diptera (i.e., Piophilidae, Sepsidae, and Phoridae) in the traps could be due, in part, to trap design but further work will be needed to confirm this. Trapping studies using small bait traps rarely document smaller Diptera, although Schoenly traps, Malaise traps, yellow pan traps, or sweeping over a bait have been successful in capturing Piophilidae and Sepsidae [47,48]. It is important to mention that previous studies of small bait traps only reported Calliphoridae, even though the other families included in our study are of forensic relevance. Nevertheless, had we done the same, our conclusions would not have changed because the Calliphoridae complex is largely responsible for the differences documented here.

4.2. Other Considerations

The potential reasons for the sampling differences of the Diptera species and families documented are numerous. First, the fact that our bait was frozen and thawed may be involved, but this remains questionable. While one study reported that frozen meat attracts Calliphoridae differently than never-frozen meat [49], another study found no difference in Calliphoridae activity between refrigerated and frozen pig carcasses [50]. Second, the two sampling methods greatly differed, as the traps are passive, and the use of a sweep net is more active. For this reason, it is more likely for the data retrieved from sweeping over the carcasses to be biased, depending on the individual carrying out the sampling procedure. However, sweep net sampling remains the most widely used method for retrieving flying insects from a corpse during forensic investigations. For this reason, despite the biases this methodology may introduce, it was necessary in this study to allow for a proper comparison.

An interesting aspect of the current study is that trap captures are affected by the proximity of decomposing animal matter. This source of bias should be considered when setting small bait traps for necrophagous Diptera in an urban setting where trashcans containing decomposing animal tissues (e.g., meat) may be nearby. Moreover, the influence of carcasses on trap catches suggests that without the presence of the carcasses to attract necrophagous insects from afar, the insect assemblages captured by the small bait traps could have been even less similar to those found on the carcasses than what were documented here. One way to test this would be to place traps at increasing distances from carcasses in a given habitat.

Finally, one of the singular aspects of this study is the statistical effect size (i.e., the magnitude of the experimental effect [31]), which tells us a lot about the disparity between what is sampled by small bait traps and on carcasses. In the present study, the effect size was so large that after performing a first fully replicated experiment in July, the results were already clear and very significant. We did however replicate the experiment a second time in August to ensure that these results were not obtained by chance and documented the same consistent differences. Thus, despite the small sample size, the effect size was so large with eight experimental units (and 9620 documented Diptera) that we did not have any reason to believe that additional experimental units were needed.

5. Conclusions

While the results of the current study are in contradiction to the claim that small bait traps could be an accurate predictor of early arriving Diptera of forensic importance [1], they are not surprising. As early as the 1940s, a descriptive study carried out in South Africa had reached the same conclusions [51]. In addition, the effects of abundance and type of necromass on colonizing insect communities have been documented beyond doubt on other decaying substrates (e.g., dung, dead wood: [52–55]). Moreover, in previous study of sapromyiophilous plants that attract fly pollinators by mimicking carrion and dung odors, antennae of houseflies showed positive dose-dependent responses to the volatile compounds released by flowers [56]. There is no rationale why these effects should be any different with carrion. It appears implausible that a time-limited sampling protocol could be comparable to method sampling around the clock. In addition, considering that traps are often baited with homogeneous tissues from internal organs, there is a strong possibility that they emit different volatile organic compounds than whole carcasses [57]. Realistically, the effect of the bait itself in small bait traps still needs to be tested using proper design, solid controls and appropriate statistics [31]. To verify the untested assumption discussed above that bait emits volatile organic compounds similar to those emitted by a cadaver/carcass, the first logical step would be to perform a chemical analysis of the volatile organic compounds released by different types of bait and carcasses or cadavers throughout decomposition. This leads us to suggest that although our results are not necessarily applicable to all trap designs or locations, if the trap in question is baited with a small amount of animal tissue, there is no reason to believe that the results will be different. In the meantime, small bait traps should be considered useful tools to provide baseline data on Diptera occurrence if cadavers or large carcasses are not readily available. In agreement with Arnaldos et al. [58], we stress that caution must be exercised when extrapolating data obtained from non-human experimental methods to forensic cases, and especially when using data from small bait traps in courtroom proceedings.

Supplementary Materials: The following are available online at https://www.mdpi.com/2075-4450/12/3/261/s1. Figure S1: Map of the study site in Cocagne, New Brunswick, Canada, with location of pigs (squares) and small bait traps (circles) in the July and August trials.

Author Contributions: Conceptualization, K.L. and G.M.; methodology, G.M.; formal analysis, G.M.; investigation, K.L. and D.R.B.; resources, G.M.; data curation, K.L.; writing—original draft preparation, K.L.; writing—review and editing, G.M. and D.R.B.; supervision, G.M.; project administration, K.L. and G.M.; funding acquisition, G.M. All authors have read and agreed to the published version of the manuscript.

Funding: This research was funded by The Natural Sciences and Engineering Research Council of Canada, grant number DDG-2020-00021.

Institutional Review Board Statement: Ethical review and approval for this study were waived by the ethical committee of our university because the study does not concern vertebrate species, carcasses being bought as meat and not live animals.

Data Availability Statement: The data presented in this study are available on request from the corresponding author. The data are not publicly available due to unfinished related ongoing further studies and manuscript.

Acknowledgments: The authors would like to thank N. Hammami, L. Tousignant and four anonymous reviewers for their comments on a previous version of this manuscript; N. Hammami and G. Kabre for assistance with fieldwork; as well as J.-P. Privé for help with experimental site.

Conflicts of Interest: The authors declare no conflict of interest. The funders had no role in the design of the study; in the collection, analyses, or interpretation of data; in the writing of the manuscript, or in the decision to publish the results.

References

1. Farinha, A.; Dourado, C.G.; Centeio, N.; Oliveira, A.R.; Dias, D.; Rebelo, M.T. Small bait traps as accurate predictors of dipteran early colonizers in forensic studies. *J. Insect Sci.* **2014**, *14*, 77. [CrossRef] [PubMed]
2. Marques, T.A.; Munger, L.; Thomas, L.; Wiggins, S.; Hildebrand, J.A. Estimating North Pacific right whale *Eubalaena japonica* density using passive acoustic cue counting. *Endanger. Species Res.* **2011**, *13*, 163–172. [CrossRef]
3. Schoenly, K. Demographic bait trap. *Environ. Entomol.* **1981**, *10*, 615–617. [CrossRef]
4. Wiener, J.G.; Smith, M.H. Relative efficiencies of four small mammal traps. *J. Mammal.* **1972**, *53*, 868–873. [CrossRef]
5. Amendt, J.; Krettek, R.; Zehner, R. Forensic entomology. *Naturwissenschaften* **2004**, *91*, 51–65. [CrossRef]
6. Benecke, M. A brief history of forensic entomology. *Forensic Sci. Int.* **2001**, *120*, 2–14. [CrossRef]
7. Gennard, D. *Forensic Entomology: An Introduction*, 1st ed.; John Wiley & Sons: Chinchester, West Sussex, UK, 2012; p. 1.
8. Hall, M.J.R. Trapping the flies that cause myiasis: Their responses to host-stimuli. *Ann. Trop. Med. Parasitol.* **1995**, *89*, 333–357. [CrossRef]
9. Ordóñez, A.; García, M.D.; Fagua, G. Evaluation of efficiency of Schoenly trap for collecting adult sarcosaprophagous dipterans. *J. Med. Entomol.* **2008**, *45*, 522–532. [CrossRef] [PubMed]
10. Schoenly, K.G.; Haskell, N.H.; Hall, R.D.; Gbur, J.R. Comparative performance and complementarity of four sampling methods and arthropod preference tests from human and porcine remains at the Forensic Anthropology Center in Knoxville, Tennessee. *J. Med. Entomol.* **2007**, *44*, 881–894. [CrossRef]
11. Hwang, C.; Turner, B.D. Spatial and temporal variability of necrophagous Diptera from urban to rural areas. *Med. Vet. Entomol.* **2005**, *19*, 379–391. [CrossRef]
12. Baz, A.; Cifrián, B.; Díaz-äranda, L.M.; Martín-Vega, D. The distribution of adult blow-flies (Diptera: Calliphoridae) along an altitudinal gradient in Central Spain. *Ann. Société Entomol. Fr.* **2007**, *43*, 289–296. [CrossRef]
13. Aak, A.; Knudsen, G.K.; Soleng, A. Wind tunnel behavioural response and field trapping of the blowfly *Calliphora vicina. Med. Vet. Entomol.* **2010**, *24*, 250–257. [CrossRef]
14. Payne, J.A. A Summer Carrion Study of the Baby Pig *Sus Scrofa* Linnaeus. *Ecology* **1965**, *46*, 592–602. [CrossRef]
15. Brundage, A.; Bros, S.; Honda, J.Y. Seasonal and habitat abundance and distribution of some forensically important blow flies (Diptera: Calliphoridae) in Central California. *Forensic Sci. Int.* **2011**, *212*, 115–120. [CrossRef]
16. Lutz, L.; Verhoff, M.A.; Amendt, J. Environmental factors influencing flight activity of forensically important female blow flies in Central Europe. *Int. J. Legal Med.* **2019**, *133*, 1267–1278. [CrossRef] [PubMed]
17. Norris, K.R. Daily patterns of flight activity of blowflies (Calliphoridae: Diptera) in the Canberra district as indicated by trap catches. *Aust. J. Zool.* **1966**, *14*, 865–853. [CrossRef]
18. Açikgöz, H.N.; Köse, S.K.; Divrak, D. Chemical meat bait traps versus basic meat bait traps in collection of adult flies. *Entomol. Res.* **2017**, *47*, 21–27. [CrossRef]
19. George, K.A.; Archer, M.S.; Toop, T. Abiotic environmental factors influencing blowfly colonisation patterns in the field. *Forensic Sci. Int.* **2013**, *229*, 100–107. [CrossRef] [PubMed]
20. Weidner, L.M.; Gemmellaro, M.D.; Tomberlin, J.K.; Hamilton, G.C. Evaluation of bait traps as a means to predict initial blow fly (Diptera: Calliphoridae) communities associated with decomposing swine remains in New Jersey, USA. *Forensic Sci. Int.* **2017**, *278*, 95–100. [CrossRef] [PubMed]
21. Schoenly, K.; Reid, W. Community structure of carrion arthropods in the Chihuahuan Desert. *J. Arid Environ.* **1983**, *6*, 253–263. [CrossRef]
22. Davies, L. Species composition and larval habitats of blowfly (Calliphoridae) populations in upland areas in England and Wales. *Med. Vet. Entomol.* **1990**, *4*, 61–68. [CrossRef]
23. Anderson, G.; VanLaerhoven, S. Initial Studies on Insect Succession on Carrion in Southwestern British Columbia. *J. Forensic Sci.* **1996**, *41*, 617–625. [CrossRef]
24. Matuszewski, S.; Hall, M.J.R.; Moreau, G.; Schoenly, K.G.; Tarone, A.M.; Villet, M.H. Pigs vs people: The use of pigs as analogues for humans in forensic entomology and taphonomy research. *Int. J. Legal Med.* **2020**, *134*, 793–810. [CrossRef]
25. Lewis, A.J.; Benbow, M.E. When entomological evidence crawls away: *Phormia regina* en masse larval dispersal. *J. Med. Entomol.* **2011**, *48*, 1112–1119. [CrossRef] [PubMed]
26. Perez, A.E.; Haskell, N.H.; Wells, J.D. Commonly used intercarcass distances appear to be sufficient to ensure independence of carrion insect succession pattern. *Ann. Entomol. Soc. Am.* **2016**, *109*, 72–80. [CrossRef]
27. Michaud, J.-P.; Moreau, G. Facilitation may not be an adequate mechanism of community succession on carrion. *Oecologia* **2017**, *183*, 1143–1153. [CrossRef]

28. Michaud, J.-P.; Moreau, G. Effect of variable rates of daily sampling of fly larvae on decomposition and carrion insect community assembly: Implications for forensic entomology field study protocols. *J. Med. Entomol.* **2013**, *50*, 890–897. [CrossRef] [PubMed]

29. Marshall, S.A.; Whitworth, T.; Roscoe, L. Blow flies (Diptera: Calliphoridae) of eastern Canada with a key to Calliphoridae subfamilies and genera of eastern North America, and a key to the eastern Canadian species of Calliphorinae, Luciliinae and Chrysomyiinae. *Can. J. Arthropod Identif.* **2011**, *11*, 11. [CrossRef]

30. Moreau, G. The pitfalls in the path of probabilistic inference in forensic entomology: A review. *Insects* **2021**, *12*, 240. [CrossRef]

31. R Development Core Team. *R: A Language and Environment for Statistical Computing*; R Foundation for Statistical Computing: Vienna, Austria, 2018.

32. Oliveira, T.C.; Vasconcelos, S.D. Insects (Diptera) associated with cadavers at the Institute of Legal Medicine in Pernambuco, Brazil: Implications for forensic entomology. *Forensic Sci. Int.* **2010**, *198*, 97–102. [CrossRef] [PubMed]

33. Davies, L. Seasonal and spatial changes in blowfly production from small and large carcasses at Durham in lowland northeast England. *Med. Vet. Entomol.* **1999**, *13*, 245–251. [CrossRef] [PubMed]

34. Grassberger, M.; Friedrich, E.; Reiter, C. The blowfly *Chrysomya albiceps* (Wiedemann) (Diptera: Calliphoridae) as a new forensic indicator in Central Europe. *Int. J. Legal Med.* **2003**, *117*, 75–81. [CrossRef] [PubMed]

35. Baz, A.; Cifrián, B.; Martín-Vega, D. Patterns of diversity and abundance of carrion insect assemblages in the Natural Park 'Hoces del Río Riaza' (Central Spain). *J. Insect Sci.* **2014**, *14*, 162. [CrossRef]

36. Goff, M.L. Comparison of insect species associated with decomposing remains recovered inside dwellings and outdoors on the Island of Oahu, Hawaii. *J. Forensic Sci.* **1991**, *36*, 748–753. [CrossRef] [PubMed]

37. Gruner, S.V.; Slone, D.H.; Capinera, J.L. Forensically important Calliphoridae (Diptera) associated with pig carrion in rural North-Central Florida. *J. Med. Entomol.* **2007**, *44*, 509–515. [CrossRef]

38. Leccese, A. Insects as forensic indicators: Methodological aspects. *Anil Aggrawals Internet J. Forensic Med. Toxicol.* **2004**, *5*, 26–32.

39. Martín-Vega, D.; Baz, A. Sex-biased captures of sarcosaprophagous Diptera in carrion-baited traps. *J. Insect Sci.* **2013**, *13*, 14. [CrossRef]

40. Hayes, E.J.; Wall, R.; Smith, K.E. Mortality rate, reproductive output, and trap response bias in populations of the blowfly *Lucilia sericata*. *Ecol. Entomol.* **1999**, *24*, 300–307. [CrossRef]

41. Sontigun, N.; Sukontason, K.L.; Klong-Klaew, T.; Sanit, S.; Samerjai, C.; Somboon, P.; Thanapornpoonpong, S.; Amendt, J.; Sukontason, K. Bionomics of the oriental latrine fly *Chrysomya megacephala* (Fabricius) (Diptera: Calliphoridae): Temporal fluctuation and reproductive potential. *Parasites Vectors* **2018**, *11*, 415. [CrossRef]

42. Nuorteva, P. The flying activity of *Phormia terrae-novae* R.-D. (Dipt, Calliphoridae) in subarctic conditions. *Ann. Zool. Fenn.* **1966**, *3*, 73–81.

43. Prinkkilá, M.-L.; Hanski, I. Complex competitive interactions in four species of *Lucilia* blowflies. *Ecol. Entomol.* **1995**, *20*, 261–272. [CrossRef]

44. Mashaly, A.; Mahmoud, A.; Ebaid, H.; Sammour, R. Effect of height to ground level on the insect attraction to exposed rabbit carcasses. *J. Environ. Biol.* **2020**, *41*, 73–78. [CrossRef]

45. Kirk, A.A. The effect of the dung pad fauna on the emergence of *Musca tempestiva* [Dipt: Muscidae] from dung pads in southern France. *Entomophaga* **1992**, *37*, 507–514. [CrossRef]

46. Vogt, W.G. Trap catches of *Musca vetustissima* Walker (Diptera: Muscidae) and other arthropods associated with cattle dung in relation to height above ground level. *Austral Entomol.* **1988**, *27*, 143–147. [CrossRef]

47. Prado e Castro, C.; Cunha, E.; Serrano, A.; García, M.D. *Piophila megastigmata* (Diptera: Piophilidae): First records on human corpses. *Forensic Sci. Int.* **2012**, *214*, 23–26. [CrossRef] [PubMed]

48. Rochefort, S.; Wheeler, T. Diversity of Piophilidae (Diptera) in northern Canada and description of a new Holarctic species of Parapiophila McAlpine. *Zootaxa* **2015**, *3925*, 229–240. [CrossRef]

49. Micozzi, M.S. Experimental study of postmortem change under field conditions: Effects of freezing, thawing, and mechanical injury. *J. Forensic Sci.* **1986**, *31*, 953–961. [CrossRef] [PubMed]

50. Bugajski, K.N.; Seddon, C.C.; Williams, R.E. A comparison of blow fly (Diptera: Calliphoridae) and beetle (Coleoptera) activity on refrigerated only versus frozen-thawed pig carcasses in Indiana. *J. Med. Entomol.* **2011**, *48*, 1231–1235. [CrossRef] [PubMed]

51. Hepburn, G.A. Sheep blowfly research I—A survey of maggot collections from live sheep and a note on the trapping of blowflies. *Onderstepoort J. Vet. Anim. Ind.* **1943**, *18*, 13–18.

52. Lobo, J.M.; Hortal, J.; Cabrero-Sañudo, F.J. Regional and local influence of grazing activity on the diversity of a semi-arid dung beetle community. *Divers. Distrib.* **2006**, *12*, 111–123. [CrossRef]

53. Macedo, R.; Audino, L.D.; Korasaki, V.; Louzada, J. Conversion of Cerrado savannas into exotic pastures: The relative importance of vegetation and food resources for dung beetle assemblages. *Agric. Ecosyst. Environ.* **2020**, *288*, 106709. [CrossRef]

54. Müller, J.; Wende, B.; Strobl, C.; Eugster, M.; Gallenberger, I.; Floren, A.; Steffan-Dewenter, I.; Linsenmair, K.E.; Weisser, W.W.; Gossner, M.M. Forest management and regional tree composition drive the host preference of saproxylic beetle communities. *J. Appl. Ecol.* **2015**, *52*, 753–762. [CrossRef]

55. Thibault, M.; Moreau, G. Enhancing bark- and wood-boring beetle colonization and survival in vertical deadwood during thinning entries. *J. Insect Conserv.* **2016**, *20*, 789–796. [CrossRef]

56. Zito, P.; Guarino, S.; Peri, E.; Sajeva, M.; Colazza, S. Electrophysiological and behavioural responses of the housefly to "sweet" volatiles of the flowers of *Caralluma europaea* (Guss.) NE Br. *Arthropod-Plant Int.* **2013**, *7*, 485–489. [CrossRef]

57. Hoffman, E.M.; Curran, A.M.; Dulgerian, N.; Stockham, R.A.; Eckenrode, B.A. Characterization of the volatile organic compounds present in the headspace of decomposing human remains. *Forensic Sci. Int.* **2009**, *186*, 6–13. [CrossRef]
58. Arnaldos, M.I.; García, M.D.; Romera, E.; Presa, J.J.; Luna, A. Estimation of postmortem interval in real cases based on experimentally obtained entomological evidence. *Forensic Sci. Int.* **2005**, *149*, 57–65. [CrossRef]

Review

Forensic Entomology in China and Its Challenges

Yu Wang †, Yinghui Wang †, Man Wang, Wang Xu, Yanan Zhang and Jiangfeng Wang *

Department of Forensic Medicine, Soochow University, Ganjiang East Road, Suzhou 215000, China; yuw@suda.edu.cn (Y.W.); ganyuchengge@163.com (Y.W.); wm15850051476@163.com (M.W.); 20194221067@stu.suda.edu.cn (W.X.); 20204221029@stu.suda.edu.cn (Y.Z.)
* Correspondence: jfwang@suda.edu.cn; Tel.: +86-181-5111-6801
† These authors contributed equally to this work.

Simple Summary: Forensic entomologists utilize sarcosaprophagous insect species to estimate the postmortem interval to aid death investigations. In this paper, we present the recent chronology of forensic entomology in China and illustrate how identification, development, and succession data are obtained and applied at the scale of such a large country. To overcome the difficulties and challenges forensic entomology faces in China, a number of countermeasures are provided.

Abstract: While the earliest record of forensic entomology originated in China, related research did not start in China until the 1990s. In this paper, we review the recent research progress on the species identification, temperature-dependent development, faunal succession, and entomological toxicology of sarcosaprophagous insects as well as common applications of forensic entomology in China. Furthermore, the difficulties and challenges forensic entomologists face in China are analyzed and possible countermeasures are presented.

Keywords: forensic entomology; postmortem interval; development; succession; species identification

Citation: Wang, Y.; Wang, Y.; Wang, M.; Xu, W.; Zhang, Y.; Wang, J. Forensic Entomology in China and Its Challenges. *Insects* **2021**, *12*, 230. https://doi.org/10.3390/insects12030230

Academic Editors: Damien Charabidze and Daniel Martín-Vega

Received: 30 December 2020
Accepted: 1 March 2021
Published: 8 March 2021

Publisher's Note: MDPI stays neutral with regard to jurisdictional claims in published maps and institutional affiliations.

1. Origin and Development of Forensic Entomology in China

The book *Washing Away of Wrongs*, published during the Song dynasty, describes the earliest case report of forensic entomology [1]. No obvious progress was achieved in forensic entomology in China until the 1990s. In the 1990s, influenced by the boom of forensic entomology across Europe, North America, and Australia, Cui Hu and Hongzhang Zhou started exploratory research in southern and northern China, respectively [2]. With the publishing of *Forensic Entomology* by Hu [3], forensic entomology began to attract extensive attention in China. An array of studies focusing on species identification, temperature-dependent development, insect succession, and entomological toxicology have been conducted since then.

1.1. Species Identification

Forensic entomology represents the application of the study of insects (and other arthropods) to legal issues [4]. The most common use of entomological evidence in medicolegal investigations is the estimation of the time that has passed since death, which is referred to as the postmortem interval (PMI) [5]. The proper identification of forensically important species constitutes the first step and also the most crucial element in forensic entomology [6]. Species identification allows the application of the proper developmental data and insect succession patterns in an investigation [7]. If species determination is incorrect, the estimated PMI will also be incorrect [5]. Without the accurate identification of forensically important insect species, basic forensic entomology research is also not possible [2]. In China, entomologists conducted comprehensive and systematic work on the identification of sarcosaprophagous insects [8–17]. Most of their peer-reviewed findings have been absorbed into books. The *Key to the Common Flies of China (2nd Edition)* by Fan [18], published in 1992, contains 1547 common fly species in nine families and 250 genera. *Flies of China* by Xue and Zhao [19], published in 1996, contains 4209 fly species of 30 families and 660 genera. *Fauna Sinica Insecta Diptera: Calliphoridae* by Fan [20], published in 1997, contains five subfamilies, 48 genera, and 232 species of blow fly. These books lay a solid foundation for both research and application of forensic entomology. *Beetles Associated with Stored Products in China* by Zhang et al. [21], *Carrion Beetles of China (Coleoptera: Silphidae)* by Ji [22], and *Atlas of Chinese beetles: Staphylinidae* by Li [23] have laid the foundation for the identification of forensically important beetle species. However, research on the larval morphology of sarcosaprophagous beetles is still insufficient.

Chinese forensic entomologists have conducted many studies on the molecular identification of forensically important insect species and a wide range of mitochondrial and nuclear DNA markers has been identified [24–31]. Many taxa have been studied including Calliphoridae [26,27], Sarcophagidae [28–30], Muscidae [32], and several Coleoptera species [25]. Because the sole reliance on single DNA fragments for defining closely related species is perilous, multiple gene regions have been used for a more reliable diagnosis of forensically important flies [29]. Single-nucleotide polymorphisms (SNPs) have been used to aid the molecular identification of sarcosaprophagous flies [24]. The availability of complete mitochondrial genomes also offers an important basis for the identification and phylogenetic analysis of forensically important species [33,34]. Moreover, researchers found that cuticular hydrocarbon composition in the puparium can also be used for taxonomic differentiation of sarcosaprophagous flies [35].

In other countries, forensic entomologists have established useful keys for the morphological identification of forensically important insects [36–39]. These identification keys facilitate more detailed and species-specific knowledge of relevant species for forensic entomology experiments and real cases [36]. Although identification keys for forensically important insects were already established for taxonomic purposes in China [18–23], a more rigorous taxonomic foundation is still required for forensic purposes. In addition, molecular identification studies still need to sample taxa from different geographical regions and take better account of genetic intra- and interspecific variabilities.

1.2. Temperature-Dependent Development of Sarcosaprophagous Insects

Accurate and reliable temperature-dependent development data of insects form the basis for PMI estimation [40]. In China, the earliest developmental studies from the perspective of forensic entomology were launched in the 1990s. Ma et al. [41] and Wang [42] studied four and five common sarcosaprophagous fly species, respectively. For estimations of larval age, changes in body length and changes of morphological features (e.g., larval cuticle, cephalopharyngeal skeleton, and posterior spiracles) have been studied [42,43]. The results suggested that larval age can be estimated not only from the number of spiracular slits and changes of body length but also from the presence, thickness, color [42], and sclerotized area of the posterior peritreme [43]. In addition, the sclerotized area of the cephalopharyngeal skeleton is also a promising indicator for the estimation of larval age [43–45] (Table 1).

The concept of the minimum PMI (PMI_{min}), as proposed by Amendt et al. [46], eliminates the uncertainty of the pre-colonization interval and provides a guideline for researchers to study the post-colonization interval of sarcosaprophagous insects. Consequently, they can establish more reliable development data for PMI_{min} estimation. To obtain more reliable and scientific results for the determination of the developmental patterns of sarcosaprophagous flies, recent studies of Chinese forensic entomologists have assessed wider temperature ranges and narrower sampling and observation intervals [47,48]. Moreover, forensic entomologists in China have also adopted more accurate temperature control devices and established additional developmental models for the estimation of PMI [49,50]. Investigated species include *Hemipyrellia ligurriens* (Wiedemann, 1830) [50], *Chrysomya megacephala* (Fabricius, 1794) [48], *Lucilia illustris* (Meigen, 1826) [47], *Parasarcophaga similis* (Meade, 1876) [51], *Boettcherisca peregrina* Robineau-Desvoidy, 1830 [52], *Calliphora grahami*

(Aldrich, 1930) [53], *Musca domestica* Linnaeus, 1758 [54], *Chrysomya pinguis* (Walker, 1858) [55], *Muscina stabulans* (Fallén, 1817) [56], *Lucilia sericata* (Meigen, 1826) [57], *Chrysomya rufifacies* (Macquart, 1842) [58], *Hydrotaea spinigera* Stein, 1910 [59], and *Sarcophaga dux* Thomson, 1869 [49].

In flies, the intrapuparial period includes quiescent developmental stages that account for about half of the development time of the life cycle, which has important implications for accurate PMI_{min} estimation [60,61]. Chinese researchers have conducted several studies on morphological changes during the intrapuparial period and have divided this period into several events based on morphological and color changes [42,51,62–68]. The aim was to improve the accuracy of age estimation when using puparial samples. The investigated flies and other species include *C. megacephala* [63], *C. rufifacies* [68], *Ca. grahami* [42], *L. sericata* [42], *M. domestica* [42], *P. similis* [51], *L. cuprina* (Wiedemann, 1830) [62], *Megaselia spiracularis* [67], *Megaselia scalaris* (Loew, 1866) [66], *Dohrniphora cornuta* (Bigot, 1857) [64], and *Hermetia illucens* (Linnaeus, 1758) [65]. In particular, Wang et al. [69] conducted a study on the morphological changes of *L. illustris* during the intrapuparial period, which further identified detailed changes in various structures. The age of intrapuparial forms can not only be estimated from overall morphological changes but also be more accurately identified by the developmental processes of compound eyes, mouthparts, antennae, thorax, legs, wings, and the abdomen. These changes offer the potential to increase the accuracy of age estimation during the intrapuparial period [69].

In recent years, researchers worldwide have broadened the scope of their research in development biology. Particularly forensically important beetles have been gradually incorporated in their studies [70–72] as these beetles can be used to extend the estimation range of PMI when primary colonizers are no longer associated with the corpse or have emerged from puparia [73]. Chinese researchers also carried out temperature-dependent developmental studies on several forensically important beetles (e.g., *Creophilus maxillosus* (Linnaeus, 1758) [73], *Necrobia rufipes* (De Geer, 1755) [74], *Omosita colon* (Linnaeus, 1758) [75], and *Dermestes tessellatocollis* Motschulsky, 1860 [76]).

In addition to morphological observations, hemolymph soluble proteins [77], cuticular hydrocarbons [78], and differential gene expression [69,79–81] have also been applied to estimate the age of immature stages. Cuticular hydrocarbons in fly puparium were found to significantly change with weathering time, indicating their potential use for PMI_{min} estimation [82,83]. The use of gene expression level changes to estimate the age of stage has proven to be an effective approach to estimate the PMI_{min}. The most prominent advantage of differential gene expression technology is that the obtained data represent quantitative results that can be subjected to error analysis, better conform to the Daubert test in forensic science, and are therefore more readily accepted by courts [84]. Notably, the combination of morphological observations with differential gene expression technology achieves higher estimation accuracy than either method alone [69,85]. Recently, researchers have studied the gene expression changes of *L. illustris* [69], *C. megacephala* [79], *Ca. grahami* [80], and *B. peregrina* [81] during the intrapuparial period and have further explored the gene expression changes of additional species at different stages. These studies lay a sound foundation for the molecular age estimation of sarcosaprophagous insects. Previous studies merely focused on establishing a connection between gene expression changes and development time. However, these studies commonly did not provide deep insight into the function of each differentially expressed gene and the molecular mechanism of phenotypic changes during growth and development [86]. In response to the rapid development of transcriptome sequencing technology and the increasing maturation of sequencing and data analysis techniques, research can now focus on developing transcriptome sequencing, gene screening, and bioinformatics analyses [86]. The generated data can be used to identify more optional, high-sensitivity genetic markers as candidates for molecular age estimation studies in forensic entomology [86]. These technologies can be expected to help establish more scientific and accurate methods for estimating the age of sarcosaprophagous insects.

The importance of a rigorous approach for the acquisition of reference development data for forensic applications has been highlighted by several researchers [46,87]. A standard approach that can be used to guide the development study is still missing in forensic entomology. Researchers in different regions or even within the same country often apply very different methods in their development studies. The divergence of these methodologies obstructs comparison of the results, thus limiting the exchange of data that may otherwise be helpful in forensic cases [88]. The studies of Bernhardt et al. [89] showed that not all tissues are similarly suitable for the gathering of sound growth data for sarcosaprophagous Diptera. Bernhardt et al. [89] suggested using minced pork as a non-human nutrition medium, since there are no developmental differences in this diet compared with human tissue. Bugelli et al. [90] found that the killing and storing methods of entomological samples can affect larval age estimation and suggested that the storing of maggots in 96% ethanol does not affect age estimation, or the specimen measurements should be done right after killing. The studies of Bernhardt et al. [89] and Bugelli et al. [90] provided important guidelines for the development of studies on forensic entomology. Further studies are needed to explore the effect of other factors with regard to the development of forensically important insects, such as fluctuating temperature and population competition between species, humidity, and photoperiod.

Table 1. Summary of studies that reported the development of sarcosaprophagous insects in China.

Order	Species	Citation	City (Province)	Temperatures (°C)	Indicators
Diptera	*Boettcherisca peregrina* Robineau-Desvoidy, 1830	Wang et al. [42]	Hangzhou (Zhejiang)	16, 20, 24, 28, 32	Dd, Lbl, T_0, K, Imc, Lmc
		Wang et al. [91]	Guangzhou (Guangdong)	15, 20, 25, 30, 35	Dd, Lbl, Lbw, Lbwi, Wp
		Wang et al. [52]	Suzhou (Jiangsu)	16, 19, 22, 25, 28, 31, 34	Dd, Ihen, Lbl, Ilen, T_0, K
		Shang et al. [85]	Changsha (Hunan)	15, 25, 35	Ige
	Calliphora grahami (Aldrich, 1930)	Ma et al. [41]	Hangzhou (Zhejiang)	12, 15, 18, 21, 24, 27, 30	Dd
		Wang et al. [42]	Hangzhou (Zhejiang)	12, 16, 20, 24, 28	Dd, Lbl, T_0, K, Imc, Lmc
		Zhao et al. [92]	Shijiazhuang (Hebei)	16, 20, 24, 28, 32	Lmc
		Wang et al. [53]	Suzhou (Jiangsu)	16, 19, 22, 25, 28, 31, 34	Dd, Ihen, Lbl, Ilen, T_0, K
		Liu et al. [84]	Changsha (Hunan)	15, 22, 27	Ige
		Chen et al. [93]	Changsha (Hunan)	Constant vs. fluctuating temperature (8 vs. 6–12; 12 vs. 10–16; 16 vs. 14–20)	Dd, Ihen, Lbl, Ilen, T_0, K
	Chrysomya megacephala (Fabricius, 1794)	Ma et al. [41]	Hangzhou (Zhejiang)	18, 21, 24, 27, 30, 33	Dd
		Wang et al. [42]	Hangzhou (Zhejiang)	16, 20, 24, 28, 32	Dd, Lbl, T_0, K, Imc, Lmc
		Zhao et al. [92]	Shijiazhuang (Hebei)	16, 20, 24, 28, 32	Lmc
		Yang et al. [48]	Chongqing (municipality)	16, 19, 22, 25, 28, 31, 34	Dd, Ihen, Lbl, Ilen, T_0, K
		Zhang et al. [63]	Suzhou (Jiangsu)	16, 19, 22, 25, 28, 31, 34	Dd, Ihen, Lbl, Ilen, T_0, K, Imc
		Wang et al. [83]	Suzhou (Jiangsu)	22.5, 27.5, 32.5	Ige
	Chrysomya nigripes Aubertin, 1932	Li et al. [94]	Guangzhou (Guangdong)	20, 24, 28, 32	Dd, Lbl, Lbw
	Chrysomya pinguis (Walker, 1858)	Zhang et al. [55]	Suzhou (Jiangsu)	16, 19, 22, 25, 28, 31, 34	Dd, Ihen, Lbl, Ilen, T_0, K
	Chrysomya rufifacies (Macquart, 1842)	Ma et al. [68]	Guangzhou (Guangdong)	20, 24, 28, 32	Imc
		Hu et al. [58]	Suzhou (Jiangsu)	16, 19, 22, 25, 28, 31, 34	Dd, Ihen, Lbl, Ilen, T_0, K
	Dohrniphora cornuta (Bigot, 1857)	Feng et al. [64]	Shenyang (Liaoning)	15, 18, 21, 24, 27, 30, 33, 36	Imc
	Hemipyrellia ligurriens (Wiedemann, 1830)	Yang et al. [50]	Chongqing (municipality)	16, 19, 22, 25, 28, 31, 34	Dd, Ihen, Lbl, Ilen, T_0, K
	Hermetia illucens Linnaeus, 1758	Li et al. [65]	Guangzhou (Guangdong)	20, 24, 28, 32	Dd, Imc
	Hydrotaea spinigera Stein, 1910	Wang et al. [59]	Suzhou (Jiangsu)	16, 19, 22, 25, 28, 31, 34	Dd, Ihen, Lbl, Ilen, T_0, K
	Lucilia cuprina (Wiedemann, 1830)	Wang et al. [62]	Shijiazhuang (Hebei)	16, 20, 24, 28, 32	Imc
	Lucilia illustris (Meigen, 1826)	Wang et al. [47]	Suzhou (Jiangsu)	15, 17.5, 20, 22.5, 25, 27.5, 30, 32.5, 35	Dd, Ihen, Lbl, Ilen, T_0, K
		Wang et al. [69]	Suzhou (Jiangsu)	20, 25, 30	Imc, Ige

Table 1. *Cont.*

Order	Species	Citation	City (Province)	Temperatures (°C)	Indicators
Diptera	*Lucilia sericata* (Meigen, 1826)	Ma et al. [41]	Hangzhou (Zhejiang)	18, 21, 24, 27, 30, 33	Dd
		Wang et al. [42]	Hangzhou (Zhejiang)	16, 20, 24, 28, 32	Dd, Lbl, T_0, K, Imc, Lmc
		Li [43]	Shijiazhuang (Hebei)	16, 20, 24, 28, 32	Lmc
		Wang et al. [57]	Suzhou (Jiangsu)	16, 19, 22, 25, 28, 31, 34	Dd, Ihen, Lbl, Ilen, T_0, K
	Megaselia scalaris (Loew, 1866)	Feng and Liu [66]	Shenyang (Liaoning)	18, 21, 24, 27, 30, 33, 36	Imc
	Megaselia spiracularis Schmitz, 1938	Feng and Liu [67]	Shenyang (Liaoning)	21, 24, 27, 30, 33, 36	Imc
		Wang et al. [95]	Suzhou (Jiangsu)	16, 19, 22, 25, 28, 31, 34	Dd, Ihen, Lbl, Ilen, T_0, K
	Musca domestica Linnaeus, 1758	Wang et al. [42]	Hangzhou (Zhejiang)	16, 20, 24, 28	Dd, Lbl, T_0, K, Imc, Lmc
		Wang et al. [54]	Suzhou (Jiangsu)	16, 19, 22, 25, 28, 31, 34	Dd, Ihen, Lbl, Ilen, T_0, K
	Muscina stabulans (Fallén, 1817)	Wang et al. [56]	Suzhou (Jiangsu)	16, 19, 22, 25, 28, 31, 34	Dd, Ihen, Lbl, Ilen, T_0, K
	Parasarcophaga crassipalpis (Macquart, 1938)	Ma et al. [41]	Hangzhou (Zhejiang)	18, 21, 24, 27, 30, 33	Dd
		Wang et al. [44]	Shijiazhuang (Hebei)	16, 20, 24, 28, 32	Lmc
	Parasarcophaga similis (Meade, 1876)	Yang et al. [51]	Suzhou (Jiangsu)	16, 19, 22, 25, 28, 31, 34	Dd, Ihen, Lbl, Ilen, T_0, K, Imc
	Sarcophaga dux Thomson, 1869	Zhang et al. [49]	Changsha (Hunan)	16, 19, 22, 25, 28, 31, 34	Dd, Ihen, Lbl, Ilen, T_0, K, Ige
Coleoptera	*Creophilus maxillosus* (Linnaeus, 1758)	Wang et al. [73]	Suzhou (Jiangsu)	17.5, 20, 22.5, 25, 27.5, 30, 32.5	Dd, Ihen, Lbl, Ilen, T_0, K
	Necrobia rufipes (De Geer, 1755)	Hu et al. [74]	Suzhou (Jiangsu)	22, 25, 28, 31, 34, 36	Dd, Ihen, Lbl, Ilen, T_0, K, LD
	Omosita colon (Linnaeus, 1758)	Wang et al. [75]	Suzhou (Jiangsu)	16, 19, 22, 25, 28, 31	Dd, Ihen, T_0, K
Hymenoptera	*Nasonia vitripennis* (Walker, 1836)	Zhang et al. [96]	Suzhou (Jiangsu)	16, 19, 22, 25, 28, 31, 34	Dd, Ihen, Lbl, T_0, K

Abbreviations: Dd: developmental duration; Ihen: isomorphen diagram; Lbl: larval body length; Ilen: isomegalen diagram; Lbw: larval body weight; Lbwi: larval body width; Wp: weight of puparia; T_0: developmental threshold temperature; K: thermal summation constant; Imc: intrapuparial morphological changes; Ige: intrapuparial gene expression; LD: larval instar determination; Lmc: larval morphological changes.

1.3. Faunal Succession of Insects

Different insect species visit and leave human remains in different orders and time sequences [97]. These regular insect activities are referred to as the faunal succession of insects and can be used to estimate the range of PMI [97]. Succession studies also provide information on the insects that occur in particular geographical regions or during particular seasons, which can be used to estimate the seasons and regions in which particular cases have taken place [2]. China has a vast territory with wide latitudinal coverage, complex geographical environments, and diverse climatic conditions. However, so far, only 20 insect succession studies have been published in 13 provinces (17 cities), accounting for about 1/3 of the total number of provincial-level administrative regions in China (Table 2).

Initially, most studies only addressed the species composition on remains in outdoor environments [98,99]. Later, researchers began to investigate the insect succession patterns under different environmental conditions [100–102]. The effects of indoor/outdoor environments [100], enclosed/unenclosed environments [103], exposure time [101], cadaver types [102], and toxicants [104] were explored with regard to body decomposition and insect succession. In particular, studies conducted in South China and the Yangtze River Delta region included not only the succession of adult insects but also the dynamic change processes of sarcosaprophagous insects on remains (such as egg laying, hatching, wandering, pupariation, eclosion, and disappearance) [102,105]. Insect succession matrixes of different seasons were obtained containing the residence times of different insects and their developmental stages [102,105].

Researchers used different types of models in insect succession studies [102,106,107]. Pig and rabbit carcasses were the most commonly used experimental remains, both of which were in nine studies. Three studies used animal tissues or organs to attract forensically important insects (Table 2). Human remains were used by four studies [98,99,102,108]. For instance, Zhou et al. [98] and Yang et al. [99] identified 38 species of sarcosaprophagous beetles and 14 species of sarcosaprophagous flies on human remains in Beijing. Chen [108] studied the body decomposition patterns and insect compositions on four human remains for four seasons under field conditions in Guizhou. Chen [108] applied the same method as used at the Anthropological Research Facility of the University of Tennessee, also known as the "Body Farm". However, such a "body farm" does not exist in China yet. A study conducted in Shenzhen found that large pig carcasses, with similar weight to human corpses, also decomposed in a similar manner to human corpses; moreover, the species composition and succession pattern of insects were similar to that of human corpses. The carcasses of small pigs decomposed faster, attracted less insect species, and had a simpler succession pattern [102]. Rabbit carcasses could not fully reflect either the body decomposition or the changes of insects that occurred on a human corpse [102].

Research on the succession patterns of insects constitutes a number of the most fundamental tasks of forensic entomology [5,102], and the significance of such research is mainly embodied in the following three aspects: First, important basic data on species composition and succession patterns of sarcosaprophagous insects can be obtained via insect succession studies, which can be used to estimate the PMI [109,110]. Second, during succession studies, live insects with forensic importance can be collected and supplied as important colony sources for developmental research [53]. Third, researchers can accumulate precious practical experience, which cannot be otherwise obtained via laboratory studies. After weeks or months of body decomposition and insect succussion observations, forensic entomologists will have gained a clearer understanding of the decomposition processes of remains and will have acquired a basic time frame of the arrival orders of insects [2]. In addition, forensic entomologists will learn how to collect insect samples and where certain species of insects can be found [3]. With such knowledge and understanding, forensic entomologists can be more efficient and confident when dealing with real cases.

Forensic entomologists in different countries have conducted comprehensive work on the faunal succession of insects. Succession patterns have been identified under different types of environments (e.g., burying [111], vehicle environments [112], and dry environments [113]) and treatments (e.g., burning [114], hanging [115], and clothing [116]). Although these studies provided an important database, the difference between the insect fauna of China and other countries, as well as differences in other factors (e.g., climate and flora) justify the need for establishing patterns of insect succession for different regions of China [102]. Currently, insect succession data are not available for most parts of China. Most existing studies only concentrated on investigating species composition while failing to meet the requirements for establishing a succession matrix, which is required for PMI estimations. Considerable opportunity still remains for studying the faunal succession of insects in all parts of China.

Table 2. Summary of studies reporting body decomposition and insect succession in China.

Citation	City (Province)	Experimental Model (Numbers of Remains)	Season/Month	Insect Species/Other Information
Zhou et al. [98]	Beijing (capital)	Human corpse (NS)	March to August	Coleoptera: 38 species
Yang et al. [99]	Beijing (capital)	Human viscera (NS)	March to November	Diptera: 14 species
Ma et al. [117]	Hangzhou (Zhejiang)	Pork meat (NS)	All seasons	Diptera: 12 species Coleoptera: 16 species Hymenoptera: 2 species
Li et al. [118]	Harbin (Heilongjiang)	Pork meat (NS)	Spring, summer, and autumn	Diptera: 8 species Coleoptera: 19 species
Chen et al. [119]	Twelve sites (Guizhou)	Pork lung (NS)	All seasons	Diptera: 27 species
Wang et al. [120]	Chengdu (Sichuan)	Rabbit (28)	1st year: May to November 2nd year: March to September	Diptera: 5 species
Chen [121]	Zhongshan (Guangdong)	Pig (4)	Autumn and winter	Total: 38 species
Chang et al. [106]	Hohhot (Inner Mongolia)	Rabbit (25), Dog (1)	July to October	Diptera: 10 species

Table 2. *Cont.*

Citation	City (Province)	Experimental Model (Numbers of Remains)	Season/Month	Insect Species/Other Information
Wang et al. [105]	Pearl River Delta (Guangdong)	Pig (18)	All seasons	Diptera: 17 species Coleoptera: 16 species Other: 9 species
Wu et al. [122]	Guangzhou (Guangdong)	Rabbit (NS)	Spring and summer	Diptera: 10 species Coleoptera: 7 species
Dong et al. [123]	Sanmenxia (Henan)	Rabbit (5)	July to October	Three families, 13 species
Chen et al. [108]	Guiyang (Guizhou)	Human corpse (4)	All seasons	Diptera: 11 species
Nie et al. [124]	Xi'an (Shanxi)	Rabbit (4)	Spring	Diptera: 10 species Coleoptera: 4 species Other: 2 species
Shi et al. [104]	Guangzhou (Guangdong)	Rabbit (4)	Summer	Effects of malathion on the insect succession
Jiang et al. [125]	Yongzhou (Hunan)	Rabbit (9)	July to September	Total: 26 species
Yin et al. [100]	Shenzhen (Guangdong)	Pig (4)	Summer	Indoor: 14 species Outdoor: 18 species
Jiang et al. [126]	Qingdao (Shandong)	Pig (12)	All seasons	Diptera: 23 species
Lv [127]	Chongqing (municipality)	Pig (11)	All seasons	Insecta: 94 species
Li et al. [103]	Guangzhou (Guangdong)	Pig (2)	Summer	Comparative study of carcasses between enclosed and open-air conditions
Yang [128]	Suzhou (Jiangsu)	Pig (22)	Summer and autumn	Diptera: 16 species Coleoptera: 12 species Other: 5 species
Wu et al. [107]	Xinxiang (Henan)	Rabbit (5), rat (6)	July to August	Diptera: 7 species
Wang et al. [101]	Guangzhou (Guangdong)	Pig (6)	Summer	Insect succession on pig carcasses using different exposure times
Wang et al. [102]	Shenzhen (Guangdong)	Human corpse (1), large pig (2), small pig (2), rabbit (2)	August to December	Total: 42 species; insect assemblages are more complex on larger carcasses, following the order of human = large pig > small pig > rabbit

NS: not specified.

1.4. Forensic Entomotoxicology in China

Chinese research on forensic entomotoxicology primarily focused on the effects of toxicants/drugs on the growth and development of insects. Tian [129] and Zhao et al. [130] found that morphine accelerated the development of flies and resulted in increased larval body length and weight [129], which caused a PMI deviation of up to 84 h [130]. Dai et al. [131] and Wang et al. [132] reported that diazepam accelerated fly development and shortened the larval development periods of *L. sericata* and *C. megacephala* by 55 and 60 h, respectively. Lv et al. [133] showed that ketamine inhibited the larval development rate of *C. megacephala* in a both dose-dependent and time-dependent manner. Zou et al. [134] showed that ketamine shortened the larval development period of *L. sericata*; the authors further showed that the development period of *L. sericata* larvae feeding on muscles and receiving ketamine at twice the lethal dose was 24 h shorter than that of control larvae. Liu et al. [135] reported that malathion inhibited the growth of larvae and puparia of *C. megacephala*, extended their developmental period by 36 h, and shortened their maximum body length by 1.1 mm. The study by Shi et al. [136] indicated that malathion and white arsenic affected the decomposition of carcasses by inhibiting insect colonization. Wang et al. [137] showed that the larval developmental time of Ca. *grahami* was significantly shorter on rabbit mince containing methamphetamine and further reported that methamphetamine can increase larval body length.

The effects of various toxicants/drugs on the growth and development of insects have also been studies by forensic entomologists in other countries. The results indicate that a number of toxicants/drugs, e.g., paracetamol [138], diazepam [139], cocaine [140], and codeine [141], can accelerate the development of forensically important insects. Other toxicants/drugs inhibited insect development, e.g., methamphetamine [142], malathion [143], alcohol [144], and amitriptyline [145]. Desmethyldiazepam [146], nandrolone [147], and gentamicin [148] did not significantly change the growth and development of sarcosaprophagous insects. The results obtained by Chinese researchers were consistent with the above results, and the same effects were found on insect development in the same toxicants/drugs. Forensic entomotoxicology was once extensively studied as a branch of forensic entomology from the end of the 20th century to the beginning of the 21st century; however, studies in this field have been declining over recent years. One of the main reasons restricting the development of forensic entomotoxicology is that although qualitative detection has been achieved, quantitative analyses still remain problematic [149]. Further studies are still required to overcome existing bottlenecks.

2. Applications of Forensic Entomology in China

The *Regulations on the Classification of the Practice of Forensic Judicial Appraisal*, released by the Chinese Ministry of Justice on 9 May 2020, stipulates that insects can serve as evidence for PMI estimation. This regulation establishes the legal status of forensic entomology. However, Chinese judicial or police departments have no full-time forensic entomology positions. Forensic entomologists are mainly researchers working at institutions of higher education or employees of research institutions, who may participate in cases as expert witnesses similar to other countries. When forensic entomologists are asked to provide expert opinions that may ultimately decide how cases are adjudicated, they will be invited to attend field investigations. In most circumstances, the police will only mail specimens or send photos of related insects to forensic entomologist, combined with brief details of the case. The resulting PMI estimated is then only used to narrow the scope of the police investigation. China has not yet established specialized academic institutions for forensic entomology research, and researchers are most commonly members of the Entomological Society of China or the Forensic Medicine Association of China. A few universities are offering forensic entomology education to M.Sc. and Ph.D. students, while undergraduate education is still not available. Because of this lack of full-time forensic entomology posts, most interested postgraduates have to consider career changes. So far, fewer than a dozen researchers are qualified to present judicial expertise on forensic entomology in the entirety of China.

Currently, forensic entomology is only applied in a few regions of China. The identification of insect species is largely based on morphological identification keys provided by entomologists [150,151]. Peer-reviewed development and succession data are used to estimate the PMI [54,150,151]. Plasticity of developmental rates has been reported between different colonies of the same species [152,153]; therefore, in applications with multiple literature reports on the development of one species, the development data obtained from the same or a nearby geographical region will be prioritized [154]. Succession data from the same or similar regions and months will be used to estimate the PMI [155].

More than 23 case reports of the application of forensic entomology in China have been published [54,58,150,151,154–164]. The most widespread application of forensic entomology is the provision of PMI clues for criminal investigators [58,150,151,154–160,163,164]. Forensic entomology also played an important role in an insurance compensation case [54]. There were 14 reported cases in the outdoor environment [58,150,151,154–157,160–164] and nine cases occurred indoors [54,151,155,158,159]. The development duration and larval body length of flies were the most commonly used indicators for the PMI estimation, both of which were utilized in 10 cases [54,58,150,151,154–156,159,161,162]. Three cases utilized the thermal summation constant of flies to estimate the PMI [158,160]. The development and succession patterns of beetles were only utilized for PMI estimation in one case [155]. Other cases used the biological characteristics of forensically important blow flies to estimate the PMI. On the basis of the seasonal distribution characteristics of *C. pinguis* and Ca. *grahami*, Li et al. [162] inferred the initial occurrence time of two fly species (from the end of February to the beginning of March) as the most likely time frame of PMI of the deceased. The actual result confirmed this inference. In a case reported by Hu et al. [164], a cadaver, found in a suitcase, contained *H. illucens*, *Me. scalaris*, and *Fannia canicularis*, while blow flies were absent. Analysis indicated that the deceased likely began to decompose during winter, and that, as the weather warmed, the cadaver was already highly decomposed and no longer attracted calliphorid species (as these prefer fresh remains). On this basis, it could be inferred that the PMI of the corpse started at the beginning of winter, which was later confirmed by investigation results.

3. Challenges for Forensic Entomology in China and Proposed Countermeasures

Forensic entomology faces many challenges in China. First, new technologies, such as video surveillance technology, DNA technology, and big data technology, have developed rapidly over the past few years [165]. These technologies contribute to the increase of police detection rates, and many of them can help to estimate the PMI [2,165]. For example, if the decedent was identified by the technology of DNA, his/her debit and credit card expenditures and telephone, accommodation, and traffic records can provide a clue for PMI. Impacted by various newly emerging case-solving technologies, the demand of criminal investigators for forensic entomology has decreased [2]. Second, forensic entomologist usually cannot provide a precise PMI result as forensic entomology is limited and can only provide a PMI_{min} or a PMI range [166]. These usually deviate by days or even months from the real PMI [151], thus making it difficult for criminal investigators to fully accept the value of forensic entomology. Third, a number of criminal investigators do not have the basic knowledge of forensic entomology and might not follow the recommended standards and practices in forensic entomology. This potentially limits the accuracy of the obtained PMI estimations. For example, the insect evidence provided may not include the first arriving species or the oldest immature insects. Without reliable and sufficient insect evidence, forensic entomologists cannot provide an accurate result. Therefore, it would be advisable if forensic entomologists could be present at the scene for specimen collection, as no one else can fully understand the importance of each piece of insect evidence.

However, forensic entomologists cannot attend every case. Consequently, education and training need to be strengthened to enhance the ability and understanding of the criminal investigators who are frequently involved in cases that would benefit from forensic entomology. Most investigators do not consider using forensic entomology when encountering death cases. The reason most often is that they are not familiar with forensic entomology or think that the application of forensic entomology is too complex. Therefore, an urgent need for education and training exists to help these investigators to understand the theory of forensic entomology, e.g., why temperature data is so important, under which circumstances forensic entomology can be utilized, and how entomological evidence is collected correctly. To achieve this, Soochow University has offered a course of 36 credit hours to undergraduate students with specialization in forensic sciences. The purpose of this course is not for students to become competent forensic entomologists but to enable them to correctly collect insect evidence or at least to induce their proactivity for asking for forensic entomologists in their future work. Forensic entomology researchers should also be trained. Some researchers cannot correctly identify the species of sarcosaprophagous insects, especially when these are immature. Some researchers have never performed basic experimental studies and/or lack the required expertise to analyze the entomological evidence associated with a case. Hence, more comprehensive training should be offered to help them overcome application barriers.

In the application, forensic entomologists should try their best to provide a scientifically informed PMI estimate. Although the PMI estimate may not absolutely coincide with the actual PMI, it is better if the real PMI falls within the scope of the estimated PMI. To improve the accuracy of PMI estimation, the following conditions must be met: (1) all entomological evidence must be collected systematically and comprehensively; (2) case information and environmental information must be understood systematically and comprehensively; furthermore, weather data must be retrieved, and accurate environmental data must be obtained through field measurements and corrections; (3) the species and stages of all insects present at the scene must be correctly identified; (4) adequate laboratory work must be conducted and accurate reference data on insect development must be obtained; (5) investigations of insect succession must be conducted in similar regions; (6) various types of cases need to be studied to accumulate experience; (7) all cases should be pre-appraised by forensic entomologists. If the insect evidence is not provided correctly and appropriately by criminal investigators, forensic entomologists should request an additional collection. Otherwise, PMI estimation should be refused. Cases that exceed the scope of forensic entomology should also be rejected to avoid mistakes.

4. Conclusions

Over nearly 30 years of exploration and development, forensic entomology has begun to play an increasingly important role in forensic practices; however, it is still far from the ultimate goal of achieving universal application in China. Faced with the challenge of various newly emerging case-solving technologies, if forensic entomology wants to further develop and progress, basic research must be strengthened. Further studies should be conducted to establish more accurate development data, based not only on morphological methods but also on techniques of differential gene expression, biochemical properties, and artificial intelligence. Moreover, basic data of succession patterns of insects should be obtained in more environments and regions, to further probe into the mechanisms of colonization of corpses by insects. Both training and education of criminal investigators should be strengthened to promote the application of forensic entomology.

Author Contributions: Writing—original draft preparation, Y.W. (Yu Wang), Y.W. (Yinghui Wang), M.W., W.X., and Y.Z.; writing—review and editing, Y.W. (Yu Wang) and J.W.; supervision, J.W.; funding acquisition, Y.W. (Yu Wang) and J.W. All authors have read and agreed to the published version of the manuscript.

Funding: This work was supported by the National Natural Science Foundation of China (grant numbers 31872258, 32070508, and 82002007) and the Priority Academic Program Development of Jiangsu Higher Education.

Institutional Review Board Statement: Not available.

Conflicts of Interest: The authors declare no conflict of interest.

References

1. Tzu, S. *Washing Away of Wrongs*; Shanghai Classics Publishing House: Shanghai, China, 1981.
2. Wang, J. *Practical Forensic Entomology*; Xi'an Jiaotong University Press: Xi'an, China, 2019.
3. Hu, C. *Forensic Entomology*; Chongqing Publishing House: Chongqing, China, 2000.
4. Goff, M.L. *A Fly for the Prosecution: How Insect Evidence Helps Solve Crimes*; Harvard University Press: Cambridge, UK, 2000.
5. Byrd, J.H.; Castner, J.L. *Forensic Entomology: The Utility of Arthropods in Legal Investigations*; CRC Press: New York, NY, USA, 2009.
6. Harvey, M.L.; Gaudieri, S.; Villet, M.H.; Dadour, I.R. A global study of forensically significant calliphorids: Implications for identification. *Forensic Sci. Int.* **2008**, *177*, 66–76. [CrossRef]
7. Chen, W.Y.; Hung, T.H.; Shiao, S.F. Molecular identification of forensically important blow fly species (Diptera: Calliphoridae) in Taiwan. *J. Med. Entomol.* **2004**, *41*, 47–57. [CrossRef] [PubMed]
8. Gan, Y. *Chrysomya* in China. *Acta Entomol. Sin.* **1958**, *8*, 340–350.
9. Fan, Z. The identification keys of common fly larvae in Shanghai. *Acta Entomol. Sin.* **1957**, *1*, 405–422.
10. Ma, D.; Fan, Z. Studies on flies in Xinjiang: Calliphoridae and Muscidae. *Bull. Dis. Control Prev.* **1998**, *13*, 33–37.
11. Fan, Z. Calliphora in China. *Acta Entomol. Sin.* **1957**, *5*, 321–348.
12. Zhang, M. Studies on the common larvae of flesh flies in China. *Entomotax* **1982**, *6*, 99–112.
13. Xue, W. One new species and three new records of the genus *Hydrotaea* (Diptera: Muscidae) from China. *Acta Entomol. Sin.* **1976**, *1*, 109–111.
14. Xue, W.; Verves, Y.G.; Du, J. A review of subtribe *Boettcheriscina* Verves 1990 (Diptera: Sarcophagidae), with descriptions of a new species and genus from China. *Ann. Soc. Ent. Fr.* **2011**, *47*, 303–329. [CrossRef]
15. Guo, F. Sarcophagidae in Shanghai. *Acta Entomol. Sin.* **1952**, *1*, 60–86.
16. Zhang, S.F.; Liu, Y.P. A study on six species of Dermestidae. *Acta Zhengzhou Grain Coll.* **1985**, *3*, 76–80.

17. Zhang, S.F.; Liu, Y.P. Histeridae associated with stored products: III. Species of genus *Saprinus* and *Gnathoncus*. *J. Jilin Grain Coll.* **1992**, *3*, 1–6.
18. Fan, Z.D. *The Keys of Common Flies of China*; Science Press: Beijing, China, 1992.
19. Xue, W.; Zhao, J. *Flies of China*; Liaoning Science and Technology Publishing House: Shenyang, China, 1996.
20. Fan, Z.D. *Fauna Sinica: Insecta. Diptera: Calliphoridae. Diptera: Calliphoridae*; Science Press: Beijing, China, 1997.
21. Zhang, S.F.; Liu, Y.P.; Wu, T.C. *Beetles Associated with Stored Products in China*; China Agriculture Science and Technology Press: Beijing, China, 1998.
22. Ji, Y. *The Carrion Beetles of China (Coleoptera: Silphidae)*; China Forestry Press: Beijing, China, 2012.
23. Li, L. *Atlas of Chinese Beetles: Staphylinidae*; Straits Publishing Group: Fuzhou, China, 2010.
24. Chen, W.; Shang, Y.; Ren, L.; Xie, K.; Zhang, X.; Zhang, C.; Sun, S.; Wang, Y.; Zha, L.; Guo, Y. Developing a MtSNP-based genotyping system for genetic identification of forensically important flesh flies (Diptera: Sarcophagidae). *Forensic Sci. Int.* **2018**, *290*, 178–188. [CrossRef] [PubMed]
25. Su, R.N.; Guo, Y.D.; Xie, D.; Peng, Y.L.; Sheng, L.H. Identification of forensically important beetles (Coleoptera: Histeridae) in China based on 16S rRNA and Cyt b. *Trop. Biomed.* **2013**, *30*, 375–387.
26. Meng, F.; Ren, L.; Wang, Z.; Deng, J.; Guo, Y.; Chen, C.; Finkelbergs, D.; Cai, J. Identification of forensically important blow flies (Diptera: Calliphoridae) in China based on COI. *J. Med. Entomol.* **2017**, *54*, 1193–1200. [CrossRef]
27. Liu, Q.L.; Cai, J.F.; Chang, Y.F.; Gu, Y.; Guo, Y.D.; Wang, X.H.; Weng, J.F.; Zhong, M.; Wang, X.; Yang, L. Identification of forensically important blow fly species (Diptera: Calliphoridae) in China by mitochondrial cytochrome oxidase I gene differentiation. *Insect Sci.* **2011**, *18*, 554–564. [CrossRef]
28. Guo, Y.D.; Cai, J.F.; Meng, F.M.; Chang, Y.F.; Gu, Y.; Lan, L.M.; Liang, L.; Wen, J.F. Identification of forensically important flesh flies based on a shorter fragment of the cytochrome oxidase subunit I gene in China. *Med. Vet. Entomol.* **2012**, *26*, 307–313. [CrossRef]
29. Guo, Y.; Zha, L.; Yan, W.; Li, P.; Cai, J.; Wu, L. Identification of forensically important sarcophagid flies (Diptera: Sarcophagidae) in China based on COI and period gene. *Int. J. Legal Med.* **2014**, *128*, 221–228. [CrossRef] [PubMed]
30. Guo, Y.; Cai, J.; Chang, Y.; Li, X.; Liu, Q.; Wang, X.; Wang, X.; Zhong, M.; Wen, J.; Wang, J. Identification of forensically important sarcophagid flies (Diptera: Sarcophagidae) in China, based on COI and 16S rDNA gene sequences. *J. Forensic Sci.* **2011**, *56*, 1534–1540. [CrossRef] [PubMed]
31. Guo, Y.D.; Cai, J.F.; Li, X.; Xiong, F.; Su, R.N.; Chen, F.L.; Liu, Q.L.; Wang, X.H.; Chang, Y.F.; Zhong, M.; et al. Identification of the forensically important sarcophagid flies *Boerttcherisca peregrina*, *Parasarcophaga albiceps* and *Parasarcophaga dux* (Diptera: Sarcophagidae) based on COII gene in China. *Trop. Biomed.* **2010**, *27*, 451–460.
32. Ren, L.; Chen, W.; Shang, Y.; Meng, F.; Zha, L.; Wang, Y.; Guo, Y. The application of COI gene for species identification of forensically important muscid flies (Diptera: Muscidae). *J. Med. Entomol.* **2018**, *55*, 1150–1159. [CrossRef]
33. Zhang, Z.; Ren, L.; Leng, J.; Shang, Y.; Zhu, G. The complete mitochondrial genome of *Chrysomya nigripes* (Diptera: Calliphoridae). *Mitochondrial DNA B* **2019**, *4*, 4012–4013. [CrossRef]
34. Shang, Y.; Ren, L.; Chen, W.; Zha, L.; Cai, J.; Dong, J.; Guo, Y. Comparative mitogenomic analysis of forensically important Sarcophagid flies (Diptera: Sarcophagidae) and implications of species identification. *J. Med. Entomol.* **2019**, *56*, 392–407. [CrossRef]
35. Ye, G.; Li, K.; Zhu, J.; Zhu, G.; Hu, C. Cuticular hydrocarbon composition in pupal exuviae for taxonomic differentiation of six necrophagous flies. *J. Med. Entomol.* **2007**, *44*, 450–456. [CrossRef]
36. Grzywacz, A.; Hall, M.J.; Pape, T.; Szpila, K. Muscidae (Diptera) of forensic importance—An identification key to third instar larvae of the western Palaearctic region and a catalogue of the muscid carrion community. *Int. J. Legal Med.* **2017**, *131*, 1–12. [CrossRef] [PubMed]
37. Lutz, L.; Williams, K.A.; Villet, M.H.; Ekanem, M.; Szpila, K. Species identification of adult African blowflies (Diptera: Calliphoridae) of forensic importance. *Int. J. Legal Med.* **2017**, *132*, 831–842. [CrossRef] [PubMed]
38. Díaz-Aranda, L.M.; Martín-Vega, D.; Baz, A.; Cifrián, B. Larval identification key to necrophagous Coleoptera of medico-legal importance in the western Palaearctic. *Int. J. Legal Med.* **2018**, *132*, 1795–1804. [CrossRef]
39. Sukontason, K.; Bunchu, N.; Chaiwong, T.; Moophayak, K.; Sukontason, K.L. Forensically important flesh fly species in Thailand: Morphology and developmental rate. *Parasitol. Res.* **2010**, *106*, 1055–1064. [CrossRef] [PubMed]
40. Amendt, J.; Richards, C.S.; Campobasso, C.P.; Zehner, R.; Hall, M.J.R. Forensic entomology: Applications and limitations. *Forensic Sci. Med. Pathol.* **2011**, *7*, 379–392. [CrossRef]
41. Ma, Y.K.; Hu, C.; Min, J.X. Effects of temperature on the growth and development of four common necrophagous flies and their significance in forensic medicine. *Chin. J. Forensic Med.* **1998**, *13*, 81–84.
42. Wang, J.F. *Fundamental Research on the Application of Morphology and Development Patterns of Some Necrophagous Flies in the Determination of Postmortem Interval*; Zhejiang University: Zhejiang, China, 1999.
43. Li, Y. *Effect of Temperature on the Growth and Development of the Larvae of Lucilia Sericata (Diptera: Calliphoridae) and its Significance in Forensic Medicine*; Hebei Medical University: Shijiazhuang, China, 2007.
44. Wang, L.; Liu, B.; Ze-Min, L.I. Chronometrical morphology changes of larval cephalopharyngeal skeleton in the different developmental stages of *Parasarcophaga crassipalpis* and its application in forensic medicine. *Chin. Bull. Entomol.* **2008**, *45*, 133–137.
45. Zhao, B.; Wang, L.; Wang, H.; Wang, W.; Qi, L.; Li, Z. Morphological changes of *Lucilia sericata* larvae and its significance in forensic medicine. *Chin. J. Vector Biol. Control* **2009**, *20*, 534–537.

46. Amendt, J.; Campobasso, C.P.; Gaudry, E.; Reiter, C.; LeBlanc, H.N., Jr.; Hall, M. Best practice in forensic entomology—Standards and guidelines. *Int. J. Legal Med.* **2007**, *121*, 90–104. [CrossRef] [PubMed]

47. Wang, Y.; Li, L.; Wang, J.; Wang, M.; Yang, L.; Tao, L.; Zhang, Y.; Hou, Y.; Chu, J.; Hou, Z. Development of the green bottle fly *Lucilia illustris* at constant temperatures. *Forensic Sci. Int.* **2016**, *267*, 136–144. [CrossRef] [PubMed]

48. Yang, Y.; Li, X.; Shao, R.; Lyu, Z.; Li, H.; Li, G.; Xu, L.; Wan, L. Developmental times of *Chrysomya megacephala* (Fabricius) (Diptera: Calliphoridae) at constant temperatures and applications in forensic entomology. *J. Forensic Sci.* **2016**, *61*, 1278–1284. [CrossRef] [PubMed]

49. Zhang, X.; Li, Y.; Shang, Y.; Ren, L.; Chen, W.; Wang, S.; Guo, Y. Development of *Sarcophaga dux* (diptera: Sarcophagidae) at constant temperatures and differential gene expression for age estimation of the pupae. *J. Therm. Biol.* **2020**, *93*, 102735. [CrossRef]

50. Yang, Y.; Lyu, Z.; Li, X.; Li, K.; Yao, L.; Wan, L. Technical note: Development of *Hemipyrellia ligurriens* (Wiedemann) (Diptera: Calliphoridae) at constant temperatures: Applications in estimating postmortem interval. *Forensic Sci. Int.* **2015**, *253*, 48–54. [CrossRef]

51. Yang, L.; Wang, Y.; Li, L.; Wang, J.; Wang, M.; Zhang, Y.; Chu, J.; Liu, K.; Hou, Y.; Tao, L. Temperature-dependent development of *Parasarcophaga similis* (Meade 1876) and its significance in estimating postmortem interval. *J. Forensic Sci.* **2017**, *62*, 1234–1243. [CrossRef]

52. Wang, Y.; Wang, J.F.; Zhang, Y.N.; Tao, L.Y.; Wang, M. Forensically important *Boettcherisca peregrina* (Diptera: Sarcophagidae) in China: Development pattern and significance for estimating postmortem interval. *J. Med. Entomol.* **2017**, *54*, 1491–1497. [CrossRef]

53. Wang, Y.; Zhang, Y.; Liu, C.; Hu, G.; Wang, M.; Yang, L.; Chu, J.; Wang, J. Development of *Aldrichina grahami* (Diptera: Calliphoridae) at constant temperatures. *J. Med. Entomol.* **2018**, *55*, 1402–1409. [CrossRef]

54. Wang, Y.; Yang, L.; Zhang, Y.; Tao, L.; Wang, J. Development of *Musca domestica* at constant temperatures and a first case report about its application in the estimation of the minimum postmortem interval. *Forensic Sci. Int.* **2018**, *285*, 172–180. [CrossRef]

55. Zhang, Y.; Wang, Y.; Sun, J.; Hu, G.; Wang, M.; Amendt, J.; Wang, J. Temperature-dependent development of the blow fly *Chrysomya pinguis* and its significance in estimating postmortem interval. *Roy. Soc. Open Sci.* **2019**, *6*, 1900039. [CrossRef]

56. Wang, Y.; Hu, G.; Zhang, Y.; Wang, M.; Amendt, J.; Wang, J. Development of *Muscina stabulans* at constant temperatures with implications for minimum postmortem interval estimation. *Forensic Sci. Int.* **2019**, *298*, 71–79. [CrossRef] [PubMed]

57. Wang, M.; Wang, Y.; Hu, G.; Wang, Y.; Xu, W.; Wu, M.; Wang, J. Development of *Lucilia sericata* (Diptera: Calliphoridae) under constant temperatures and its significance for the estimation of time of death. *J. Med. Entomol.* **2020**, *57*, 1373–1381. [CrossRef]

58. Hu, G.; Wang, Y.; Sun, Y.; Zhang, Y.; Wang, M.; Wang, J. Development of *Chrysomya rufifacies* (Diptera: Calliphoridae) at constant temperatures within its colony range in Yangtze River Delta Region of China. *J. Med. Entomol.* **2019**, *56*, 1215–1224. [CrossRef]

59. Wang, Y.; Zhang, Y.; Wang, M.; Hu, G.; Fu, Y.; Zhi, R.; Wang, J. Development of *Hydrotaea spinigera* at constant temperatures and its significance for estimating postmortem interval. *J. Med. Entomol.* **2021**, *58*, 56–63. [CrossRef]

60. Martín-Vega, D.; Simonsen, T.J.; Wicklein, M.; Hall, M.J. Age estimation during the blow fly intra-puparial period: A qualitative and quantitative approach using micro-computed tomography. *Int. J. Legal Med.* **2017**, *131*, 1429–1448. [CrossRef] [PubMed]

61. Brown, K.; Thorne, A.; Harvey, M. *Calliphora vicina* (Diptera: Calliphoridae) pupae: A timeline of external morphological development and a new age and PMI estimation tool. *Int. J. Legal Med.* **2015**, *129*, 835–850. [CrossRef]

62. Wang, H.; Wang, L. Pupal morphogenesis of *Lucilia cuprina* in different constant temperatures and its significance in forensic medicine. *Chin. J. Forensic Med.* **2002**, *23*, 324–326.

63. Zhang, Y.; Wang, Y.; Yang, L.; Tao, L.; Wang, J. Development of *Chrysomya megacephala* at constant temperatures within its colony range in Yangtze River Delta region of China. *Forensic Sci. Res.* **2018**, *3*, 74–82. [CrossRef] [PubMed]

64. Feng, D.; Wu, J.; Sun, D. Intrapuparial age estimation of forensically important *Dohrniphora cornuta* (Diptera: Phoridae). *J. Med. Entomol.* **2020**. [CrossRef]

65. Li, L.; Wang, Y.; Wang, J. Intra-puparial development and age estimation of forensically important *Hermetia illucens* (L.). *J. Asia-Pac. Entomol.* **2016**, *19*, 233–237. [CrossRef]

66. Feng, D.; Liu, G. Pupal age estimation of forensically important *Megaselia scalaris* (Loew) (Diptera: Phoridae). *Forensic Sci. Int.* **2014**, *236*, 133–137. [CrossRef] [PubMed]

67. Feng, D.; Liu, G. Pupal age estimation of forensically important *Megaselia spiracularis* Schmitz (Diptera: Phoridae). *Forensic Sci. Int.* **2013**, *231*, 199–203. [CrossRef]

68. Ma, T.; Huang, J.; Wang, J. Study on the pupal morphogenesis of *Chrysomya rufifacies* (Macquart) (Diptera: Calliphoridae) for postmortem interval estimation. *Forensic Sci. Int.* **2015**, *253*, 88–93. [CrossRef] [PubMed]

69. Wang, Y.; Gu, Z.; Xia, S.; Wang, J.; Zhang, Y.; Tao, L. Estimating the age of intra-puparial period of *Lucilia illustris* using two approaches: Morphological changes and differentially expressed genes. *Forensic Sci. Int.* **2018**, *287*, 1–11. [CrossRef]

70. Gruszka, J.; Matuszewski, S. Estimation of physiological age at emergence based on traits of the forensically useful adult carrion beetle *Necrodes littoralis* L (Silphidae). *Forensic Sci. Int.* **2020**, *314*, 110407. [CrossRef] [PubMed]

71. Martín-Vega, D.; Díaz-Aranda, L.M.; Baz, A.; Cifrián, B. Effect of temperature on the survival and development of three forensically relevant *Dermestes* species (Coleoptera: Dermestidae). *J. Med. Entomol.* **2017**, *54*, 1140–1150. [CrossRef] [PubMed]

72. Frątczak-Łagiewska, K.; Grzywacz, A.; Matuszewski, S. Development and validation of forensically useful growth models for Central European population of *Creophilus maxillosus* L. (Coleoptera: Staphylinidae). *Int. J. Legal Med.* **2020**, *134*, 1531–1545. [CrossRef]

73. Wang, Y.; Yang, J.B.; Wang, J.F.; Li, L.L.; Wang, M.; Yang, L.J.; Tao, L.Y.; Chu, J.; Hou, Y.D. Development of the forensically important beetle *Creophilus maxillosus* (Coleoptera: Staphylinidae) at constant temperatures. *J. Med. Entomol.* **2017**, *54*, 281–289.

74. Hu, G.; Wang, M.; Wang, Y.; Tang, H.; Chen, R.; Zhang, Y.; Zhao, Y.; Jin, J.; Wang, Y.; Wu, M.; et al. Development of *Necrobia rufipes* (De Geer, 1775) (Coleoptera: Cleridae) under constant temperatures and its implication in forensic entomology. *Forensic Sci. Int.* **2020**, *311*, 110275. [CrossRef] [PubMed]

75. Wang, Y.; Wang, M.; Hu, G.; Xu, W.; Wang, Y.; Wang, J. Temperature-dependent development of *Omosita colon* at constant temperature and its implication for PMI$_{min}$ estimation. *J. Forensic Leg. Med.* **2020**, *72*, 101946. [CrossRef]

76. Wang, Y.; Hu, G.; Liu, N.; Wang, M.; Chen, R.; Zhu, R.; Wang, Y.; Ren, X.; Wang, Y.; Xu, W.; et al. Development of *Dermestes tessellatocollis* Motschulsky under different constant temperatures and its implication in forensic entomology. *Forensic Sci. Int.* **2021**, *321*, 110723. [CrossRef] [PubMed]

77. Zhu, G.; Ye, G.; Hu, C. Development profile of hemolymph soluble proteins during the larval and pupal stages in *Aldrichina grahami*. *J. Zhejiang Univ.* **2004**, *30*, 69–72.

78. Xu, H.; Ye, G.; Xu, Y.; Hu, C.; Zhu, G. Age-dependent changes in cuticular hydrocarbons of larvae in *Aldrichina grahami* (Aldrich) (Diptera: Calliphoridae). *Forensic Sci. Int.* **2014**, *242*, 236–241. [CrossRef]

79. Wang, Y.; Gu, Z.; Li, L.; Wang, J. Gene expression during the intra-puparial stage of *Chrysomya megacephala*: Implications for postmortem interval estimation. *J. Asia-Pac. Entomol.* **2019**, *22*, 841–846. [CrossRef]

80. Liu, Z.; Han, H.; Chen, W.; Wang, S.; Meng, F.; Cai, J.; Guo, Y. Evaluation of reference genes and age estimation of forensically useful Aldrichina grahami (Diptera: Calliphoridae) during intrapuparial period. *J. Med. Entomol.* **2021**, *58*, 47–55.

81. Shang, Y.; Ren, L.; Yang, L.; Wang, S.; Chen, W.; Dong, J.; Ma, H.; Qi, X.; Guo, Y. Differential gene expression for age estimation of forensically important *Sarcophaga peregrina* (Diptera: Sarcophagidae) intrapuparial. *J. Med. Entomol.* **2020**, *57*, 65–77. [CrossRef] [PubMed]

82. Zhu, G.; Jia, Z.; Yu, X.; Wu, K.; Chen, L.; Lv, J.; Benbow, M.E. Predictable weathering of puparial hydrocarbons of necrophagous flies for determining the postmortem interval: A field experiment using *Chrysomya rufifacies*. *Int. J. Legal Med.* **2017**, *131*, 885–894. [CrossRef] [PubMed]

83. Zhu, G.H.; Xu, X.H.; Xiao, J.Y.; Zhang, Y.; Wang, J.F. Puparial case hydrocarbons of *Chrysomya megacephala* as an indicator of the postmortem interval. *Forensic Sci. Int.* **2007**, *169*, 1–5. [CrossRef]

84. Baqué, M.; Amendt, J.; Verhoff, M.A.; Zehner, R. Descriptive analyses of differentially expressed genes during larval development of *Calliphora vicina* (Diptera: Calliphoridae). *Int. J. Legal Med.* **2015**, *129*, 891–902. [CrossRef]

85. Tarone, A.M.; Foran, D.R. Gene expression during blow fly development: Improving the precision of age estimates in forensic entomology. *J. Forensic Sci.* **2011**, *56*, S112–S122. [CrossRef]

86. Zajac, B.K.; Amendt, J.; Horres, R.; Verhoff, M.A.; Zehner, R. De novo transcriptome analysis and highly sensitive digital gene expression profiling of *Calliphora vicina* (Diptera: Calliphoridae) pupae using MACE (Massive Analysis of cDNA Ends). *Forensic Sci. Int. Gen.* **2015**, *15*, 137–146. [CrossRef]

87. Catts, E.P.; Goff, M.L. Forensic entomology in criminal investigations. *Annu. Rev. Entomol.* **1992**, *37*, 253–272. [CrossRef] [PubMed]

88. Tomberlin, J.K.; Benbow, M.E. *Forensic Entomology: International Dimensions and Frontiers*; CRC Press: Boca Raton, FL, USA, 2015.

89. Bernhardt, V.; Schomerus, C.; Verhoff, M.A.; Amendt, J. Of pigs and men—Comparing the development of *Calliphora vicina* (Diptera: Calliphoridae) on human and porcine tissue. *Int. J. Legal Med.* **2017**, *131*, 847–853. [CrossRef]

90. Bugelli, V.; Campobasso, C.P.; Verhoff, M.A.; Amendt, J. Effects of different storage and measuring methods on larval length values for the blow flies (Diptera: Calliphoridae) *Lucilia sericata* and *Calliphora vicina*. *Sci. Justice* **2016**, *57*, 159–164. [CrossRef]

91. Wang, H.; Shi, Y.; Liu, X.; Zhang, R. Growth and development of *Boettcherisca peregrine* under different temperature conditions and its significance in forensic entomology. *J. Environ. Entomol.* **2010**, *32*, 166–172.

92. Zhao, B.; Wang, H.; Wang, L.; Wang, W.; Li, Z. Chronometric morphological change of larvae of two species of necrophagous flies and its implication in forensic medicine. *Chin. Bull. Entomol.* **2010**, *47*, 360–367.

93. Chen, W.; Yang, L.; Ren, L.; Shang, Y.; Wang, S.; Guo, Y. Impact of constant versus fluctuating temperatures on the development and life history parameters of *Aldrichina grahami* (Diptera: Calliphoridae). *Insects* **2019**, *10*, 184. [CrossRef] [PubMed]

94. Li, L.; Wang, Y.; Wang, J.; Ma, M.; Lai, Y. Temperature-dependent development and the significance for estimating postmortem interval of *Chrysomya nigripes* Aubertin, a new forensically important species in China. *Int. J. Legal Med.* **2016**, *130*, 1363–1370. [CrossRef]

95. Wang, Y.; Zhang, Y.; Hu, G.; Wang, M.; Zhu, R.; Zhai, Y.; Sun, J.; Li, X.; Wang, L.; Wu, M.; et al. Development of *Megaselia spiracularis* (Diptera: Phoridae) at different constant temperatures. *J. Therm. Biol.* **2020**, *93*, 102722. [CrossRef] [PubMed]

96. Zhang, Y.; Wang, Y.; Liu, C.; Wang, J.; Hu, G.; Wang, M.; Yang, L.; Chu, J. Development of *Nasonia vitripennis* (Hymenoptera: Pteromalidae) at constant temperatures in China. *J. Med. Entomol.* **2018**, *56*, 368–377. [CrossRef] [PubMed]

97. Smith, K.G.V. *A Manual of Forensic Entomology*; Cornell University Press: Ithaca, NJ, USA, 1986.

98. Zhou, H.; Yang, Y.; Ren, J.; Lu, L.; Wang, S.; Yan, R.; Li, Y. Studies on forensic entomology in Beijing district I: Sarcosaprophagous beetles and their local specificity. *Acta Entomol. Sin.* **1997**, *40*, 62–70.

99. Yang, Y. Study on necrophagous mucoid flies living on human corpses in Beijing and its application on forensic medicine practice. *Chin. J. Forensic Med.* **1998**, *13*, 159–162.

100. Yin, X.; Ma, M.; Zhou, H.; Lai, Y.; Wang, J. The community succession of sarcosaphagous insects on pig carcasses in summer indoor and outdoor environment in Shenzhen area. *Fa Yi Xue Za Zhi* **2014**, *30*, 172–177.

101. Wang, Y.; Wang, J.; Wang, Z.; Tao, L. Insect succession on pig carcasses using different exposure time—A preliminary study in Guangzhou, China. *J. Forensic Leg. Med.* **2017**, *52*, 24–29. [CrossRef] [PubMed]

102. Wang, Y.; Ma, M.; Jiang, X.; Wang, J.; Li, L.; Yin, X.; Wang, M.; Lai, Y.; Tao, L. Insect succession on remains of human and animals in Shenzhen, China. *Forensic Sci. Int.* **2017**, *271*, 75–86. [CrossRef]

103. Li, L.; Wang, J.; Wang, Y. A comparative study of the decomposition of pig carcasses in a methyl methacrylate box and open air conditions. *J. Forensic Leg. Med.* **2016**, *42*, 92–95. [CrossRef] [PubMed]

104. Shi, Y.; Liu, X.; Wang, H.; Zhang, R. Seasonality of insect succession on exposed rabbit carrion in Guangzhou, China. *Insect Sci.* **2009**, *16*, 425–439. [CrossRef]

105. Wang, J.; Li, Z.; Chen, Y.; Chen, Q.; Yin, X. The succession and development of insects on pig carcasses and their significances in estimating PMI in south China. *Forensic Sci. Int.* **2008**, *179*, 11–18. [CrossRef] [PubMed]

106. Chang, Y.F.; Zhang, J.L.; Zhou, X.R.; Liao, Z.G.; Cai, J.F.; Deng, Z.H.; Lan, L.M.; Wang, Z.X.; Liang, W.B.; Deng, R.L. The constitution and succession of sarcosaphagous flies community in hohhot. *Forensic Sci.Technol.* **2006**, *4*, 15–18.

107. Wu, H.; Yang, L.; Wang, B.; Yuan, H.; Zhai, Y. Study on the community succession of sarcosaphagous insects at summer in Xinxiang area. *Chin. J. Forensic Med.* **2017**, *32*, 13–15.

108. Chen, L.; Qiu, L.; Guo, J.; Lin, Y. Initial research on process of necrophagous flies participating in decomposing corpse parenchyma in four seasons. *Chin. J. Forensic Med.* **2009**, *24*, 188–190.

109. Matuszewski, S.; Bajerlein, D.; Konwerski, S.; Szpila, K. Insect succession and carrion decomposition in selected forests of Central Europe. Part 2: Composition and residency patterns of carrion fauna. *Forensic Sci. Int.* **2010**, *195*, 42–51. [CrossRef]

110. Matuszewski, S.; Bajerlein, D.; Konwerski, S.; Szpila, K. Insect succession and carrion decomposition in selected forests of Central Europe. Part 3: Succession of carrion fauna. *Forensic Sci. Int.* **2011**, *207*, 150–163. [CrossRef]

111. Pastula, E.C.; Merritt, R.W. Insect arrival pattern and succession on buried carrion in Michigan. *J. Med. Entomol.* **2013**, *50*, 432–439. [CrossRef]

112. Voss, S.C.; Forbes, S.L.; Dadour, I.R. Decomposition and insect succession on cadavers inside a vehicle environment. *Forensic Sci. Med. Pathol.* **2008**, *4*, 22–32. [CrossRef] [PubMed]

113. Aballay, F.H.; Murua, A.F.; Acosta, J.C.; Centeno, N.D. Succession of carrion fauna in the arid region of San Juan Province, Argentina and its forensic relevance. *Neotrop. Entomol.* **2012**, *41*, 27–31. [CrossRef]

114. Heo, C.C.; Mohamad, A.M.; Ahmad, F.M.; Jeffery, J.; Kurahashi, H.; Omar, B. Study of insect succession and rate of decomposition on a partially burned pig carcass in an oil palm plantation in Malaysia. *Trop. Biomed.* **2008**, *25*, 202–208. [PubMed]

115. Lynch-Aird, J.; Moffatt, C.; Simmons, T. Decomposition rate and pattern in hanging pigs. *J. Forensic Sci.* **2015**, *60*, 1155–1163. [CrossRef] [PubMed]

116. Voss, S.C.; Cook, D.F.; Dadour, I.R. Decomposition and insect succession of clothed and unclothed carcasses in Western Australia. *Forensic Sci. Int.* **2011**, *211*, 67–75. [CrossRef]

117. Ma, Y.; Hu, C. A preliminary study on the species and biological characters of necrophagous insects in Hangzhou area. *J. Zhejiang Agric. Univ.* **1997**, *23*, 375–380.

118. Li, L.; Zhang, Y.; Ma, Y. The major necrophagous insects and their regular activity on carcass in Harbin. *J. Northeast Forest. Univ.* **2002**, *30*, 93–96.

119. The species and distribution of the necrophagous flies in Guizhou Province. *Acta Entomol. Sin.* **2004**, *47*, 849–853.

120. Wang, Y.; Liu, M.; Sun, D.H. A study on sarcosaphagous insects species variety with seasons in Chengdu. *Fa Yi Xue Za Zhi* **2003**, *19*, 86–87. [PubMed]

121. Chen, Q. *Study on the Characteristics of Cadaver Decomposition and the Disciplinarian of Arthropod Successional Patterns in Cadaver at Suburban Districts of Zhongshan*; Sun Yat-sen University: Guangzhou, China, 2006.

122. Wu, D.; Mao, R.; Guo, M.; Zhou, J.; Jia, F.; Ou, G.; Zhang, R. Species and succession of necrophagous insect community in spring and summer seasons in Guangzhou. *Acta Sci. Nat. Univ. Sunyatseni* **2008**, *47*, 56–60.

123. Dong, Y.; Zhu, Y.; Wang, L. Primary study on common species sarcosaphagous flies. *Forensic Sci.Technol.* **2009**, *2*, 14–16.

124. Nie, T.; Wei, Z.; Tian, Y.; Tian, F.; Lian, Z. Primary study on sarcosaphagous insects on rabbit carcass in Xi' an. *Chin. Bull. Entomol.* **2010**, *47*, 587–591.

125. Jiang, Y.; Cai, J.; Yang, L.; Yi, W.; Lan, L.; Li, X.; Li, J. A study of sarcosaphagous insects from arthropod in Yongzhou district of Hunan Province. *Chin. J. Applied Entomol.* **2011**, *48*, 191–196.

126. Jiang, D.; Zhan, X.; Sun, K.; Gu, Y.; Ren, W.; Li, B.; Zhang, P.; Li, W. Study on the common species and seasonal succession of necrophagous flies in Qingdao region. *Chin. J. Forensic Med.* **2014**, *29*, 460–463.

127. Zhou, L. *Insect Succession on Carcasses in Southern Chongqing City and Its Forensic Application*; Chongqing Medical University: Chongqing, China, 2015.

128. Yang, L. *Succession and Development of Sarcosarprophagous Insects and Their Significances in Estimating PMI in Yangtze River Delta Region*; Soochow University: Suzhou, China, 2017.

129. Tian, J. *Effect of Morphine in Tissues on Development of Chrysomya megacephala (Diptera) and Implication of This Effect on Estimation of Postmortem Intervals Using Arthropod Development Patterns*; Hebei Medical University: Shijiazhuang, China, 2004.

130. Zhao, W.; Wang, B.; Hu, S.; Feng, X. The relationship between the postmortem interval and effects of morphine on the growth stage of *Lucilia sericata*. *J. Pathog. Biol.* **2008**, *3*, 612–615.

131. Dai, J. *Effect of Diazepam on the Growth and Development of Lucilia Sericata (Diptera Calliphoridae) and Its Forensic Importance*; Hebei Medical University: Shijiazhuang, China, 2005.
132. Wang, L. *Effect of Diazepam in Tissues on Development of Chrysomya Megacephala (Diptera) and Implication of This Effect on Estimation of Postmortem Intervals Using Arthropod Development Patterns*; Hebei Medical University: Shijiazhuang, China, 2006.
133. Lv, Z.; Zhai, X.D.; Zhou, H.M.; Li, P.; Ma, J.; Guan, L.; Mo, Y. Effect of ketamine on the development of *Chrysomya megacephala* (Diptera:Calliphoridae). *Chin. J. Parasitol. Parasitic Dis.* **2012**, *30*, 361–366.
134. Zou, Y.; Huang, M.; Huang, R.; Wu, X.; You, Z.; Lin, J.; Huang, X.; Qiu, X.; Zhang, S. Effect of ketamine on the development of *Lucilia sericata* (Meigen) (Diptera: Calliphoridae) and preliminary pathological observation of larvae. *Forensic Sci. Int.* **2013**, *226*, 273–281. [CrossRef] [PubMed]
135. Liu, X.; Shi, Y.; Wang, H.; Zhang, R. Determination of Malathion levels and its effect on the development of *Chrysomya megacephala* (Fabricius) in South China. *Forensic Sci. Int.* **2009**, *192*, 14–18. [CrossRef] [PubMed]
136. Shi, Y.; Liu, X.; Wang, H.; Zhang, R. Effects of malathion on the insect succession and the development of *Chrysomya megacephala* (Diptera: Calliphoridae) in the field and implications for estimating postmortem interval. *Am. J. Forensic Med. Pathol.* **2010**, *31*, 46–51.
137. Wang, S.; Zhang, C.; Chen, W.; Ren, L.; Ling, J.; Shang, Y.; Guo, Y. Effects of methamphetamine on the development and its determination in *Aldrichina grahami* (Diptera: Calliphoridae). *J. Med. Entomol.* **2020**, *57*, 691–696. [CrossRef] [PubMed]
138. Brien, C.; Turner, B. Impact of paracetamol on *Calliphora vicina* larval development. *Int. J. Legal Med.* **2004**, *118*, 188–189.
139. Carvalho, L.M.; Linhares, A.X.; Trigo, J.R. Determination of drug levels and the effect of diazepam on the growth of necrophagous flies of forensic importance in southeastern Brazil. *Forensic Sci. Int.* **2001**, *120*, 140–144. [CrossRef]
140. De Carvalho, L.M.L.; Linhares, A.X.; Palhares, F.A.B. The effect of cocaine on the development rate of immatures and adults of *Chrysomya albiceps* and *Chrysomya putoria* (Diptera: Calliphoridae) and its importance to postmortem interval estimate. *Forensic Sci. Int.* **2012**, *220*, 27–32. [CrossRef] [PubMed]
141. Kharbouche, H.; Augsburger, M.; Cherix, D.; Sporkert, F.; Giroud, C.; Wyss, C.; Champod, C.; Mangin, P. Codeine accumulation and elimination in larvae, pupae, and imago of the blowfly *Lucilia sericata* and effects on its development. *Int. J. Legal Med.* **2008**, *122*, 205–211. [CrossRef]
142. Goff, M.L.; Brown, W.A.; Omori, A.I. Preliminary observations of the effect of methamphetamine in decomposing tissues on the development rate of *Parasarcophaga ruficornis* (Diptera: Sarcophagidae) and implications of this effect on the estimations of postmortem intervals. *J. Forensic Sci.* **1992**, *37*, 867–872. [PubMed]
143. Rashid, R.A.; Osman, K.; Ismail, M.I.; Zuha, R.M.; Hassan, R.A. Determination of malathion levels and the effect of malathion on the growth of *Chrysomya megacephala* (Fabricius) in malathion-exposed rat carcass. *Trop. Biomed* **2008**, *25*, 184–190.
144. Tabor, K.L.; Fell, R.D.; Brewster, C.C.; Pelzer, K.; Behonick, G.S. Effects of antemortem ingestion of ethanol on insect successional patterns and development of *Phormia regina* (Diptera: Calliphoridae). *J. Med. Entomol.* **2005**, *42*, 481–489. [CrossRef] [PubMed]
145. Goff, M.L.; Brown, W.A.; Omori, A.I.; LaPointe, D.A. Preliminary observations of the effects of amitriptyline in decomposing tissues on the development of *Parasarcophaga ruficornis* (Diptera: Sarcophagidae) and implications of this effect to estimation of postmortem interval. *J. Forensic Sci.* **1993**, *38*, 316–322. [CrossRef]
146. Pien, K.; Laloup, M.; Pipeleers-Marichal, M.; Grootaert, P.; De Boeck, G.; Samyn, N.; Boonen, T.; Vits, K.; Wood, M. Toxicological data and growth characteristics of single post-feeding larvae and puparia of *Calliphora vicina* (Diptera: Calliphoridae) obtained from a controlled nordiazepam study. *Int. J. Legal Med.* **2004**, *118*, 190–193. [CrossRef]
147. Souza, C.M.; Thyssen, P.J.; Linhares, A.X. Effect of nandrolone decanoate on the development of three species of *Chrysomya* (Diptera: Calliphoridae), flies of forensic importance in Brazil. *J. Med. Entomol.* **2014**, *48*, 111–117. [CrossRef] [PubMed]
148. Ferraz, A.C.; Dallavecchia, D.L.; Silva, D.C.; Figueiredo, A.L.; Proença, B.; Silva-Filho, R.G.; Aguiar, V.M. Effects of the antibiotics Gentamicin on the postembryonic development of *Chrysomya putoria* (Diptera: Calliphoridae). *J. Insect Sci.* **2014**, *14*, 1–5. [CrossRef]
149. Da Silva, E.I.T.; Wilhelmi, B.; Villet, M.H. Forensic entomotoxicology revisited—Towards professional standardisation of study designs. *Int. J. Legal Med.* **2017**, *13*, 1399–1412. [CrossRef]
150. Guo, Y.; Cai, J.; Tang, Z.; Feng, X.; Lin, Z.; Yong, F.; Li, J.; Chen, Y.; Meng, F.; Wen, J. Application of *Aldrichina grahami* (Diptera, Calliphoridae) for forensic investigation in central-south China. *Rom J. Leg. Med.* **2011**, *19*, 55–58.
151. Wang, J.F.; Chang, P.; Liao, M.Q.; Li, Z.G.; Ma, T.; Gui, L.F.; Yu, L.C. Application of forensic entomology in indoor, outdoor and poisoning death cases in Guangdong Province. *J. Poltical Sci. Law* **2012**, *29*, 116–121.
152. Owings, C.G.; Spiegelman, C.; Tarone, A.M.; Tomberlin, J.K. Developmental variation among *Cochliomyia macellaria* Fabricius (Diptera: Calliphoridae) populations from three ecoregions of Texas, USA. *Int. J. Legal Med.* **2014**, *128*, 709–717. [CrossRef] [PubMed]
153. Zhou, F.; Tomberlin, J.K.; Zheng, L.; Yu, Z.; Zhang, J. Developmental and waste reduction plasticity of three black soldier fly strains (Diptera: Stratiomyidae) raised on different livestock manures. *J. Med. Entomol.* **2013**, *50*, 1224–1230. [CrossRef] [PubMed]
154. Liao, M.Q.; Li, Z.G.; Wang, J.F.; Luo, S.M.; Ma, T. Insects can also be found on bodies that are severely burned in winter. *J. Poltical Sci. Law* **2012**, *29*, 112–115.
155. Wang, M.; Chu, J.; Wang, Y.; Li, F.; Liao, M.; Shi, H.; Zhang, Y.; Hu, G.; Wang, J. Forensic entomology application in China: Four case reports. *J. Forensic Leg. Med.* **2019**, *63*, 40–47. [CrossRef]
156. Peng, J.F.; Xia, Y.F. The applicaiton of forensic entomology in three cases. *Fa Yi Xue Za Zhi* **1999**, *15*, 122.

157. Wang, C.Y. Application of forensic entomology in solving a case. *Forensic Sci. Technol.* **2003**, *5*, 60.

158. Chen, L.S. Estimating the postmortem interval using the accumulated degree days. *Forensic Sci. Technol.* **2007**, *5*, 236–237.

159. Yu, L.C.; Lv, G.L.; Wang, J.F. Using entomology to estimate the time of death: A case report. *J. Trop. Med.* **2008**, *8*, 519–520.

160. Gu, J.M.; Zhang, J.Z.; Sun, Z.; Zhang, W.J.; Qin, B.L.; Cao, S.Y.; Zhou, L. Estimation of the discarding time of a dismembered corpse. *Forensic Sci. Technol.* **2009**, *4*, 75–76.

161. Liu, Y.; Chen, Y.Q.; Guo, Y.D.; Zha, L.; Li, L.J. Estimation of post-mortem interval for a drowning case by using flies (Diptera) in Central-South China: Implications for forensic entomology. *Rom. J. Leg. Med.* **2013**, *21*, 293–298.

162. Li, X.B.; Gong, Q.; Wan, L.H.; Jiang, D.; Jiao, Y.G. A case of estimating the time of death by necrophagous flies. *Chin. J. Forensic Med.* **2013**, *3*, 248–249.

163. Wei, Z.W.; Huang, A.H. Estimating the postmortem interval of a composed body by regression equation of maggot. *Chin. J. Forensic Med.* **2016**, *31*, 316–317.

164. Hu, G.; Wang, M.; Wang, Y.; Liao, M.; Hu, J.; Zhang, Y.; Yu, Y.; Wang, J. Estimation of post-mortem interval based on insect species present on a corpse found in a suitcase. *Forensic Sci. Int.* **2020**, *306*, 110046. [CrossRef] [PubMed]

165. Wang, Q.; Lin, H.; Xu, J.; Huang, P.; Wang, Z. Current research and prospects on postmortem interval estimation. *Fa Yi Xue Za Zhi* **2018**, *34*, 459–467. [PubMed]

166. Amendt, J.; Campobasso, C.P.; Grassberger, M. *Current Concepts in Forensic Entomology*; Springer: Dordrecht, The Netherlands, 2010.

 *insects*

Case Report

Unusual Application of Insect-Related Evidence in Two European Unsolved Murders

Francesco Introna [1], Cristina Cattaneo [2], Debora Mazzarelli [2], Francesco De Micco [3] and Carlo Pietro Campobasso [3,*]

1   Institute of Legal Medicine, Department of Interdisciplinar Medicine, University of Bari "Aldo Moro", Policlinico Hospital, 70124 Bari, Italy; francesco.introna@uniba.it
2   Institute of Legal Medicine, Department of Health and Biomedical Sciences, University of Milan, 20133 Milan, Italy; cristina.cattaneo@unimi.it (C.C.); debora.mazzarelli@guest.unimi.it (D.M.)
3   Institute of Legal Medicine, Department of Experimental Medicine, University of Campania "Luigi Vanvitelli", 80138 Naples, Italy; francesco.demicco@gmail.com
*   Correspondence: carlopietro.campobasso@unicampania.it; Tel.: +39-81-5666019

Simple Summary: Proper collection and analysis of physical evidence including insects, bloodstains, or any other material can be of probative value in a court of law. This is the first casework where hairs were involved as insect-related evidence. Hairs constitute important categories of trace evidence as they can provide useful information for an association between a suspect and a crime scene or a suspect and a victim. Two "cold cases" occurred in two different European countries in which the trace evidence relating to insects was the last piece of a complex puzzle useful for the conviction of the perpetrator.

Abstract: Insect-related evidence must be considered of probative value just as bloodstains, fingerprints, fibers, or any other materials. Such evidence if properly collected and analyzed can also provide useful details in the reopening of old unsolved murders, also called "cold cases". This paper presents the case of two murders that occurred in two different European countries and remained unsolved for years. The remains of a girl found in Italy 17 years after her disappearance helped to solve a murder that occurred in Britain 8 years prior. The cases were unexpectedly linked together because of the similarities in the ritualistic placing of strands of hair and connections with the suspect. The trace evidence relating to insects and hairs played a relevant role in the conviction of the perpetrator. In Italy, the defense raised the doubt that the strands of hair found nearby the skeletal remains could be the result of insect feeding activity and not the result of a cut by sharp objects. Therefore, it was fundamental to distinguish between sharp force lesions and insect feeding activity on hair. This unusual application of insect-related evidence clearly emphasizes the importance of an appropriate professional collection and analysis of any physical evidence that could be of robust probative value.

Keywords: forensic entomology; cold cases; insect evidence; hair evidence

 check for updates

Citation: Introna, F.; Cattaneo, C.; Mazzarelli, D.; De Micco, F.; Campobasso, C.P. Unusual Application of Insect-Related Evidence in Two European Unsolved Murders. *Insects* **2021**, *12*, 444. https://doi.org/10.3390/insects12050444

Academic Editors: Damien Charabidze and Daniel Martín-Vega

Received: 24 February 2021
Accepted: 7 May 2021
Published: 13 May 2021

Publisher's Note: MDPI stays neutral with regard to jurisdictional claims in published maps and institutional affiliations.

1. Introduction

The major goal of medicocriminal entomology is the determination of time, cause, and manner of an investigated death [1]. The most frequently requested task is the estimation of the time since death, better known as the minimum postmortem interval (PMI min) [2]. However, new insights into and achievements of carrion entomology and ecology have increased the opportunities for the use of insects in death investigations and their accuracy [3–5].

Several useful elements can be inferred from the study of insects found on the cadaver or nearby. The correct analysis and interpretation of insect evidence can be crucial in any death investigation, especially when dealing with badly decomposed or skeletonized human remains. Insect-related evidence can provide valuable information to answer questions concerning the use of drugs, child and elderly neglect, sexual abuse, cadaver transfer and concealment of the remains following death, victim identification, determination of specific sites of antemortem trauma, and postmortem artefacts on a body and at the death scene [6,7]. The final aim of any criminal investigation is to provide a strong basis for an association between a suspect and a crime scene or a suspect and a victim.

Therefore, insect-related evidence must be considered of probative value just as bloodstains, fingerprints, hairs, fibers, or any other biological materials. In the recent past, several papers have highlighted the crucial role of correct procedures in sampling and storing entomological evidence [8–10]. If collected properly, such evidence can also provide useful details in the reopening of old unsolved murders and in the presentation of important evidence to the court [11].

This paper presents the case of two murders that occurred in two different European countries and remained unsolved for years. The murders were unexpectedly linked together because of the similarities in the ritualistic placing of strands of hair and connections with the suspect. Hair evidence and its postmortem infestation by insects played a relevant role in the conviction of the perpetrator. The remains of a girl found in Italy 17 years after her disappearance helped to solve a murder that occurred in Britain 8 years prior.

2. Case History

On 17 March 2010, the mummified remains of a 16 year old female were discovered hidden under curved tiles in the darkest corner of a church's loft in Southern Italy. The victim was identified as that of a teenage girl who disappeared from her home 17 years before, on 12 September 1993. The cause of death was massive blood loss due to multiple stab wounds. At autopsy, the signs of multiple stab wounds (15 in total) were found on the skeletal remains. Large amounts of *Diptera puparia* (mostly parasitized) and cast pupal skin of clothes moths (Lepidoptera) were present on the body and nearby the corner of the church's loft where the victim was dragged and concealed. Among the *Diptera* species, Sarcophagids, Phoridae, Fanniidae, Muscids of *Muscina stabulans*, and Calliphorids of *Calliphora vicina*, *Lucilia sericata*, and *Chrysomya albiceps*, on which the clothes moths fed, were found (Figure 1). Among the clothes moths, *Tinea pellionella* and *Tinea bisseliella* were found. The insect species collected were those that were active during the fall, when the victim disappeared. Physical evidence collected at the death scene was some light-brown strands of hair, perfectly and squarely cut, near the skeletonized hands (Figure 2).

(a)   (b)

Figure 1. (**a**) Diptera puparia mostly parasitized; (**b**) clothes moth cases and hairs.

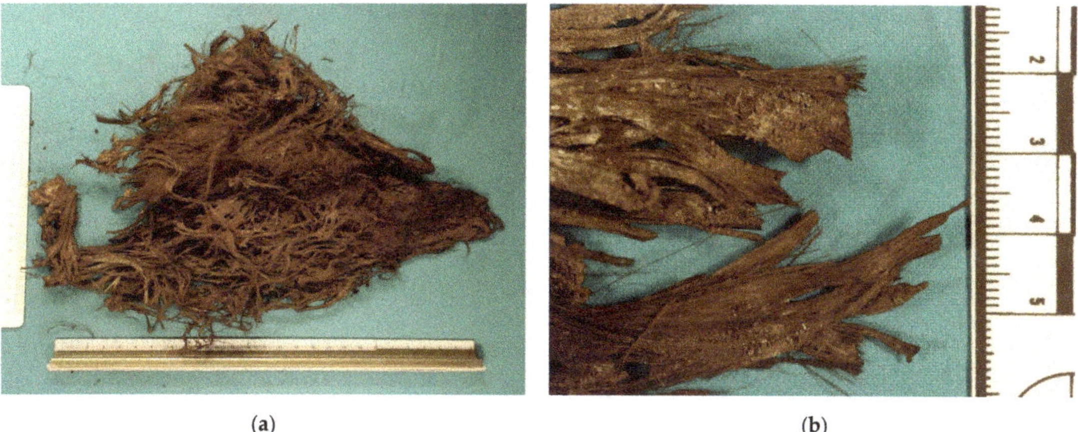

Figure 2. (**a**) Strands of hair found nearby the skeletal remains of a 16 year old female found in Italy 17 years after death; (**b**) close up of the lock of hair clearly cut by sharp force trauma (scissors).

The murder case showed some coincidences with a similar case that occurred in England 8 years prior. A 48 year old seamstress was found dead in her apartment on 12 November 2002. The woman was beaten with a hammer-type object before her throat was cut and body mutilated. She was also stabbed several times and strands of hair were placed in both of her hands. The lock of hair in the victim's left hand was cut from her own head, but that in her right hand was not hers. The main suspect in this murder was the victim's neighbor. He was a 39 year old Italian male who arrived in England several years before from a small town in southern Italy, where the teenage girl went missing. Although the neighbor initially became the main suspect, the British police was unable to collect sufficient evidence for prosecution. After the discovery of the Italian girl in 2010, the cases were linked together because of the similarities in the ritualistic placing of strands of hair and connections with the suspect. The manner in which the English woman was murdered was considered a trademark of the suspect, and the killing was linked to the murder of the adolescent girl in Italy. Two months after the discovery of the Italian girl in 2010, the suspect was arrested in England and charged with the murder of the British woman. In June 2011, he was found guilty by the English jury and sentenced to spend the rest of his life in prison. On October 2014, he was also condemned by the Italian Supreme Court to 30 years in prison for the murder of the 17 year old teenage girl.

3. Insect-Related Evidence

At the trial, the perpetrator consistently denied any involvement in both cases. However, under cross-examination he admitted to having a hair fetish and cutting the hair of girls occasionally met on the street. He liked to touch and smell the hair of girls without planning cutting episodes. According to such admission, the two murders were referred as hair fetish murders. Hair fetishism or trichophilia is a paraphilia, which is a sexual perversion, a condition characterized by abnormal sexual desires. In the case of hair fetishism, seeing or touching hair is particularly erotic and sexually arousing for a person [12]. During the trial, several women reported having their hair cut by an unknown man. One of them described finding something white and sticky in her hair afterwards. Another man described seeing the suspect sitting behind a woman on a bus with her hair in his hands.

In Italy, the defense raised the doubt that the strands of hair found nearby the skeletal remains were not cut by sharp objects but could be the result of insect feeding activity on hair. Therefore, it was fundamental to distinguish between sharp force lesions and insect feeding activity on hair. In order to study if the features in hair lesions were caused by traumatic/sharp forces or by insects, hair samples were subjected to blunt and sharp force trauma and used as pabulum for common clothes moth (*Tineola bisselliella* Lepidoptera, Tineidae) and carpet beetles (*Anthrenus* sp., Colepoptera, Dermestidae) [13]. These species colonize bodies in the late stages of decay when the food source is completely dry [14,15], feeding not only on natural fiber such as hairs, but also on clothing [16]. Artefacts on hairs caused by the feeding activity of moths and beetles were studied by two of the coauthors in a published manuscript [13].

The hair samples examined by stereomicroscopy and scanning electron microscopy (SEM) showed clear differences in hair lesions depending on the type of trauma. They can be summarized as follows: (1) irregular edges without striations and concavities, but with hair elements on different levels in hair locks manually broken; (2) regular edges with parallel striations (similar to marks from sharp weapons observed on bone and cartilage) and hair elements on the same plane in hairs cut by a knife, with the same features but of oval shape due to the compression of the two edges of scissors in hairs cut by scissors; (3) irregular edges with concave shape lesions and cocoons in hair locks used as pabulum for insects [13].

The strands of hair found close to the hands of the victim showed regular edges of the hair lesions with parallel striations of oval shape consistent with hairs cut by scissors. These findings and the results of the experimental study were considered of probative value. The hair lesions distributed on the same plane were clearly cut and not the result of gnawing activity easily distinguishable from breaking and tearing. In the Italian case, the hypothesis of insect feeding on hair was ruled out.

4. Discussion

Hair evidence is one of the most common pieces of physical evidence applied in criminal investigations [17]. The forensic analysis of hair evidence can be extremely valuable in demonstrating that there may have been an association between a suspect and a crime scene, a weapon, or a victim [18,19]. Comparing strands of hair under a microscope can provide robust corroborative evidence to establish these associations as hairs constitute important categories of trace evidence [19]. Microscopical analysis of hairs can give significant information regarding the ancestry, the body area the hair came from, and whether or not the hair was artificially treated or damaged. Toxicological and genetic analysis of hair can also provide useful information from the identification of the hair donor to the detection of drugs and pharmaceuticals and their chronology of intake, especially in drug-facilitated crimes and sexual assaults [20,21]. To the best of our knowledge, this is the first forensic case where hair evidence was related to insect artefacts due to feeding activity on hair.

At a death scene, fly and larval activity can produce several different types of artefacts: from fly specks due to regurgitation and defecation to floor and wall stripes due to post-feeding larval dispersal [22]. Other modifications of the bloodstain pattern are represented by transfer stains due to the migration of insects from a blood pooling [23]. However, *Diptera* larvae can also produce relevant modifications of superficial injuries on the human body. Stab wounds and gunshot injuries can be easily enlarged, distorted, or made unrecognizable by the feeding larval activity at the injury site [24]. Larvae actively feeding on soft tissues can destroy the residual part of the physical evidence still present on the body. Therefore, corpses heavily colonized by larval masses should be considered a matter for urgent forensic pathology examination, and the autopsy should be performed as soon as possible [6,24].

Artefacts made by insects that arrive late on the body are also reported. Dominant taxa in the late stages of decomposition are commonly species of Coleoptera and Lepidoptera such as the clothes moths (Tineidae), very common in mummies and skeletonized bodies [14,15]. These species colonize bodies when the remnants of soft tissues are really dry, feeding on them, as well as on hairs and other natural fibers such as those of clothing [16]. Coleoptera and Lepidoptera can also feed on *Diptera* puparia.

Puparia consist largely of chitin, a long, complex polysaccharide polymer composed of *n*-acetylglucosamine that is similar to keratin in human hair. Keratin and keratin-associated proteins are the major structural components of human hair fiber. Human hairs are proteinaceous shafts with circular or elliptical cross-sections with three morphologically distinct regions in each hair (medulla, cortex, and cuticle) [19]. They can be considered as fiber-reinforced composites consisting of crystalline intermediate filaments embedded in an amorphous protein matrix [25]. The strength and robustness of keratin is derived from tightly packed filaments containing a high degree of disulfide bonding, which confers rigidity and chemical resistance.

The structure and mechanical properties of puparia and human hair explain why sclerotized puparia and hairs do not degrade much, which allow samples to survive for thousands of years [19,26]. Fly puparia and human hairs have been found in ancient graves from Egypt to South America, long after soft tissues have disappeared [27,28]. *Diptera* puparia may represent useful evidence for reconstructing postmortem events in both forensic and archaeo-funerary contexts [29]. However, they may contaminate the forensic entomological evidence if they originate not from human remains but from animal cadavers or other decomposing organic material [30].

The taphonomic degradation process for teeth and bones is well studied, but the same cannot be said for human hair as very little is known about how environmental conditions may alter hair morphology [19]. A recent review [31] addressed the main factors that correlate with decompositional changes of hair. Temperature, environment, and microbes are the major factors affecting the decay of hair. In particular, humidity/aqueous conditions and warm temperatures appear to have a significant effect on the rate of hair decomposition [32,33].

Fungi and bacteria can break down hair; however, insects such as clothes moths and carpet beetles can also easily destroy such evidence as they feed on dried food sources such as natural fibers [34–36]. Many moths and beetles use keratin as a nutrient source. Moth larvae are common sources of damage of woolen textiles, and many Dermestidae larvae also consume hair. The gnawing insect activity usually leaves microscopical signs of erosions and concave lesions on the surface of the hair, but these signs were not observed in the hair samples found close to the skeletonized hands of the victim. In the Italian case, the absence of such microscopical features related to feeding insect activity were considered enough evidence to exclude hair damages due to postmortem infestation by insects as suggested by the defense. Therefore, the hypothesis of insect feeding on hair was ruled out. The hair samples showed only the marks produced from sharp force weapons such as those produced by the compression of the two edges of scissors. Such evidence was just a piece of the big puzzle presented by this international death investigation of unsolved and heinous crimes that occurred in England and Italy, but no less important and crucial for the indictment and final conviction of the suspect.

5. Conclusions

The mutually beneficial relationship between research and casework was largely discussed by Hall [37]. The unusual application of insect-related evidence to hair evidence clearly emphasizes the importance of an appropriate professional collection and analysis of any physical evidence that could be of robust probative value [38]. Differentiation of entomological activity, taphonomy, and sharp force trauma on hair can be crucial in the recovery and analysis of hair evidence.

The take-home message is to call the attention of forensic entomologists to hair examination in skeletonized or mummified bodies or in an advanced stage of decay. Forensic experts should cover every corner of the crime scene and treat each and every piece of evidence as vital. Investigative research has contributed to advancing knowledge of the basic biological and ecological processes of human and animal decomposition, increasing the chances for the admissibility of entomological evidence [39,40]. Because criminal investigations are context-dependent, any trace evidence can provide useful information about the event under investigation, and proper collection and preservation of physical evidence is mandatory [41]. The final aim of any forensic examination must be to provide statements based on objective scientific observation that can be of value in a court of law or for any interested party involved in criminal investigations [18].

Author Contributions: Conceptualization, F.I. and C.P.C.; methodology, C.C. and D.M.; validation, F.I., C.C., and C.P.C.; formal analysis, F.I., C.C., and D.M.; investigation, F.I., C.C., and D.M.; resources, C.P.C. and F.D.M.; writing—original draft preparation, C.P.C. and F.D.M.; writing—review and editing, F.I., C.C., and C.P.C.; visualization, F.D.M.; supervision, C.P.C. All authors have read and agreed to the published version of the manuscript.

Funding: This research received no external funding.

Institutional Review Board Statement: Not applicable.

Data Availability Statement: The data presented in this study are available on request from the corresponding author.

Conflicts of Interest: The authors declare no conflict of interest.

References

1. Amendt, J.; Goff, M.L.; Campobasso, C.P.; Grassberger, M. Preface. In *Current Concepts in Forensic Entomology*; Amendt, J., Goff, M.L., Campobasso, C.P., Grassberger, M., Eds.; Springer: Dordrecht, The Netherlands, 2010; pp. v–vi.
2. Amendt, J.; Campobasso, C.P.; Gaudry, E.; Reiter, C.; Leblanc, H.N.; Hall, M.J.R. Best practice in forensic entomology—standards and guidelines. *Int. J. Leg. Med.* **2006**, *121*, 90–104. [CrossRef]
3. Forbes, S.L.; Carter, D.O. Processes and Mechanisms of Death and Decomposition of Vertebrate Carrion. In *Carrion Ecology, Evolution, and Their Applications*; Benbow, M.E., Tomberlin, J.K., Tarone, A.M., Eds.; CRC Press: Boca Raton, FL, USA, 2016; pp. 13–30.
4. Benbow, M.; Tomberlin, J.; Tarone, A. *Introduction to Carrion Ecology, Evolution, and Their Applications*; Apple Academic Press: Palm Bay, FL, USA, 2015; pp. 3–12.
5. Bugelli, V.; Campobasso, C.P. Basic research and applied science in forensic entomology. *Sci. Justice* **2017**, *57*, 157–158. [CrossRef] [PubMed]
6. Amendt, J.; Richards, C.S.; Campobasso, C.P.; Zehner, R.; Hall, M.J.R. Forensic entomology: Applications and limitations. *Forensic Sci. Med. Pathol.* **2011**, *7*, 379–392. [CrossRef]
7. Campobasso, C.P.; Introna, F. The forensic entomologist in the context of the forensic pathologist's role. *Forensic Sci. Int.* **2001**, *120*, 132–139. [CrossRef]
8. Amendt, J.; Anderson, G.; Campobasso, C.P.; Dadour, I.; Gaudry, E.; Hall, M.J.R.; Moretti, T.C.; Sukontason, K.L.; Villet, M.H. Standard Practices. In *Forensic Entomology—International Dimensions and Frontiers*; Tomblerin, J.K., Benbow, M.E., Eds.; Taylor & Francis Group: Florence, KY, USA, 2015; pp. 381–398.
9. Bugelli, V.; Campobasso, C.P.; Verhoff, M.A.; Amendt, J. Effects of different storage and measuring methods on larval length values for the blow flies (Diptera: Calliphoridae) *Lucilia sericata* and *Calliphora vicina*. *Sci. Justice* **2017**, *57*, 159–164. [CrossRef]
10. Bugelli, V.; Campobasso, C.P.; Zehner, R.; Amendt, J. How should living entomological samples be stored? *Int. J. Leg. Med.* **2019**, *133*, 1985–1994. [CrossRef]
11. Magni, P.A.; Harvey, M.L.; Saravo, L.; Dadour, I.R. Entomological evidence: Lessons to be learnt from a cold case review. *Forensic Sci. Int.* **2012**, *223*, e31–e34. [CrossRef] [PubMed]
12. Janssen, D.F. Rape of the lock: Note on nineteenth-century hair fetishists. *Hist. Psychiatry* **2019**, *30*, 469–479. [CrossRef] [PubMed]
13. Mazzarelli, D.; Vanin, S.; Gibelli, D.; Maistrello, L.; Porta, D.; Rizzi, A.; Cattaneo, C. Splitting hairs: Differentiating between entomological activity, taphonomy, and sharp force trauma on hair. *Forensic Sci. Med. Pathol.* **2014**, *11*, 104–110. [CrossRef]
14. Introna, F.; De Donno, A.; Santoro, V.; Corrado, S.; Romano, V.; Porcelli, F.; Campobasso, C.P. The bodies of two missing children in an enclosed underground environment. *Forensic Sci. Int.* **2011**, *207*, e40–e47. [CrossRef] [PubMed]
15. Buckland, P.C.; Smith, K.G.V. A Manual of Forensic Entomology. *Am. J. Archaeol.* **1988**, *92*, 287. [CrossRef]
16. Robinson, C.P.; Ciccotosto, S.; Sparrow, L.G. Identification of a Key Enzyme in the Digestion of Wool by Larvae of the Webbing Clothes Moth, *Tineola bisselliella*. *J. Text. Inst.* **1993**, *84*, 39–48. [CrossRef]
17. Kolowski, J.C.; Petraco, N.; Wallace, M.M.; De Forest, P.R.; Prinz, M. A Comparison Study of Hair Examination Methodologies. *J. Forensic Sci.* **2004**, *49*, 1–3. [CrossRef]
18. Oien, C.T. Forensic Hair Comparison: Background Information for Interpretation. Forensic Science Communications. Available online: https://archives.fbi.gov/archives/aboutus/lab/forensicsciencecommunications/fsc/april2009/review/2009_04_review02.htm (accessed on 18 February 2021).
19. Rowe, W.F. Biodegradation of Hairs and Fibers. In *Forensic Taphonomy: The Postmortem Fate of Human Remains*; Haglund, W.D., Sorg, M.H., Eds.; CRC Press: New York, NY, USA, 1997; pp. 337–351.

20. Wang, X.; Johansen, S.S.; Nielsen, M.K.K.; Linnet, K. Hair analysis in toxicological investigation of drug-facilitated crimes in Denmark over a 8-year period. *Forensic Sci. Int.* **2018**, *285*, e1–e12. [CrossRef]

21. Cooper, G.A.; Kronstrand, R.; Kintz, P. Society of Hair Testing guidelines for drug testing in hair. *Forensic Sci. Int.* **2012**, *218*, 20–24. [CrossRef] [PubMed]

22. Viero, A.; Montisci, M.; Pelletti, G.; Vanin, S. Crime scene and body alterations caused by arthropods: Implications in death investigation. *Int. J. Leg. Med.* **2019**, *133*, 307–316. [CrossRef] [PubMed]

23. Rivers, D.; Geiman, T. Insect Artifacts are More than Just Altered Bloodstains. *Insects* **2017**, *8*, 37. [CrossRef]

24. Campobasso, C.P.; Di Vella, G.; Introna, F. Factors affecting decomposition and Diptera colonization. *Forensic Sci. Int.* **2001**, *120*, 18–27. [CrossRef]

25. McKittrick, J.; Chen, P.-Y.; Bodde, S.G.; Yang, W.; Novitskaya, E.; Meyers, M.A. The Structure, Functions, and Mechanical Properties of Keratin. *JOM* **2012**, *64*, 449–468. [CrossRef]

26. Koch, S.L.; Michaud, A.L.; Mikell, C.E. Taphonomy of Hair-A Study of Postmortem Root Banding. *J. Forensic Sci.* **2013**, *58*, S52–S59. [CrossRef]

27. Huchet, J.-B.; Greenberg, B. Flies, Mochicas and burial practices: A case study from Huaca de la Luna, Peru. *J. Archaeol. Sci.* **2010**, *37*, 2846–2856. [CrossRef]

28. Marchant, J. Ancient Egyptians used hair gel. Mummy analysis finds that fat-based product held styles in place. *J. Archaeol. Sci.* **2011**, *38*, 3432–3434.

29. Pradelli, J.; Tuccia, F.; Giordani, G.; Vanin, S. Puparia Cleaning Techniques for Forensic and Archaeo-Funerary Studies. *Insects* **2021**, *12*, 104. [CrossRef] [PubMed]

30. Archer, M.S.; Elgar, M.A.; Briggs, C.A.; Ranson, D.L. Fly pupae and puparia as potential contaminants of forensic entomology samples from sites of body discovery. *Int. J. Leg. Med.* **2005**, *120*, 364–368. [CrossRef] [PubMed]

31. Donfack, J.; Castillo, H.S. A Review and Conceptual Model of Factors Correlated with Postmortem Root Band Formation. *J. Forensic Sci.* **2018**, *63*, 1628–1633. [CrossRef]

32. Serowik, J.M.; Rowe, W.F. Biodeterioration of Hair in a Soil Environment. In *Biodeterioration Research 1*; Llewellyn, G.C., O'Rear, C.R., Eds.; Plenum Press: New York, NY, USA, 1987; pp. 87–93.

33. Kundrat, J.A.; Rowe, W.F. A Study of Hair Degradation in Agricultural Soil. In *Biodeterioration Research 2: General Biodeterioration, Degradation, Mycotoxins, Biotins, and Wood Decay*; Llewellyn, G.C., O'Rear, C.R., Eds.; Plenum Press: New York, NY, USA, 1989; pp. 91–98.

34. DeGaetano, D.H.; Kempton, J.B.; Rowe, W.F. Fungal tunneling of hair from a buried body. *J. Forensic Sci.* **1992**, *37*, 1048–1054. [CrossRef]

35. Allsop, D.; Seal, K.J. *Introduction to Biodeteriorarion*; Edward Arnold: London, UK, 1986.

36. Janaway, R.C. Degradation of Clothing and Other Dress Materials Associated with Buried Bodies of Both Archaeological and Forensic Interest. In *Advances in Forensic Taphonomy. Method, Theory, and Archaeological Perspectives*; Haglund, W.D., Sorg, M.H., Eds.; CRC Press: New York, NY, USA, 2002; pp. 379–402.

37. Hall, M.J.R. The Relationship between Research and Casework in Forensic Entomology. *Insects* **2021**, *12*, 174. [CrossRef] [PubMed]

38. Lutz, L.; Verhoff, M.; Amendt, J. To Be There or Not to Be There, That Is the Question—On the Problem of Delayed Sampling of Entomological Evidence. *Insects* **2021**, *12*, 148. [CrossRef]

39. Greenberg, B.; Kunich, J.C. *Entomology and the Law*; Cambridge University Press: Cambridge, UK, 2002.

40. Hall, R.D. The Forensic Entomologist as Expert Witness. In *Forensic Entomology: The Utility of Arthropods in Legal Investigations*; Byrd, J.H., Castner, J.L., Eds.; CRC Press: Boca Raton, FL, USA, 2001; pp. 379–400.

41. Holder, E.H.; Robinson, L.O.; Laub, J.H. *Death Investigation: A Guide for the Scene Investigator*; US Department of Justice, National Institute of Justice: Washington, DC, USA, 2011.

Article

Investigations on Arthropods Associated with Decay Stages of Buried Animals in Italy

Teresa Bonacci [1], Federica Mendicino [1], Domenico Bonelli [1], Francesco Carlomagno [1], Giuseppe Curia [2], Chiara Scapoli [3] and Marco Pezzi [3,*]

[1] Department of Biology, Ecology and Earth Sciences, University of Calabria, Via P. Bucci, 87036 Arcavacata di Rende, Cosenza, Italy; teresa.bonacci@unical.it (T.B.); federica.mendicino@gmail.com (F.M.); domenico.bonelli@unical.it (D.B.); francescocarlomagno@gmail.com (F.C.)
[2] Azienda Sanitaria Provinciale di Cosenza, Servizio Veterinario, Via J. F. Kennedy, Traversa II, 87036 Roges di Rende, Cosenza, Italy; peppinocuria@libero.it
[3] Department of Life Sciences and Biotechnology, University of Ferrara, Via L. Borsari 46, 44121 Ferrara, Italy; chiara.scapoli@unife.it
* Correspondence: marco.pezzi@unife.it

Simple Summary: The burial of corpses may interfere with the succession of sarcosaprophagous fauna and forensic evaluation of post-mortem interval. For the first time in Italy, an experimental study was conducted on arthropods associated with buried pig carcasses in a rural area near Cosenza (Southern Italy). One carcass was left above the ground and five were buried: one of the buried ones was periodically exhumed to evaluate the effects of disturbance on decay processes and on arthropod fauna, and the other four were exhumed only once at given time intervals. The results revealed differences in taxa and colonization of arthropod fauna in the above ground and periodically exhumed carcasses. No arthropod colonization was detected in the carcasses exhumed only once, showing that a burial at about 25 cm depth could be sufficient to prevent colonization by sarcosaprophagous taxa.

Abstract: Burial could be used by criminals to conceal the bodies of victims, interfering with the succession of sarcosaprophagous fauna and with the evaluation of post-mortem interval. In Italy, no experimental investigation on arthropods associated with buried remains has been conducted to date. A first experimental study on arthropods associated with buried carcasses was carried out in a rural area of Arcavacata di Rende (Cosenza), Southern Italy, from November 2017 to May 2018. Six pig carcasses (*Sus scrofa* Linnaeus) were used, five of which were buried in 60-cm deep pits, leaving about 25-cm of soil above each carcass, and one was left above ground. One of the buried carcasses was periodically exhumed to evaluate the effects of disturbance on decay processes and on arthropod fauna. The other four carcasses were exhumed only once, respectively after 43, 82, 133, and 171 days. As expected, the decay rate was different among carcasses. Differences in taxa and colonization of arthropod fauna were also detected in the above ground and periodically exhumed carcasses. In carcasses exhumed only once, no arthropod colonization was detected. The results showed that a burial at about 25 cm depth could be sufficient to prevent colonization by sarcosaprophagous taxa and these data could be relevant in forensic cases involving buried corpses.

Keywords: arthropods; burial; decay; forensic entomology; insects; pig

check for updates

Citation: Bonacci, T.; Mendicino, F.; Bonelli, D.; Carlomagno, F.; Curia, G.; Scapoli, C.; Pezzi, M. Investigations on Arthropods Associated with Decay Stages of Buried Animals in Italy. *Insects* **2021**, *12*, 311. https://doi.org/10.3390/insects12040311

Academic Editors: Damien Charabidze and Daniel Martín-Vega

Received: 19 February 2021
Accepted: 26 March 2021
Published: 1 April 2021

Publisher's Note: MDPI stays neutral with regard to jurisdictional claims in published maps and institutional affiliations.

1. Introduction

Diptera and Coleoptera are the most studied insect orders in forensic entomology because they are associated to dead bodies for both trophic and reproductive activities [1–3]. Insects are well known to colonize a corpse immediately after death [4] although intrinsic and extrinsic factors may affect their arrival and activities on the remains [5]. Temperature, humidity, and environmental conditions around the dead body affect the decay rate and the species composition [6,7]. Criminals sometimes use burial to conceal the bodies of their victims: this practice favors a reduction of the decay rate [8] and often prevents the arrival of sarcosaprophagous fauna on the body. When burial prevents access of insects to a body, entomological methods to estimate post-mortem interval (PMI) are not sufficiently reliable [9,10]. When colonization occurs, investigators may use the discovered specimens for forensic analyses, but insect relative abundance and diversity are often different between buried and exposed corpses [11]. Previous studies on buried corpses and flesh pieces reported the presence of many necrophagous species, mostly belonging to families of Diptera, such as Phoridae, Muscidae, and Sarcophagidae [11–23].

Forensic entomology practitioners may also have trouble in correctly evaluating PMI in the case of buried corpses because of lack of reference data (as occurring in Italy) and interference of local factors. These challenges may be met by a correct integration of field casework and laboratory studies, also requiring experimental studies on animal models in scenarios relevant for forensic entomology. These experimental studies may provide useful data for future applications to forensic investigations.

In order to identify arthropod taxa and succession on buried remains, we carried out for the first time in Italy an investigation using an animal model, namely pig carcasses buried in a rural area in Calabria (Southern Italy).

2. Materials and Methods

2.1. Study Site

The investigation was carried out from 28 November 2017, to 17 May 2018 in a rural area in Southern Italy, located within the campus of the University of Calabria (Arcavacata di Rende, Cosenza, 39°21′35.31″ N and 16°13′53.48″ W). The site was about 220 m a.s.l. and consisted of a grassy area with juvenile trees of *Quercus pubescens* Wild. (Fagales: Fagaceae), *Olea europaea* L. (Scrophulariales: Oleaceae) and *Populus alba* L. (Salicales: Salicaceae) (Figure 1). The soil granulometry, analyzed by the Laboratory of Geodynamics and Earth Surface Processes of the Department of Ecology, Biology and Earth Sciences of the University of Calabria, revealed a composition of 54% sand, 23% silt, 14% clay, and 9% gravel. The soil was therefore identified as "sandy clay loam", according to the classification by the United States Department of Agriculture. Based on previous studies [21], six 120-kg female pigs, *Sus scrofa* Linnaeus (Artiodactyla: Suidae), were used. The experimental animals were purchased at a farm adjacent to the sampling site and the animals were sacrificed at the local slaughterhouse under veterinary control, according to Regulation (EU) 2017/625 of the European Parliament on animal health and welfare (Document 32017R0625), published in Official Controls Regulation (EU) L95 (ISSN 1977-0707) on 7 April 2017. Five pits, each 1.2 × 1.5 m large and 60-cm deep were mechanically excavated two days before the beginning of the experiment. The pits were spaced at least 15 m from each other on all sides (Figure 1) to prevent interference in the succession fauna. All animals were set inside the pits on the left side over a 25-mm mesh chicken wire. Since the pig carcasses laid on the left side were about 35-cm high at their maximum, the thickness of soil above each carcass was about 25 cm. Four carcasses were respectively labelled as B1, B2, B3, and B4 (Figure 1). For comparison, a fifth carcass, labelled SC, was positioned at the same time and location on the soil surface and left to open air, protected by the chicken wire which was removable for inspections. A sixth carcass, labelled BE, was placed in another pit in similar conditions to those of B1, B2, B3, and B4, but was periodically exhumed to evaluate the effects of disturbance on decay processes and on the associated arthropod fauna (Figure 1). Carcasses B1, B2, B3, and B4 were exhumed only once during the experimental period, respectively after 43, 82, 133, and 171 days. Carcass BE was exhumed every 10 or 15 days according to weather conditions. Data of air temperature for each day during the experimental period were provided by Centro Funzionale Multirischi—ARPACAL (Agenzia Regionale per la Protezione dell'Ambiente della Calabria) (Catanzaro, Italy), referring to the station Cosenza 118. Figure 2 shows the average values of air temperature between 9 a.m. and 5 p.m. for each day of the experimental period. Soil temperature was measured with a digital thermometer with a probe (IHM Moineau Instruments, Boutonne, France) every day, during the experimental period, each hour from 9 a.m. to 5 p.m., at 10-cm depth. The average of data of soil temperature are shown in Figure 2. Daily data about mm of rainfall were provided by ARPACAL pluviometer station (website http://www.arpacal.it/; accessed on 7 August 2020), referring to station Cosenza 118 (cod. 1017) (Figure 3).

Figure 1. Satellite view (Google Earth) of the experimental site showing the rural area where the pig carcasses were positioned (Flags). SC, position of carcass set above the soil; B1–B4, positions of buried carcasses exhumed only once; BE, position of periodically exhumed carcass.

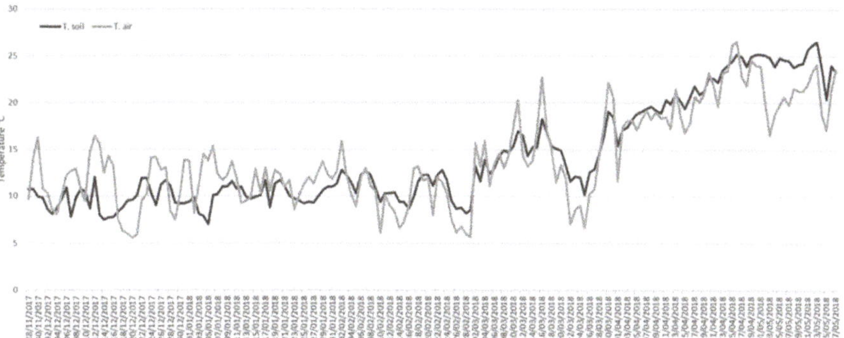

Figure 2. Data of soil (T. soil) and air temperature (T. air) and recorded at the experimental site from 28 November 2017, to 17 May 2018. Data are expressed as average of daily measurements at each hour from 9 a.m. to 5 p.m. Soil temperature data were measured with a manual probe every day, each hour from 9 a.m. to 5 p.m., at 10-cm depth above carcass B4. Air temperature data were provided by Centro Funzionale Multirischi—ARPACAL (Catanzaro, Italy), station Cosenza 118.

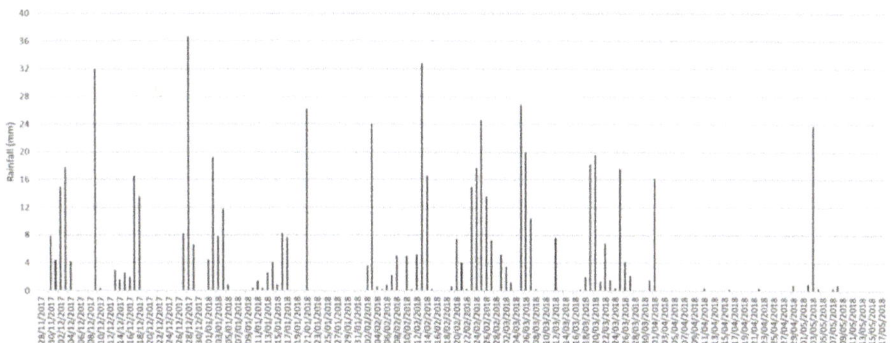

Figure 3. Daily rainfall data (mm) provided by ARPACAL pluviometer station (website http://www.arpacal.it/; accessed on 7 August 2020), referring to station Cosenza 118 (cod. 1017).

2.2. Arthropod Sampling and Carcass Decay

Every day, carcass SC and the surfaces of all pits were examined for arthropod presence twice a day (from 9 a.m. to 1 p.m., and from 3 p.m. to 5 p.m.). The decay process of carcasses SC and BE was also examined according to a previous protocol [21]. For each carcass, the decomposition rate and the duration of each decomposition stage were recorded. Sampling of arthropods for carcass BE was carried out every 10–15 days: the carcass was exhumed and carefully observed only for 20 min, to avoid any interference on arthropod colonization. Adults were captured by an entomological net and tweezers and larvae were hand collected by tweezers during the observation time.

The same sampling procedure of arthropod fauna was carried out at the exhumations of carcasses B1, B2, B3, and B4. At each exhumation of BE and of the other buried carcasses, soil samples from the sides, above and (when possible) under the carcass, in positions corresponding to the head, anus, dorsum, back and abdomen (for a total of about 10 kg of soil) were collected and examined by a Berlese funnel in the Applied and Forensic Entomology Laboratory of the University of Calabria [24].

The adult arthropods collected during the observation time and by the Berlese funnel on soil samples were stored in 60% ethanol and taxonomically identified at the maximum possible level using keys [25–41]. Concerning the dipteran larvae collected by tweezers, they were reared to adults in plastic boxes containing a layer of sand covered with about 100 g bovine liver, at a temperature similar to that of field collection. Once the larvae reached the pupal stage, the liver residuals were removed, leaving only the sand where the larvae reached the pupal stage. The emerged Diptera adults were stored in 60% ethanol and identified at the maximum possible level using keys [35,38]. Concerning the larvae of Dermestidae, those collected by tweezers were reared to adults in plastic boxes containing a layer of sand covered with about 100 g dry pork skin, at a temperature similar to that of field collection. Once the larvae reached the pupal stage, the skin residuals were removed, leaving only the sand where the larvae reached the pupal stage. The emerged Dermestidae adults were stored in 60% ethanol and identified at the maximum possible level using keys [41]. Concerning the arthropods collected by the Berlese funnel, that causes the death of arthropods, preserving them in a liquid, the collected larvae were identified using taxonomical keys for larvae at the maximum possible level [42,43].

3. Results

3.1. Decay Rate

All along the experimental period, for air temperature the daily average ranged from a minimum of 5.6 to a maximum of 26.5 °C, and for soil temperature the daily average ranged from a minimum of 7.0 to a maximum of 26.5 °C (Figure 2). Rainfall throughout the duration of the study was frequent and abundant (Figure 3), with a total of 600.6 mm, according to rain data provided by ARPACAL pluviometer station Cosenza 118 (cod. 1017).

The decay rate of the buried carcasses compared to carcass SC was much slower. The depth of pit, the soil compactness (due to abundant rainfall) and the average low temperature recorded at 25-cm depth may have affected the decomposition process.

The decay of carcass BE lasted 171 days, while that of carcass SC required 82 days. Four decomposition stages were observed for carcass SC (fresh, bloated, decay/advanced decay and dry), and for BE (fresh, bloated, adipocere/advanced decay and dry).

Carcasses B1 and B2, respectively exhumed after 43 and 82 days, were observed at the adipocere stage, while B3 and B4, respectively exhumed after 133 and 171 days, were observed at the advanced decay stage.

3.2. Insect Succession

From carcass SC, a total of 6162 arthropod specimens were collected and identified as belonging to five orders and 25 families (Tables 1–3). The most abundant order was that of Diptera, with 12 families (Table 1).

Table 1. Taxa of Diptera found as adults (A) and larvae (L) in carcasses SC and BE, with number of individuals in brackets, at each of the four decay stages. The decay/advanced decay stage of carcass SC corresponds to the adipocere/advanced decay stage of carcass BE. Specimens were collected by an entomological net and/or tweezers, or by a Berlese funnel (*) or by mixed methods (**). SC, carcass set above the soil; BE, periodically exhumed carcass.

Infraorder/ Family	Genus/ Species	Stage of Decomposition							
		Fresh		Bloated		Decay/Advanced Decay		Dry	
		SC	BE	SC	BE	SC	BE	SC	BE
Calliphoridae	*Calliphora vicina*	A(3)	–	A(129); L(6)	L(92)	A(38); L(1)	–	L(4)	–
	Calliphora vomitoria	A(16)	–	A(258); L(14)	–	A(93); L(534)	L(617 **)	L(3695)	–
	Chrysomya albiceps	–	–	A(74)	–	A(17)	–	–	–
	Lucilia caesar	A(7)	–	A(103)	–	A(10)	–	–	–
	Lucilia sericata	–	–	A(107); L(4)	–	A(18)	L(7)	–	–
Fanniidae	*Fannia canicularis*	–	–	A(26)	–	A(24)	L(26)	–	–
	Fannia lineata	–	–	–	–	A(1)	L(5)	–	–
Muscidae	*Hydrotaea aenescens*	–	–	–	–	–	L(123)	–	–
	Hydrotaea capensis	–	–	–	–	A(1)	L(1373)	–	–
	Hydrotaea dentipes	A(1)	–	A(149)	–	A(147); L(50)	–	L(87)	–
	Hydrotaea ignava	–	–	–	–	A(1)	L(19)	–	–
	Musca domestica	A(10)	–	A(235)	–	A(45)	–	–	–
	Muscina levida	–	–	A(1)	–	A(1)	–	–	–
	Muscina prolapsa	–	–	A(6)	–	–	L(26)	–	–
	Muscina stabulans	–	–	A(1)	–	–	L(118)	–	–
	Phaonia subventa	–	–	A(7)	–	A(4)	–	–	–
	Polietes meridionalis	–	–	A(5)	–	–	–	–	–
	Azelia sp.	–	–	–	–	A(1)	–	–	–
	Helina sp.	–	–	A(3)	–	–	–	–	–
	Thricops sp.	–	–	A(10)	–	–	–	–	–
Phoridae	*Megaselia scalaris*	–	–	A(3)	A(1)	–	A(3)	–	A(1)
	Conicera tibialis	–	–	–	–	–	A(3 **)	–	–
	Chaetopleurophora sp.	–	–	A(2)	–	–	–	–	–
Piophilidae	*Piophila casei*	–	–	A(7)	–	A(6)	–	–	–
	Stearibia nigriceps	–	–	A(46)	–	A(10)	–	–	–
	Prochyliza sp.	–	–	A(3)	–	–	–	–	–
Sepsidae	*Sepsis duplicata*	–	–	A(11)	–	A(3)	–	–	–
	Sepsis cynipsea	–	–	A(2)	–	–	–	–	–
Anthomyiidae	*Anthomyia* sp.	–	–	A(2)	–	A(1)	–	–	–
Sarcophagidae	*Sarcophaga* sp.	–	–	A(1)	–	–	–	–	–
Cecidomyidae	–	–	–	–	–	–	A(2)	–	–
Drosophilidae	–	–	–	–	A(1)	–	A(4)	–	–
Dryomyzidae	–	–	–	A(4)	–	–	–	–	–
Ephydridae	–	–	–	–	A(2 *)	–	–	–	–
Heleomyzidae	–	A(1)	–	–	–	–	–	–	–
Lauxaniidae	–	A(1)	–	–	–	–	–	–	–
Psychodidae	–	–	–	–	A(1)	–	A(12 **)	–	–
Scatopsidae	–	–	–	–	A(2)	–	A(2)	–	–
Sciaridae	–	–	–	–	A(7 *)	–	A(30 **)	–	A(13 *)
Simulidae	–	–	–	A(2 *)	–	A(8 *)	A(1)	A(1 *)	–
Sphaeroceridae	–	–	–	–	A(1 *)	–	A(36 *)	–	–
Tipulomorpha	–	–	–	–	–	–	L(13 *)	–	–

Table 2. Taxa of Coleoptera found as adults (A) and larvae (L) in carcasses SC and BE, with number of individuals in brackets, at each of the four decay stages. The decay/advanced decay stage of carcass SC corresponds to the adipocere/advanced decay stage of carcass BE. Specimens were collected by an entomological net and/or tweezers, or by a Berlese funnel (*) or by mixed methods (**). SC, carcass set above the soil, BE, periodically exhumed carcass.

Family	Genus/Species	Fresh		Bloated		Decay/Advanced Decay		Dry	
		SC	BE	SC	BE	SC	BE	SC	BE
Cleridae	*Necrobia ruficollis*	–	–	–	–	–	A(2)	–	–
	Necrobia violacea	–	–	–	–	–	A(4)	–	–
Dermestidae	*Dermestes frischi*	–	–	–	–	–	A(9); L(10)	–	A(2)
Geotrupidae	*Anoplotrupes stercorosus*	–	–	A(1)	–	–	–	–	–
Histeridae	*Margarinotus brunneus*	–	–	–	–	–	A(6)	–	A(3)
	Margarinotus ventralis	–	–	–	–	–	A(1)	–	A(1)
	Saprinus semistriatus	–	–	–	–	–	A(10)	–	A(2)
Hydrophilidae	*Sphaeridium lunatum*	–	–	–	–	–	A(1)	–	–
Nitidulidae	*Nitidula flavomaculata*	–	–	–	–	A(11)	–	A(19)	–
Silphidae	*Necrodes littoralis*	–	–	–	–	A(1)	–	–	–
	Thanatophilus rugosus	–	–	–	–	–	A(1)	–	–
Staphylinidae	*Creophilus maxillosus*	–	–	–	–	A(14)	A(5)	A(3)	–
	Anotylus sp.	–	–	–	A(3 **)	–	A(3)	–	–
	Carpelimus sp.	A(1)	–	A(10)	–	–	–	–	–
	Heterothops sp.	–	–	–	–	–	A(4)	–	–
	Paederus sp.	–	–	–	–	A(1)	–	A(2)	–
	Platystethus sp.	–	–	–	–	–	A(9)	–	–
	Philonthus sp.	–	–	–	A(1 *)	–	A(1)	–	–
	Quedius sp.	–	–	–	A(1)	–	A(4)	–	A(1)
	Aleocharinae	–	–	–	A(16)	–	A(10)	–	–
	Piestinae	–	–	–	–	–	A(1)	–	–
Carabidae	*Abax* sp.	–	–	–	A(1)	–	–	–	–
	Acinopus sp.	–	–	–	A(1)	–	–	–	–
	Ophonus cribricollis	–	–	–	A(1)	–	–	–	–
	Ophonus sp.	–	–	–	A(1)	–	–	–	–
	Trechus quadristriatus	–	–	–	A(1)	–	–	–	–
Buprestidae	–	–	–	A(1)	–	–	–	–	–
Chrysomelidae	–	–	–	A(1)	–	–	–	–	–
Curculionidae	–	–	–	A(1)	–	A(1)	–	–	–
Scarabeidae	–	–	–	A(1)	–	–	A(3)	–	–

During the fresh stage (about 18 h from positioning) the taxa of adults collected were *Calliphora vicina* Robineau-Desvoidy, *Calliphora vomitoria* (Linnaeus), *Lucilia sericata* (Meigen) (Diptera: Calliphoridae), *Hydrotaea dentipes* (Fabricius), *Musca domestica* Linnaeus (Diptera: Muscidae), *Homoneura* sp. van der Wulp (Diptera: Lauxaniidae), Heleomyzidae (Diptera) and *Carpelimus* sp. Leach (Coleoptera: Staphylinidae).

The first oviposition by Diptera was observed as early as five minutes after the positioning of carcass SC on the soil, inside an eye.

The first instar of dipteran larvae was observed 96 h after egg laying, due to the low temperatures on the day of positioning and on the following days. After 96 h, the carcass SC was in the early bloated stage.

Table 3. Other arthropod taxa found as adults (A) in carcasses SC and BE, with number of individuals in brackets, at each of the four decay stages. The decay/advanced decay stage of carcass SC corresponds to the adipocere/advanced decay stage of carcass BE. Specimens were collected by an entomological net and/or tweezers, or by a Berlese funnel (*) or by mixed methods (**). SC, carcass set above the soil; BE periodically exhumed carcass.

Order	Family	Genus/Species	Stage of Decomposition							
			Fresh		Bloated		Decay/Advanced Decay		Dry	
			SC	BE	SC	BE	SC	BE	SC	BE
Araneae	Amaurobiidae	–	–	–	–	–	A(2)	–	–	–
	Lycosidae	–	–	–	–	A(7)	–	–	–	–
	Thomisidae	–	–	–	–	–	A(1)	–	–	–
	Salticidae	–	–	–	–	–	A(1)	–	–	–
Dermaptera	Forficulidae	*Forficula auricularia*	–	–	–	A(1)	–	–	–	–
Hymenoptera	Braconidae	–	–	–	A(18)	A(1 *)	A(8)	A(1)	–	–
	Formicidae	*Messor* sp.	–	–	–	A(1)	A(1)	–	–	A(1)
		Camponotus sp.	–	–	–	A(2)	–	–	–	–
		Pheidole sp.	–	–	–	–	–	A(36)	–	–
	Pteromalidae	*Nasonia vitripennis*	–	–	–	–	–	A(19 **)	–	A(2)
Rhynchota	Cicadellidae	*Eupteryx* sp.	–	–	A(1)	–	–	–	–	–
	Delphacidae	–	–	–	A(1)	–	–	–	–	–
Geophilomorpha	–	–	–	–	–	A(1)	–	–	–	–

During the bloated stage (lasting about 13 days), the taxa of adults were *C. vomitoria*, *C. vicina*, *Chrysomya albiceps* Wiedemann, *Lucilia caesar* (Linnaeus), *L. sericata* (Diptera: Calliphoridae), *Fannia canicularis* (Linnaeus) (Diptera: Fanniidae), *H. dentipes*, *M. domestica*, *Muscina prolapsa* (Harris), *Phaonia subventa* (Harris), *Polietes meridionalis* Peris & Llorente, *Thricops* sp. Róndani (Diptera: Muscidae), *Piophila casei* (Linnaeus), *Stearibia nigriceps* (Meigen) and, *Prochyliza* sp. Walker (Diptera: Piophilidae), and *Sepsis duplicata* Haliday (Diptera: Sepsiidae). During this stage, adults belonging to the families Dryomizidae, Sarcophagidae (Diptera) and Braconidae (Hymenoptera) and to the genus *Carpelimus* sp. were also identified.

During this stage, the larvae detected belonged to the species *C. vicina*, *C. vomitoria* and *L. sericata* (Table 1).

In the decay/advanced decay stage (lasting about 38 days) the most abundant taxa of adults collected on carcass SC were *C. vomitoria*, *C. vicina*, *Ch. albiceps*, *F. canicularis*, *H. dentipes*, *L. sericata*, *L. caesar*, *M. domestica*, *Pi. casei*, and *S. nigriceps*.

Adults belonging to the family Braconidae and adults of *Nitidula flavomaculata* Rossi (Coleoptera: Nitidulidae) were also collected. Among larvae, the most abundant taxa were *C. vomitoria* and *H. dentipes*.

In the dry stage (lasting about 27 days), the collected taxa of adults were *C. vicina*, *C. vomitoria*, *H. dentipes* and Simulidae for Diptera, and *Creophilus maxillosus* (Linnaeus), *Paederus* sp. Fabricius (Staphylinidae) and *N. flaveomaculata* for Coleoptera. The only larvae collected belonged to *C. vicina*, *C. vomitoria*, and *H. dentipes*.

Based on the immature stages collected, the only species breeding on carcass SC were respectively *C. vomitoria*, *C. vicina*, *H. dentipes*, and *L. sericata*.

Concerning the other carcasses exhumed only once (B1, B2, B3, and B4, respectively exhumed at 43, 82, 133, and 171 days after the date of positioning), carcass B1, exhumed 43 days after the positioning, was in the decay/adipocere stage. Carcass B2, exhumed 82 days after the positioning, was also in the decay/adipocere stage. Carcass B3, exhumed 133 days after the positioning, was in the advanced decay stage, as well as carcass B4, exhumed 171 days after the positioning.

After a careful examination of all carcasses, no signs of colonization by arthropods was detected on them at the date of exhumation. No arthropods were also detected or collected during the daily inspections of the surface of all pits of B1–B4.

Concerning the carcass BE, which was exhumed every 10–15 days to evaluate the effects of disturbance on decay processes and associated arthropod fauna, during daily inspections on the pit surface a total of nine taxa of adults were collected: *C. vomitoria*, *C. vicina*, *Helina* sp. Robineau-Desvoidy, *Hydrotaea meteorica* (Linnaeus), *M. domestica*, *P. subventa* (Diptera: Muscidae), *Sarcophaga tibialis* Macquart (Diptera: Sarcophagidae), *Messor* sp. Forel (Hymenoptera: Formicidae), and *Scolopendra cingulata* Latreille (Scolopendromorpha: Scolopendridae).

Carcass BE appeared in the bloated stage from 10 to 65 days from positioning (respectively corresponding to the first and sixth exhumation), thus this stage lasted about 56 days. In this time interval, during the third exhumation (43 days after positioning), carcass BE was mostly colonized by larvae of *C. vicina* and of the coleopteran families Staphilinidae and Carabidae.

The same carcass appeared in the adipocere/advanced decay stage from 74 to 163 days after positioning (respectively corresponding from the seventh to the twelfth exhumation), thus this stage lasted about 90 days. In this time interval, the most frequent orders were Diptera and Coleoptera (Tables 1 and 2), and the order with the highest number of taxa detected at the larval stage was Diptera.

The taxa recorded at the larval stage were *C. vomitoria*, *L. sericata*, *F. canicularis* and *Fannia lineata* (Stein) (Diptera: Fanniidae), *Hydrotaea aenescens* (Wiedemann), *H. capensis*, *H. ignava*, *Mu. prolapsa*, and *Muscina stabulans* (Fallén) (Diptera: Muscidae).

Concerning Coleoptera, the only species found at the larval stage was *Dermestes frischi* (Kugelann) (Coleoptera: Dermestidae). Larvae belonging to the infraorder Tipulomorpha (Diptera) were also detected. At the dry stage, occurring 171 days after positioning (corresponding to the thirteenth exhumation) no larval stage was detected. Concerning adults, for Diptera, only individuals of *Megaselia scalaris* (Loew) (Phoridae) and other individuals belonging to the family Sciaridae were found.

Some Coleoptera of the families Histeridae, Dermestidae and Staphylinidae were also found. From dipteran larvae collected from carcasses and reared in laboratory the parasitoid *Nasonia vitripennis* (Walker) (Hymenoptera: Pteromalidae) emerged and this species was also detected in soil samples analyzed by the Berlese funnel.

4. Discussion

The size and instar of dipteran larvae and/or patterns of postmortem insect succession are relevant to estimate the interval of arthropod colonization on a corpse [2]. Although biotic and abiotic factors may affect both decay and insect successional pattern, the entomological method used to estimate PMI could be applied when corpses colonized by insects are discovered in both outdoor and indoor environments. Burial is a common method used by criminals to conceal a corpse, but burial depth and soil hardness are physical barriers that significantly affect temperature and insect succession [13]. In our study, in which pig carcasses weighing about 120 kg each were buried at a depth of 60 cm, a layer of about 25 cm was on the top of each carcass exhumed only once (B1–B4). According to our data, this layer of soil was sufficient to prevent the arrival of arthropods and slow down the decay rate of carcasses.

The decay rate was therefore different among the carcass left above ground (SC), the carcass periodically exhumed (BE) and those exhumed only once (B1–B4). No arthropods were observed on the surface of the pits of carcasses exhumed only once, probably because the soil compactness at 25-cm depth prevented the spread of odors arising from decaying remains. It is also possible that anoxic conditions may have developed though high saturation of the soil by water associated to low evaporation. This hypothesis is supported by the low temperatures and frequent rains detected during the experimental period and by the observed adipocere stages.

Concerning the periodically exhumed carcass, the effects of the disturbance on the pit surface were the reduced soil compactness and the mixing of the surface with the deeper soil in contact with the carcass, thus spreading odors, attracting arthropods and increasing the number of colonizing taxa.

Although the above ground and buried carcasses were set simultaneously in their positions, a different insect successional pattern was found on the exposed carcass in comparison to the buried ones. Dipteran species such as *C. vicina*, *C. vomitoria* and *L. sericata* arrived and bred on both carcasses SC and BE. On carcass SC, *Ch. albiceps* arrived but did not breed, and *H. dentipes* was the only one of its genus breeding on it. Contrary to what occurred in the carcasses exhumed only once (B1–B4), in carcass BE the larvae, whose presence signaled breeding activity, belonged to the species *F. canicularis*, *F. lineata*, *H. aenescens*, *H. capensis*, *H. ignava*, *Mu. prolapsa* and *Mu. stabulans*. Some colonization by Tipulomorpha was also found. The highest number of larvae detected belonged to *H. capensis*. The presence of *H. ignava* and other unidentified species of the genus *Hydrotaea* was previously reported in buried remains at 30-cm and 60-cm depth [21]. Colonization by *Mu. prolapsa* and *Mu. stabulans* on buried animal remains was also reported [44]. The relatively low insect diversity on carcass SC in comparison to carcass BE is interesting and could be related to the different decay rate, but this aspect requires more in-depth studies. It is interesting to notice that in carcasses exhumed only once (B1-B4) no colonization by arthropods was detected, not even that due to common forensic indicator species reported in buried remains, such as *Conicera tibialis* Schmitz (Diptera: Phoridae) and *Me. scalaris* [2,16,21,23,45]. However, these two dipteran species were found at the adult stage in carcasses SC and BE: *Me. scalaris* was found in both carcasses, but *Co. tibialis* only on BE. The absence of these two species is intriguing: probably the soil texture or some other factor may be involved, and this point requires further investigation. In Italy, *Me. scalaris* was reported as the only species colonizing a human corpse buried in a wooden coffin at 30–40 cm depth [46]. This species, also reported in pig carcasses buried at 60-cm depth [21], has medical and forensic interest because it is known to cause myiasis [47]. The species *Co. tibialis*, commonly known as "coffin fly", is usually detected in exhumed human corpses [16,20,48]. In our investigation, this was observed visiting carcass BE but no larvae were detected. Because the decay of buried carcasses usually occurs at a much slower rate than air exposed ones [13], dipteran larvae normally associated with the earlier stages of decomposition were detected later on carcass BE in comparison to carcass SC. It is also interesting to notice that species considered "thermophilic", such as *Ch. albiceps*, *L. caesar* and *L. sericata* [49] were found in carcasses SC and BE in winter months. Moreover, according to a previous study conducted in Region Calabria, *C. vomitoria* is a species abundant in *Fagus* sp. L. (Fagales: Fagaceae) woodlands and infrequent in rural and urban areas [50]. However, in our study this species resulted abundant both as adults and larvae, although the investigation was carried out in a rural area near urban settlements.

Concerning Coleoptera, in carcass BE a higher number of taxa of the family Staphylinidae was found in comparison to the other carcasses. The species of this family colonizing carcasses are characterized by predatory habits towards adults and immature stages of sarcosaprophagous insects [1,2]. A species that was found on both carcasses SC and BE is *Cr. maxillosus*, commonly reported as a carrion colonizer [51–53]. Its role as a forensic bioindicator of PMI is under evaluation [54]. In Italy, *Cr. maxillosus* was previously reported on two human bodies found in rural areas in Region Veneto [55] and on another mummified body found in a woodland area of Northwestern Italy [56].

A possible explanation of the high number of taxa of Staphylinidae on carcass BE, periodically exhumed and reburied, was the fact that the soil was mixed, becoming more humid, less compact and carrying attractive odors, thus more suitable for colonization by hypogean Coleoptera. Moreover, on carcass BE the necrophagous coleopteran *D. frischi* and the coleopteran species with necrophilus habits *Margarinotus brunneus* (Fabricius), *Margarinotus ventralis* (Marseul), *Saprinus semistriatus* (Scriba) (Histeridae), *Necrobia ruficollis* (Fabricius) and *Necrobia violacea* Linnaeus (Cleridae) were found. Among Coleoptera, the only species found breeding on BE was *D. frischi*. Adults and preimaginal stages of this species were previously detected in Northwestern Italy on the mummified human body found outdoor in woodlands [56]. In the region of Calabria, preimaginal stages of *D. frischi* were previously reported in another mummified human body, found outdoors in a rural area [57].

The hypothesis that periodical exhumation and reburial of carcass BE has a role in making soil more suitable to colonization by predator Coleoptera may be supported by the presence of some adults belonging the family Carabidae, such as *Abax* sp. Bonelli, *Acinopus* sp. Dejean, *Ophonus* sp. Dejean, *Ophonus cribricollis* (Dejean) and *Trechus quadristriatus* (Schrank), and an adult of the water beetle, *Sphaeridium lunatum* Fabricius (Coleoptera: Hydrophilidae), usually found in very humid environments [58].

Concerning the effects of the soil granulometry on arthropod colonization, in a previous study on a similar soil type a large colonization was detected on pig carcasses buried at depths of 30 and 60 cm [21]. However, in our study no arthropod colonization was detected in carcasses B1–B4, exhumed only once, probably because the soil became too compact due to winter rains and low temperatures. On the contrary, on carcass BE the arthropod colonization was present because of soil reworking during periodical exhumation and reburial.

Overall, the results show that in our experimental conditions a burial at about 25-cm depth is sufficient to prevent colonization by sarcosaprophagous taxa. These findings, obtained for the first time in Italy on pig carcasses, have forensic relevance because they provide useful data when the depth of burial, the type of soil and the weather conditions interfere with a correct evaluation of PMI by entomological methods in cases involving buried corpses. In our experimental conditions, in buried carcasses the delay in the decay process was affected by the lack of necrophagous insects, in turn affected by the compactness of the soil. A whole range of factors should be considered when entomological evidence is used for forensic purposes. The rate of decomposition and the regular sequence of sarcosaprophagous fauna related to decomposition stages is affected by many factors, in turn related to the location of the corpse and/or its burial [59].

Our results represent an experimental investigation of the effects of burial on arthropod colonization of corpses, which should be extended to different environments, such as urban, suburban, and forest ones, and to different seasons, in order to support future forensic investigations.

Author Contributions: Conceptualization, T.B. and M.P.; data curation, T.B., F.M., D.B., F.C., and M.P.; funding acquisition, T.B.; investigation, T.B., F.M., D.B., F.C., and M.P.; methodology, G.C. and C.S.; supervision, G.C. and C.S.; validation, T.B. and M.P.; writing—original draft, T.B., F.M., D.B., F.C., and M.P.; writing—review & editing, T.B., F.M., G.C., C.S., and M.P. All authors have read and agreed to the published version of the manuscript.

Funding: This research was funded by a grant (ex60%-DiBEST Unical-2017) from the Department of Biology, Ecology and Earth Sciences, University of Calabria, Arcavacata di Rende (Cosenza, Italy) assigned to Professor Teresa Bonacci.

Institutional Review Board Statement: Not applicable.

Data Availability Statement: All data presented in this study are available in the article.

Acknowledgments: The authors owe thanks to Giuseppe Antonio Biagio, Museum of Natural History of Calabria and Botanical Garden, University of Calabria, for technical assistance during the experiments. The authors also wish to thank the Centro Funzionale Multirischi—ARPACAL (Agenzia Regionale per la Protezione dell'Ambiente della Calabria) (Catanzaro, Italy) for kindly providing the air temperature data.

References

1. Smith, K.G.V. *A Manual of Forensic Entomology*, 1st ed.; Trustees of the British Museum (Natural History): London, UK, 1986; p. 205.
2. Byrd, J.H.; Castner, J.L. *Forensic Entomology: The Utility of Arthropods in Legal Investigations*, 2nd ed.; CRC Press: Boca Raton, FL, USA, 2010; p. 681.
3. Amendt, J.; Richards, C.S.; Campobasso, C.P.; Zehner, R.; Hall, M.J.R. Forensic entomology: Applications and limitations. *Forensic Sci. Med. Pathol.* **2011**, *7*, 379–392. [CrossRef]
4. Catts, E.P.; Goff, M.L. Forensic entomology in criminal investigations. *Ann. Rev. Entomol.* **1992**, *37*, 253–272. [CrossRef]
5. George, K.A.; Archer, M.S.; Toop, T. Abiotic environmental factors influencing blowfly colonization patterns in the field. *Forensic Sci. Int.* **2013**, *229*, 100–107. [CrossRef]
6. Gilbert, B.M.; Bass, W.M. Seasonal dating of burials from the presence of fly pupae. *Am. Antiq.* **1967**, *32*, 534–535. [CrossRef]
7. Shean, B.S.; Messinger, L.; Papworth, M. Observations of differential decomposition on sun exposed v. shaded pig carrion in coastal Washington state. *J. Forensic Sci.* **1993**, *38*, 938–949. [CrossRef]
8. Moses, R.J. Experimental adipocere formation: Implications for adipocere formation on buried bone. *J. Forensic Sci.* **2012**, *57*, 589–595. [CrossRef] [PubMed]
9. Rodriguez, W.C.; Bass, W.M. Decomposition of buried bodies and methods that may aid in their location. *J. Forensic Sci.* **1985**, *30*, 836–852. [CrossRef] [PubMed]
10. Simmons, T.; Cross, P.A.; Adlam, R.E.; Moffatt, C. The influence of insects on decomposition rate in buried and surface remains. *J. Forensic Sci.* **2010**, *55*, 889–892. [CrossRef] [PubMed]
11. Gaudry, E. The insects colonization of buried remains. In *Current Concepts in Forensic Entomology*, 1st ed.; Amendt, J., Goff, M.L., Campobasso, C.P., Grassberger, M., Eds.; Springer: New York, NY, USA, 2010; pp. 273–311.
12. Lundt, H. Ökologische Untersuchungen über die tierische Besiedlung von Aas im Boden. *Pedobiologia* **1964**, *4*, 158–180.
13. Payne, J.A.; King, E.W.; Beinhart, G. Arthropod succession and decomposition of buried pigs. *Nature* **1968**, *219*, 180–181. [CrossRef]
14. Goff, M.L. Estimation of postmortem interval using arthropod development and successional patterns. *Forensic Sci. Rev.* **1993**, *5*, 81–94. [PubMed]
15. Van Laerhoven, S.L.; Anderson, G.S. Insect succession on buried carrion in two biogeoclimatic zones of British Columbia. *J. Forensic Sci.* **1999**, *44*, 32–43.
16. Bourel, B.; Tornel, G.; Hédouin, V.; Grosset, D. Entomofauna of buried bodies in northern France. *Int. J. Legal Med.* **2004**, *118*, 215–220. [CrossRef] [PubMed]
17. Huchet, J.B.; Greenberg, B. Flies, Mochicas and burial practices: A case study from Huaca de la Luna, Peru. *J. Archaeol. Sci.* **2010**, *37*, 2846–2856. [CrossRef]
18. Szpila, K.; Voss, J.G.; Pape, T. A new dipteran forensic indicator in buried bodies. *Med. Vet. Entomol.* **2010**, *24*, 278–283. [CrossRef] [PubMed]
19. Gunn, A.; Bird, J. The ability of the blowflies *Calliphora vomitoria* (Linnaeus), *Calliphora vicina* (Rob-Desvoidy) and *Lucilia sericata* (Meigen) (Diptera: Calliphoridae) and the muscid flies *Muscina stabulans* (Fallen) and *Muscina prolapsa* (Harris) (Diptera: Muscidae) to colonise buried remains. *Forensic Sci. Int.* **2011**, *207*, 198–204.
20. Martin-Vega, D.; Gomez-Gomez, A.; Baz, A. The "coffin fly" *Conicera tibialis* (Diptera: Phoridae) breeding on buried human remains after a postmortem interval of 18 years. *J. Forensic Sci.* **2011**, *56*, 1654–1656. [CrossRef]
21. Pastula, E.C.; Merritt, R.W. Insect arrival pattern and succession on buried carrion in Michigan. *J. Med. Entomol.* **2013**, *50*, 432–439. [CrossRef]
22. Huchet, J.B. Insect remains and their traces: Relevant fossil witnesses in the reconstruction of past funerary practices. *Anthropologie* **2014**, *52*, 329–346.
23. Botham, J.L. Decomposition and arthropod succession on buried remains during winter and summer in Central South Africa: Forensic implications and predictive analyses. In Proceedings of the 19th Entomological Congress of the Entomological Society of Southern Africa, Grahamstown, South Africa, 12–15 July 2015.
24. Bano, R.; Roy, S. Extraction of soil microarthropods: A low cost Berlese-Tullgren funnels extractor. *Int. J. Fauna Biol. Stud.* **2016**, *3*, 14–17.
25. Le Quesne, W.J.; Payne, K.R. *Handbooks for the Identification of British Insects—Cicadellidae (Typhlocybinae) with a Checklist of the British Auchenorhyncha (Hemiptera, Homoptera)*; Royal Entomological Society of London: London, UK, 1981; Volume 2, Part 2(c); p. 95.
26. Disney, R.H.L. *Handbooks for the Identification of British Insects. Scuttle Flies Diptera, Phoridae (except Megaselia)*; Royal Entomological Society of London: London, UK, 1983; Volume 10, Part 6; p. 81.
27. Disney, R.H.L. *Handbooks for the Identification of British Insects—Scuttle Flies Diptera, Phoridae (genus Megaselia)*; Royal Entomological Society of London: London, UK, 1989; Volume 10, Part 8; p. 155.
28. Goulet, H.; Huber, J.T. *Hymenoptera of the World: An Identification Guide to Families*; Centre for Land and Biological Resources Research: Ottawa, ON, Canada, 1993; p. 668.
29. Rozkosny, R.; Gregor, F.; Pont, A.C. The european fanniidae (Diptera). *Acta Sci. Nat. Brno* **1997**, *31*, 1–80.
30. Thiele, H.U. *Carabid Beetles in Their Environments. a Study on Habitat Selection by Adaptations in Physiology and Behavior*; Springer: New York, NY, USA, 1977; p. 369.

31. Pont, A.C.; Meier, R. *The Sepsidae (Diptera) of Europe*; Fauna entomologica Scandinavica; Brill: Leiden, The Netherlands, 2002; Volume 37, p. 196.

32. Brandmayr, P.; Zetto, T.; Pizzolotto, R. *I coleotteri Carabidi per la valutazione ambientale e la conservazione della biodiversità*; APAT (Agenzia per la protezione dell'ambiente e per i servizi tecnici): Rome, Italy, 2005; Volume 34, p. 240, Manuali e Linee Guida.

33. Oesterbroek, P. *The European Families of Diptera—Identification, Diagnosis, Biology*; KNNV Publishing: Utrech, Olanda, The Netherlands, 2006; p. 210.

34. Latella, L.; Gobbi, M. *La Fauna Del Suolo. Tassonomia Ecologia E Metodi DI Studio Dei Principali Gruppi DI Invertebrati Terrestri Italiani*; MUSE—Museo delle Scienze: Trento, Italy, 2008; Volume 3, p. 191, Quaderni del Museo Tridentrino di Scienze Naturali.

35. Szpila, K. Key for identification of European and Mediterranean blowflies (Diptera, Calliphoridae) of medical and veterinary importance—adult flies. In *Forensic Entomology, an Introduction*, 2nd ed.; Gennard, D., Ed.; Wiley-Blackwell: West Sussex, UK, 2012; pp. 77–81.

36. Rochefort, S.; Giroux, M.; Savage, J.; Wheeler, T.A. Key to forensically important Piophilidae (Diptera) in the Nearctic Region. *Can. J. Arthropod Identif.* **2015**, 27. [CrossRef]

37. Chinery, M. *Guida Degli Insetti D'Europa*; Franco Muzzio Editore: Rome, Italy, 2016; p. 431.

38. Gregor, F.; Rozkosny, R.; Barták, M.; Vanhara, J. *Manual of Central European Muscidae (Diptera). Morphology, Taxonomy, Identification and Distribution*; E. Schweizerbart Science Publisher: Stuttgart, Germany, 2016; Volume 162, p. 219.

39. Ants (Hymenoptera: Formicidae) Collected in Cowling Arboretum and McKnight Prairie. Available online: https://acad.carleton.edu/curricular/BIOL/resources/ant/index.html (accessed on 18 December 2017).

40. Ballerio, A.; Rey, A.; Uliana, M.; Rastelli, M.; Rastelli, S.; Romano, M.; Colacurcio, L. Coleotteri Scarabeoidea d'Italia. Available online: http://www.societaentomologicaitaliana.it/Coleotteri%20Scarabeoidea%20d\T1\textquoterightItalia%202014/scarabeidi/home.htm (accessed on 11 December 2017).

41. Mike's Insect Keys. Available online: https://sites.google.com/view/mikes-insect-keys/mikes-insect-keys (accessed on 18 December 2017).

42. Szpila, K. Key for the identification of third instars of European blowflies (Diptera: Calliphoridae) of forensic importance. In *Current Concepts in Forensic Entomology*; Amendt, J., Goff, M.L., Campobasso, C.P., Grassberger, M., Eds.; Springer: Dordrecht, The Netherlands, 2010; pp. 43–56.

43. Glime, J.M. Terrestrial insects: Holometabola—Diptera nematocera: Tipuloidea. Chapter 12–18. In *Bryophyte ecology, Volume 2: Bryological Interaction*; Glime, J.M., Ed.; International Association of Bryologists and Michigan Technological University: Houghton, MI, USA, 2016.

44. Gunn, A. The colonisation of remains by the muscid flies *Muscina stabulans* (Fallén) and *Muscina prolapsa* (Harris) (Diptera: Muscidae). *Forensic Sci. Int.* **2016**, *266*, 349–356. [CrossRef]

45. Turner, R.; Wiltshire, P. Experimental validation of forensic evidence: A study of the decomposition of buried pigs in heavy clay soil. *Forensic Sci. Int.* **1999**, *101*, 113–122. [CrossRef]

46. Campobasso, C.P.; Henry, R.; Disney, L.; Introna, F. A case of *Megaselia scalaris* (Loew) (Dipt., Phoridae) breeding in a human corpse. *Anil Aggrawal's Internet J. Forensic Med. Toxicol.* **2004**, *5*, 3–5.

47. Francesconi, F.; Lupi, O. Myiasis. *Clin. Microbiol. Rev.* **2012**, *25*, 79–105. [CrossRef] [PubMed]

48. Merritt, R.W.; Snider, R.; de Jong, J.L.; Benbow, M.E.; Kimbirauskas, R.K.; Kolar, R.E. Collembola of the grave: A cold case history involving arthropods 28 years after death. *J. Forensic Sci.* **2007**, *52*, 1359–1361. [CrossRef] [PubMed]

49. Martin-Vega, D.; Baz, A. Sarcosaprophagous Diptera assemblages in natural habitats in central Spain: Spatial and seasonal changes in composition. *Med. Vet. Entomol.* **2013**, *27*, 64–76. [CrossRef]

50. Greco, S.; Brandmayr, P.; Bonacci, T. Synanthropy and temporal variability of Calliphoridae living in Cosenza (Calabria, Southern Italy). *J. Insect Sci.* **2014**, *14*, 216. [CrossRef]

51. Matuszewski, S.; Bajerlein, D.; Konwerski, S.; Szpila, K. An initial study of insect succession and carrion decomposition in various forest habitats of Central Europe. *Forensic Sci. Int.* **2008**, *180*, 61–69. [CrossRef] [PubMed]

52. Matuszewski, S.; Bajerlein, D.; Konwerski, S.; Szpila, K. Insect succession and carrion decomposition in selected forests of Central Europe. Part 2: Composition and residency patterns of carrion fauna. *Forensic Sci. Int.* **2010**, *195*, 42–51. [CrossRef]

53. Dekeirsschieter, J.; Frederick, C.; Verheggen, F.J.; Drugmand, D.; Haubruge, E. Diversity of forensic rove beetles (Coleoptera, Staphylinidae) associated with decaying pig carcass in a forest biotope. *J. Forensic Sci.* **2013**, *58*, 1032–1040. [CrossRef] [PubMed]

54. Matuszewski, S. Estimating the preappearance interval from temperature in *Creophilus maxillosus* L. (Coleoptera: Staphylinidae). *J. Forensic Sci.* **2012**, *57*, 136–145. [CrossRef] [PubMed]

55. Turchetto, M.; Vanin, S. Forensic entomology and climate change. *Forensic Sci. Int.* **2001**, *146* (Suppl.), 207–209. [CrossRef] [PubMed]

56. Bugelli, V.; Gherardi, M.; Focardi, M.; Pinchi, V.; Vanin, S.; Campobasso, C.P. Decomposition pattern and insect colonization in two cases of suicide by hanging. *Forensic Sci. Res.* **2018**, *3*, 94–102. [CrossRef]

57. Bonacci, T.; Vercillo, V.; Benecke, M. *Dermestes frischii* and *D. undulatus* (Coleoptera: Dermestidae) on a human corpse in Southern Italy: First report. *Rom. J. Leg. Med.* **2017**, *25*, 180–184. [CrossRef]

58. Chin, H.C.; Hurahashi, H.; Marwi, M.A.; Jeffery, J.; Omar, B. Opportunistic insects associated with pig carrions in Malaysia. *Sains Malays.* **2011**, *40*, 601–604.

59. Turner, B. Forensic entomology: Insects against crime. *Sci. Prog.* **1987**, *71*, 133–144.

Communication

New Species of Soldier Fly—*Sargus bipunctatus* (Scopoli, 1763) (Diptera: Stratiomyidae), Recorded from a Human Corpse in Europe—A Case Report

Marek Michalski [1,*], Piotr Gadawski [2], Joanna Klemm [3] and Krzysztof Szpila [4]

[1] Department of Experimental Zoology and Evolutionary Biology, Faculty of Biology and Environmental Protection, University of Lodz, Banacha Street 12/16, 90-237 Łódź, Poland
[2] Department of Invertebrate Zoology and Hydrobiology, Faculty of Biology and Environmental Protection, University of Lodz, Banacha Street 12/16, 90-237 Łódź, Poland; piotr.gadawski@biol.uni.lodz.pl
[3] Facility of Forensic Medicine Barzdo i Żydek, Franciszkańska Street 104/112, 91-845 Łódź, Poland; joanna.monika.klemm@gmail.com
[4] Department of Ecology and Biogeography, Faculty of Biological and Veterinary Sciences, Nicolaus Copernicus University, Lwowska Street 1, 87-100 Toruń, Poland; szpila@umk.pl
* Correspondence: marek.michalski@biol.uni.lodz.pl

Simple Summary: In the current study, we present the first record of twin-spot centurion fly larvae, *Sargus bipunctatus* (Scopoli, 1763), feeding on a human corpse. The morphology of collected imagines and larvae of *S. bipunctatus* was documented, and a standard COI barcode sequence was obtained. Morphology- and DNA-based methods were used to distinguish the larvae of *S. bipunctatus* and its relative, *Hermetia illucens* (Linnaeus, 1758). The potential of *S. bipunctatus* for practical applications in forensic entomology is currently difficult to assess.

Citation: Michalski, M.; Gadawski, P.; Klemm, J.; Szpila, K. New Species of Soldier Fly—*Sargus bipunctatus* (Scopoli, 1763) (Diptera: Stratiomyidae), Recorded from a Human Corpse in Europe—A Case Report. *Insects* **2021**, *12*, 302. https://doi.org/10.3390/insects12040302

Academic Editors: Damien Charabidze and Daniel Martín-Vega

Received: 27 February 2021
Accepted: 26 March 2021
Published: 30 March 2021

Publisher's Note: MDPI stays neutral with regard to jurisdictional claims in published maps and institutional affiliations.

Abstract: The only European Stratiomyidae species known for feeding on human corpses was the black soldier fly *Hermetia illucens* (Linnaeus, 1758). Analysis of fauna found on a human corpse, discovered in central Poland, revealed the presence of feeding larvae of another species from this family: the twin-spot centurion fly *Sargus bipunctatus* (Scopoli, 1763). The investigated corpse was in a stage of advanced decomposition. The larvae were mainly observed in the adipocere formed on the back and lower limbs of the corpse, and in the mixture of litter and lumps of adipocere located under the corpse. Adult specimens and larvae were identified based on morphological characters, and final identification was confirmed using DNA barcoding. Implementing a combination of morphological and molecular methods provided a reliable way for distinguishing the larvae of *S. bipunctatus* and *H. illucens*. The potential of *S. bipunctatus* for practical applications in forensic entomology is currently difficult to assess. Wide and reliable use of *S. bipunctatus* in the practice of forensic entomology requires further studies of the bionomy of this fly.

Keywords: carrion; larva; first record; barcoding DNA; integrative taxonomy

1. Introduction

The family Stratiomyidae (soldier flies), representing the suborder of Brachycera Orthorrhapha, includes more than 2600 described species [1]. Large stratiomyiids are often characterized by their mimicry of wasps or bees (Hymenoptera: Aculeata). Larvae of these flies are dorso-ventrally flattened, with strongly sclerotized integument. Their cuticle has a polygonal pattern due to numerous calcareous incrustations. The larvae of the majority of species live in terrestrial, humid environments, with the exception of a few typically aquatic species. Terrestrial species usually feed on dead organic matter, e.g., humus, decaying parts of the plants and fungi, and the faeces of vertebrates and invertebrates. Stratiomyidae utilize a mode of pupation unique among the Orthorrhapha, pupating inside the cuticle of the last larval instar [2,3].

The only species of Stratiomyidae with confirmed forensic importance is the black soldier fly *Hermetia illucens* (Linnaeus, 1758). This species, probably native to Central America, can currently be found in warmer regions around the world. In the northern part of North America, it reaches the province of Ontario [4]. In Central Europe, it is distributed from the south up to the Czech Republic [5]. The larvae of *H. illucens* are polyphagous, being able to feed on almost any type of decaying plant or animal matter. Since the 1970s, these larvae have been often used to accelerate the decomposition of organic waste, and bred as food for poultry, pigs, fish, terrarium animals, or as fishing baits [6]. As a result, the species is continuously transferred to new territories [4]. Larvae of black soldier fly have been known to be forensically important since 1915 [7]. They feed on carcasses in a phase of advanced decomposition, and are active only at temperatures exceeding 20 °C. At an optimal temperature of 30 °C, the full development cycle from hatch to maturity takes 43 days [8]. Larvae are also used to estimate the postmortem interval (PMI) in cases where several weeks have passed since death [9]. However, the use of *H. illucens* larvae to determine PMI based on the development approach is quite problematic, due to their presence on highly decomposed corpses, when the time between death and oviposition is long and difficult to estimate [9]. Moreover, specimens from different populations of this species may differ significantly in the rate of development, so that the broad use of developmental models established for particular populations is not valid [8].

Entomological material collected from a recent case from Poland indicates that *H. illucens* is not the only species of Stratiomyidae that can successfully develop on human corpses. Stratiomyiid larvae were collected feeding on human remains and identified, using DNA barcoding and morphological characters, to be *Sargus bipunctatus* (Scopoli, 1763). The trophic relationship of *S. bipunctatus* with dead organic matter of animal origin has already been mentioned by Chick [10]. However, this study marks the first record of larvae of this species feeding on human remains, thereby extending the list of European fly species potentially important for medico–legal purposes.

2. Case Description

An unidentified human corpse in an advanced stage of decomposition was found in the City of Lodz (central Poland) on the evening of 21 April 2019, in Jozef Pilsudski park (51°776278′ N, 19°400248′ E). It was located in a small clearing surrounded by dense vegetation, 10 m from a small open river channel. The body was lying in an anatomical position on the ground covered with creeping vegetation, mainly blackberry (*Rubus* L.) (See Figure 1).

Figure 1. The site of the corpse disclosure, 22 April 2019. Photo—M. Michalski.

The corpse was dressed in an undershirt and long denim trousers. The head and torso were almost completely skeletonized. Soft tissues of the upper back and the proximal parts of the upper limbs were mummified. The lower parts of the back and the tissues of the lower limbs, covered by denim trousers, had changed to adipocere. Gnaw marks on the feet phalanges and the presence of faeces indicated the activity of vertebrate scavengers. Based on the morphological features of the skeleton examined by the forensic physician, it was initially estimated that the human remains belonged to a woman aged 25–45 years. During the examination, as well as during the subsequent autopsy, no antemortem injuries were found. Therefore, it was impossible to establish the circumstances and cause of death.

After the body examination, the remains were taken to the morgue. The entomological material was collected at the site of their disclosure the next morning. The supplementary material was collected during the autopsy and subsequent body examination on the 23 April 2019. The collected material was preserved in 75% ethyl alcohol.

Imagines of predatory beetles, from the families Staphylinidae and Histeridae, were most dominant on the corpse. A few specimens of *Omosita* spp. (Nitidulidae), *Necrobia violacea* Linnaeus, 1758 (Cleridae), *Thanatophilus sinuatus* (Fabricius, 1775), and *Oiceoptoma thoracicum* (Linnaeus, 1758) (Silphidae) were also found. Among the flies, the most abundant were larvae of the Piophilidae family feeding in the adipocere. Numerous larvae and pupae of *Fannia* sp. were collected from the folds of the clothes, and a few Muscid larvae belonging to the genera *Hydrotaea* Robineau-Desvoidy, 1830 and *Muscina* Robineau-Desvoidy, 1830 were collected from the soft tissue residue. Three empty puparia of *Chrysomya albiceps* (Wiedemann, 1819), attached to clothing, were the only indicators of the presence of blow flies.

Several larvae, ~1 cm long, were collected from the folds of the clothes covered with litter and adipocere formed on the back and lower limbs of the corpse. The specimens, covered with moist soil mixed with organic matter, were preliminarily identified as larvae of *H. illucens*. After cleaning in an ultrasonic cleaner, the surface of the larvae was re-examined and showed a clear striped pattern, uncharacteristic for *H. illucens*.

On the 1 October 2019, numerous mature specimens of *Sargus bipunctatus* were found in the compost and manure dumping place in the Łódź Zoo, located approximately 1.1 km away from the site where the body was discovered (Figure 2). This is likely the primary origin of *S. bipunctatus* specimens in the area.

Figure 2. Imago of *Sargus bipunctatus*. Łódź—Zoo, 1 October 2019. Photo—M. Michalski.

3. Material and Methods

Larval specimens were identified first as the genus *Sargus* Fabricius, 1798 [=*Geosargus* Bezzi, 1907] based on available literature [11], then to the species level based on the keys provided by Rozkošny [12]. Species-level identification and subsequent photographic documentation was performed using a Leica M205 FA stereo microscope (Leica Microsystems GmbH, Wetzlar, Germany) with imaging software provided by the manufacturer. FOCUS Projects 4 Professional software (Franzis Verlag GmbH, Haar, Germany) was used to perform photo stacking. To confirm species identifications, the barcode region of *cytochrome oxidase unit I* (COI) was amplified from DNA extracted from three specimens, and compared to data stored in online repositories. DNA extraction was performed in the Molecular Laboratory of the Department of Invertebrate Zoology and Hydrobiology at the University of Lodz, Poland. DNA was extracted from tissue dissected from the anal segments of the larvae using a GeneMATRIX Tissue DNA Purification Kit (EURx, Gdansk, Poland), following manufacturer protocol. Dissected tissue was incubated overnight in lysis buffer with Proteinase K. The 658 bp barcode region of COI was then amplified for each specimen using a Polymerase Chain Reaction (PCR) and the standard barcode primer pair, LCO1490/HCO2198 (Biomers.net GmbH, Ulm, Germany, [13]). A PCR was performed in a final volume of 11 µL reaction mix, containing 5 µL of DreamTaq reaction Buffer (ThermoFisher Scientific, Waltham, MA, USA), 0.8 µL of LCO1490 primer, 0.8 µL of HCO2198 primer, 2.4 µL of ultrapure water, and 2 µL of DNA template. The PCR conditions consisted of 94 °C for 1 min followed by 5 cycles of 30 s at 94 °C, 1 min 30 s at 45 °C and 1 min at 72 °C; 36 cycles at 94 °C for 30 s, 51 °C for 1 min 30 s, and 72 °C for 1 min; with the final extension of 5 min at 72 °C. Amplification success was confirmed via visualisation using agarose gel electrophoresis. PCR products were purified using a mix of FastAP (1 U/µL, ThermoFisher Scientific, Waltham, MA, USA) and Exonuclease I (20 U/µL ThermoFisher Scientific, Waltham, MA, USA). Direct sequencing of the PCR product with the marker-specific primers was outsourced to Macrogen Europe (Amsterdam, The Netherlands). The obtained COI sequences were edited and primers removed using Geneious Pro 11 (Biomatters Ltd., Auckland, New Zealand [14]). From three analysed specimens, only one DNA extraction was successful, and provided a good quality sequence (617 bp). The identity of the obtained COI sequence was verified using the Barcode of Life Data System (BOLD) Identification Engine [15]. As a result, a list of twenty top matches was obtained with 99.82% similarity, with all sequences belonging to *Sargus bipunctatus* (Scopoli, 1763) (BIN URI: BOLD:ACI9008). Nineteen records were collected in Vancouver, Canada, and one from Frankfurt, Germany. The obtained COI sequence was then deposited in the BOLD v4 and GenBank online repositories under accession numbers: BOLD Process ID: DPTPL001-21 (Sample ID: DptPL_Lodz_LA_1); GenBank: MW661345 (dx.doi.org/10.5883/DS-DIPTPL) to make it available for future studies [15,16].

4. Discussion

The twin-spot centurion fly, *Sargus bipunctatus* (Scopoli, 1763) [=*Chrysochroma bipunctatum* (Scopoli, 1763); *Sargus bipunctatus* Costa, 1844], is widespread in Europe and the northwest of North America [1,11,17]. Single records are also known from the mountainous areas in Iran, Tunisia, and Turkey [1,18–20]. According to Nartshuk [21], it is a Euro–Caucasian species associated with temperate deciduous forests. The larvae of this fly feed in terrestrial environments on various substrates, such as decaying plant debris, compost, and the faeces of vertebrates [11,22–24]. According to Dušek and Rozkošny [25], larvae can also be found in egg sacs of the Moroccan locust *Dociostaurus moroccanus* (Thunberg, 1815), and Oldroyd [26] recorded larvae in decaying mushrooms of the species, *Cerioporus squamosus* (Huds.) Quélet (1886). The variety of substrates from which larvae of *S. bipunctatus* has been collected indicates a broad feeding spectrum. Despite this, only Chick [10] has reported the relationship between larvae of *S. bipunctatus* and decaying animal remains. Numerous females of the species were observed on the carcass of a domestic pig on the 24th day of decomposition. Nineteen days later, larvae were found feeding in a mixture of soil, mulch, and the putrefactive liquid exuding from the carcass. Sukontason et al. [27] recorded single cases of *Sargus* sp. larvae feeding on human bodies found in the forests of Thailand from 2000–2006.

Larvae of *S. bipunctatus* are very distinctive and easy to distinguish from the specimens of other necrophagous flies by their dorsoventral flattening. The only species possible to misidentify it with is *Hermetia illucens*, which belongs to the same family Stratiomyidae. The geographic distributions of both species overlap [17,28–30] (see Figure 3). Moreover, it is expected that the northern range of *H. illucens* will expand through both natural processes and the unintentional release of mature flies from black soldier fly breeding farms. Larvae of *S. bipunctatus* and *H. illucens* can be easily distinguished based on morphological features alone. *Sargus bipunctatus* has a distinct colouration pattern, with six longitudinal stripes on its abdominal segments, and very short dorsal, dorsolateral, lateral and ventral bristles (Figure 4).

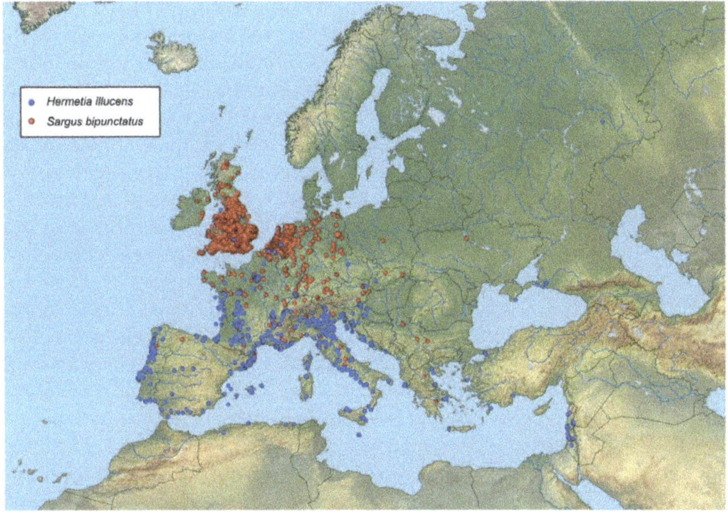

Figure 3. Distribution of *Hermetia illucens* and *Sargus bipunctatus* in Western Palearctic, based on data coming from GBIF and iNaturalist.

Figure 4. Third instar larvae of *Sargus bipunctatus*. (**a**): upperside, (**b**): underside and *Hermetia illucens*, (**c**): upperside, (**d**): underside. Scale bar: 5 mm. Photos—M. Michalski.

Due to the characteristic morphology of both larvae (Figure 4) and adult forms (Figure 2), this species is difficult to overlook when observing its food substrate. Despite this, larvae of this species have not been recorded on the corpses of large vertebrates in any succession experiments conducted in Central Europe to date [31–36]. As such, the presence of *S. bipunctatus* in this case, may only be an incidental colonization of human remains. At the current time, it is difficult to evaluate the potential of this species in forensic applications, due to the narrow range of environments included in the studies of insect succession on carcasses in Central Europe [31–33]. In the discussed case, there is a significant correlation between the presence of larvae and the stage of advanced decomposition. This relation indicates that the species could be used for the estimation of postmortem interval, using a method based on successional patterns rather than development rate.

Imagines of *S. bipunctatus* are active from July to November, with maximum abundances recorded from September to October [28]. In the analysed case, the presence of fully developed larvae of *S. bipunctatus* in spring may indicate that the corpse was in the advanced decomposition stage during the period of imagines activity, probably in the fall of the previous year. This hypothesis may also be confirmed by the observed coexistence of fully grown larvae of Muscidae and empty puparia of *Chrysomya albiceps* [35]. More detailed conclusions are not currently possible due to the lack of precise data on the development of *S. bipunctatus*. Access to such information is a crucial issue for the practical use of forensic entomology [37]. Based on the well-planned field and laboratory experiments, even relatively rare insects may be considered highly important for medico–legal purposes. A good example is the beetle, *Necrodes littoralis* (Linnaeus, 1758), included in the red lists of threatened animals in Central European countries [38,39]. Analysis of specimens collected in real cases and during insect succession studies has shown its frequent presence on large vertebrate carrion, including human corpses [33,40]. Such knowledge stimulated extensive studies of *N. littoralis* development, conducted under laboratory conditions [41,42]. Finally, the presence of immature stages of this beetle was used for the estimation of the time of death [43,44]. We hope that this model path from the laboratory studies to casework will be, at least partly, successfully replicated for *S. bipunctatus*. Further studies of this species should have a precise forensic profile and cover field studies of environmental preferences, preferred food sources and habitats, role in necrophagous insect community, and activity period during successional changes of carrion, as well as laboratory experiments concerning thermal requirements during the development of immature life stages.

Author Contributions: Conceptualization, M.M. and K.S.; methodology, P.G., J.K., M.M. and K.S.; software, P.G. and M.M.; validation, P.G., J.K., M.M. and K.S.; formal analysis, P.G., M.M. and K.S.; investigation, P.G., J.K., M.M. and K.S.; resources, P.G. and M.M.; data curation, P.G. and M.M.; writing—original draft preparation, P.G., J.K., M.M. and K.S.; writing—review and editing, P.G., J.K., M.M. and K.S.; visualization, M.M.; supervision, M.M. and K.S.; project administration, P.G., M.M. and K.S.; funding acquisition, P.G. and K.S. All authors have read and agreed to the published version of the manuscript.

Funding: This work was supported by the National Science Center of Poland [grant number 2016/23/NZ8/02123 to P.G. and 2018/31/B/NZ8/02113 to K.S.].

Institutional Review Board Statement: Not applicable.

Informed Consent Statement: Not applicable.

Data Availability Statement: The obtained COI sequence is available at: Barcode of Life Data System (BOLD) (http://boldsystems.org), Process ID: DPTPL001-21 (Sample ID: DptPL_Lodz_LA_1). GenBank: (https://www.ncbi.nlm.nih.gov/genbank/) Genbank Accession number: MW661345 (dx.doi.org/10.5883/DS-DIPTPL).

Conflicts of Interest: The authors declare no conflict of interest. The funders had no role in the design of the study; in the collection, analyses, or interpretation of data; in the writing of the manuscript, or in the decision to publish the results.

References

1. Woodley, N.E. A World Catalog of the Stratiomyidae (Insecta: Diptera). *Myia* **2001**, *11*, 1–473.
2. James, M.T. Stratiomyidae. Chapter 36. In *Manual of Nearctic Diptera. Vol. 1. Monograph No. 27*; McAlpine, J.F., Peterson, B.V., Shewell, G.E., Teskey, H.J., Vockeroth, J.R., Wood, D.M., Eds.; Research Branch, Agriculture Canada: Ottawa, Canada, 1981; pp. 497–511.
3. Kovac, D.; Rozkošný, R. Insecta: Diptera, Stratiomyidae. In *Freshwater Invertebrates of the Malaysian Region*; Yule, C.M., Yong, H.S., Eds.; Academy of Sciences Malaysia: Kuala Lumpur, Malaysia, 2005; pp. 798–804.
4. Marshall, S.A.; Woodley, N.E.; Hauser, M. The historical spread of the Black Soldier Fly, *Hermetia illucens* (L.) (Diptera, Stratiomyidae, Hermetiinae), and its establishment in Canada. *J. Entomol. Soc. Ont.* **2015**, *146*, 51–54.
5. Roháček, J.; Hora, M.A. Northernmost European Record of the Alien Black Soldier Fly *Hermetia illucens* (Linnaeus, 1758) (Diptera: Stratiomyidae). *Acta Musei Sil. Sci. Nat.* **2013**, *62*, 101–106. [CrossRef]
6. Caruso, D.; Devic, E.; Subamia, I.; Talamond, P.; Baras, E. *Technical Handbook of Domestication and Production of Diptera Black Soldier Fly (BSF) Hermetia Illucens, Stratiomyidae*; Kampus IPB Taman Kencana; PT Penerbit IPB Press: Bogor, Indonesia, 2013; pp. 1–141.
7. Dunn, L.H. *Hermetia illucens* breeding in a human cadaver. *Entomol. News* **1916**, *27*, 59–61.
8. Tomberlin, J.K.; Adler, P.H.; Myers, H.M. Development of the Black Soldier Fly (Diptera: Stratiomyidae) in Relation to Temperature. *Environ. Entomol.* **2009**, *38*, 930–934. [CrossRef]
9. Lord, W.D.; Goff, M.L.; Adkins, T.R.; Haskell, N.H. The Black Soldier Fly *Hermetia illucens* (Diptera: Stratiomyidae) As a Potential Measure of Human Postmortem Interval: Observations and Case Histories. *J. Forensic Sci.* **1994**, *39*, 215–222. [CrossRef]
10. Chick, A.I.R. *Sargus bipunctatus* (Scopoli) (Diptera, Stratiomyidae) on Carrion in Nottinghamshire, and Some Considerations for Forensic Entomology. *Dipter. Dig. Second Ser.* **2012**, *19*, 162.
11. McFadden, M.W. Soldier Fly Larvae in America North of Mexico. *Proc. U. S. Natl. Mus.* **1967**, *121*, 1–72. [CrossRef]
12. Rozkošný, R. *A Biosystematic study of the European Stratiomyidae (Diptera). Vol. 1*; W. Junk: Hague, The Netherlands; Boston, MA, USA; London, UK, 1982; 401p.
13. Folmer, O.; Black, M.; Hoeh, W.; Lutz, R.; Vrijenhoek, R. DNA primers for amplification of mitochondrial cytochrome c oxidase subunit I from diverse metazoan invertebrates. *Mol. Mar. Biol. Biotechnol.* **1994**, *3*, 294–299.
14. Kearse, M.; Moir, R.; Wilson, A.; Stones-Havas, S.; Cheung, M.; Sturrock, S.; Buxton, S.; Cooper, A.; Markowitz, S.; Duran, C.; et al. Geneious Basic: An integrated and extendable desktop software platform for the organization and analysis of sequence data. *Bioinformatics* **2012**, *28*, 1647–1649. [CrossRef]
15. Ratnasingham, S.; Hebert, P.D.N. BOLD: The Barcode of Life Data System. *Mol. Ecol. Notes* **2007**, *7*, 355–364. [CrossRef]
16. Clark, K.; Karsch-Mizrachi, I.; Lipman, D.J.; Ostell, J.; Sayers, E.W. GenBank. *Nucleic Acids Res.* **2016**, *41*, D67–D72. [CrossRef]
17. de Jong, Y.; Verbeek, M.; Michelsen, V.; Bjørn, P.; Los, W.; Steeman, F.; Bailly, N.; Basire, C.; Chylarecki, P.; Stloukal, E.; et al. Fauna Europaea – all European animal species on the web. *Biodivers. Data J.* **2014**, *2*, e4034. [CrossRef] [PubMed]
18. Demirözer, O.; Üstüner, T.; Hayat, R.; Uzun, A. Contribution to the Knowledge of the Stratiomyidae (Diptera) Fauna of Turkey. *Entomol. News* **2017**, *126*, 252–273. [CrossRef]
19. Yimlahi, D.; Üstüner, T.; Zinebi, S.; Belqat, B. New records of the soldier flies of Morocco with a bibliographical inventory of the North African fauna (Diptera, Stratiomyidae). *ZooKeys* **2017**, *709*, 87–125. [CrossRef] [PubMed]
20. Kazerani, F.; Farashiani, M.E.; Karimidoost, A.; Maghsudloo, M.K. New Data on the Subfamily Sarginae (Diptera: Stratiomyidae) from Iran. *Biharean Biol.* **2019**, *13*, 1–3.
21. Nartshuk, E.P. The Character of Soldier Fly Distribution (Diptera, Stratiomyidae) in Eastern Europe. *Entomol. Rev.* **2009**, *89*, 46–55. [CrossRef]
22. Roberts, M.J. Structure of the Mouthparts of the Larvae of the Flies Rhagio and Sargus in Relation to Feeding Habits. *J. Zool.* **1969**, *159*, 381–398. [CrossRef]
23. McFadden, M.W. The soldier flies of Canada and Alaska (Diptera: Stratiomyidae). *Can. Entomol.* **1972**, *104*, 531–562. [CrossRef]
24. Barendregt, A. Het Voorkomen van *Sargus bipunctatus* (Scopoli, 1763) (Diptera, Stratiomyidae) in Nederland. *Entomol. Ber.* **1980**, *40*, 33–37.
25. Dušek, J.; Rozkošny, R. Revision Mitteleuropäischer Arten Der Familie Stratiomyidae (Diptera) Mit Besonderer Berücksichtigung Der Fauna ČSSR II. *Časopis Ceskoslov. Společnosti Entomol.* **1964**, *61*, 360–373.
26. Oldroyd, H. *Handbooks for the Identification of British Insects. Diptera Brachycera Section (a) Tabanoidea and Asiloidea IX. (4)*; Pemberley Natural History Books: Iver, UK, 1969; pp. 1–132.
27. Sukontason, K.; Narongchai, P.; Kanchai, C.; Vichairat, K.; Sribanditmongkol, P.; Bhoopat, T.; Kurahashi, H.; Chockjamsai, M.; Piangjai, S.; Bunchu, N.; et al. Forensic Entomology Cases in Thailand: A Review of Cases from 2000 to 2006. *Parasitol. Res.* **2007**, *101*, 1417–1423. [CrossRef]
28. GBIF.org. GBIF Occurrence Download. Available online: https://doi.org/10.15468/dl.jed6qb (accessed on 15 March 2021).
29. GBIF.org. GBIF Occurrence Download. Available online: https://doi.org/10.15468/dl.xqkvdb (accessed on 15 March 2021).
30. iNaturalist. Available online: https://www.inaturalist.org (accessed on 15 March 2021).
31. Grassberger, M.; Frank, C. Initial study of arthropod succession on pig carrion in a central european urban habitat. *J. Med Entomol.* **2004**, *41*, 511–523. [CrossRef] [PubMed]
32. Anton, E.; Niederegger, S.; Beutel, R.G. Beetles and flies collected on pig carrion in an experimental setting in Thuringia and their forensic implications. *Med Vet. Entomol.* **2011**, *25*, 353–364. [CrossRef] [PubMed]

33. Matuszewski, S.; Bajerlein, D.; Konwerski, S.; Szpila, K. Insect succession and carrion decomposition in selected forests of Central Europe. Part 3: Succession of carrion fauna. *Forensic Sci. Int.* **2011**, *207*, 150–163. [CrossRef]

34. Matuszewski, S.; Frątczak, K.; Konwerski, S.; Bajerlein, D.; Szpila, K.; Jarmusz, M.; Szafałowicz, M.; Grzywacz, A.; Mądra, A. Effect of body mass and clothing on carrion entomofauna. *Int. J. Leg. Med.* **2016**, *130*, 221–232. [CrossRef] [PubMed]

35. Mądra, A.; Frątczak, K.; Grzywacz, A.; Matuszewski, S. Long-term study of pig carrion entomofauna. *Forensic Sci. Int.* **2015**, *252*, 1–10. [CrossRef]

36. Jarmusz, M.; Grzywacz, A.; Bajerlein, D. A comparative study of the entomofauna (Coleoptera, Diptera) associated with hanging and ground pig carcasses in a forest habitat of Poland. *Forensic Sci. Int.* **2020**, *309*, 110212. [CrossRef] [PubMed]

37. Amendt, J.; Richards, C.S.; Campobasso, C.P.; Zehner, R.; Hall, M.J.R. Forensic entomology: Applications and limitations. *Forensic Sci. Med. Pathol.* **2011**, *7*, 379–392. [CrossRef]

38. Głowaciński, Z. (Ed.) *Red List of Threatened Animals in Poland*; Institute of Nature Conservation PAS: Kraków, Poland; 155p.

39. Esser, J. Rote Liste und Gesamtartenliste der Kurzflügelkäferartigen und Stutzkäfer (Coleoptera: Staphylinoidea und Histeridae) von Berlin. In *Rote Listen der gefährdeten Pflanzen, Pilze und Tiere von Berlin*; Der Landesbeauftragte für Naturschutz und Landschaftspflege/Senatsverwaltung für Umwelt, Verkehr und Klimaschutz: Berlin, Germany, 2017; 57p. [CrossRef]

40. Charabidze, D.; Vincent, B.; Pasquerault, T. The biology and ecology of *Necrodes littoralis*, a species of forensic interest in Europe. *Int. J. Leg. Med.* **2016**, *130*, 273–280. [CrossRef]

41. Matuszewski, S. Estimating the pre-appearance interval from temperature in *Necrodes littoralis* L. (Coleoptera: Silphidae). *Forensic Sci. Int.* **2011**, *212*, 180–188. [CrossRef] [PubMed]

42. Gruszka, J.; Matuszewski, S. Estimation of physiological age at emergence based on traits of the forensically useful adult carrion beetle *Necrodes littoralis* L. (Silphidae). *Forensic Sci. Int.* **2020**, *314*, 110407. [CrossRef] [PubMed]

43. Bajerlein, D.; Taberski, D.; Matuszewski, S. Estimation of postmortem interval (PMI) based on empty puparia of *Phormia regina* (Meigen) (Diptera: Calliphoridae) and third larval stage of *Necrodes littoralis* (L.) (Coleoptera: Silphidae) – Advantages of using different PMI indicators. *J. Forensic Leg. Med.* **2018**, *55*, 95–98. [CrossRef] [PubMed]

44. Matuszewski, S.; Mądra-Bielewicz, A. Post-mortem interval estimation based on insect evidence in a quasi-indoor habitat. *Sci. Justice* **2019**, *59*, 109–115. [CrossRef] [PubMed]

insects

Review

The Forensic Entomology Case Report—A Global Perspective

Zanthé Kotzé [1,*], Sylvain Aimar [2], Jens Amendt [3], Gail S. Anderson [4], Luc Bourguignon [5], Martin J.R. Hall [6] and Jeffery K. Tomberlin [1]

1. Department of Entomology, Texas A&M University, 400 Bizzell St, College Station, TX 77843, USA; jktomberlin@tamu.edu
2. Forensics Fauna and Flora Unit, Forensic Sciences Laboratory of the French Gendarmerie, 95000 Pontoise, France; sylvain.aimar@gendarmerie.interieur.gouv.fr
3. Institute of Legal Medicine, University Hospital Frankfurt/Main, Goethe-University, 60323 Frankfurt, Germany; amendt@em.uni-frankfurt.de
4. Centre for Forensic Research, School of Criminology, Simon Fraser University, Burnaby, BC V5A 1S6, Canada; gail_anderson@sfu.ca
5. National Institute for Criminalistics and Criminology, 1120 Brussels, Belgium; Luc.Bourguignon@just.fgov.be
6. Department of Life Sciences, Natural History Museum, London SW7 5BD, UK; m.hall@nhm.ac.uk
* Correspondence: z.kotze@gmail.com

Simple Summary: Forensic entomologists are most often tasked with determining when arthropods colonized living or deceased vertebrates. In most cases, this estimation involves humans; however, pets, livestock, and other domesticated animals can also be illegally killed or victims of neglect. Globally, there is no standard format for the case report, and much of the content is based on the personal preferences of the analyst or standards set within a country. The article below proposes a general overview of sections to be considered when drafting a case report.

Citation: Kotzé, Z.; Aimar, S.; Amendt, J.; Anderson, G.S.; Bourguignon, L.; Hall, M.J.R.; Tomberlin, J.K. The Forensic Entomology Case Report—A Global Perspective. *Insects* **2021**, *12*, 283. https://doi.org/10.3390/insects12040283

Academic Editors: Damien Charabidze and Daniel Martín-Vega

Received: 11 March 2021
Accepted: 22 March 2021
Published: 25 March 2021

Publisher's Note: MDPI stays neutral with regard to jurisdictional claims in published maps and institutional affiliations.

Abstract: Forensic practitioners analyzing entomological evidence are faced with numerous challenges when presenting their findings to law practitioners, particularly in terms of terminology used to describe insect age, what this means for colonization time of remains, and the limitations to estimates made. Due to varying legal requirements in different countries, there is no standard format for the entomological case report prepared, nor any guidelines as to the sections that are required, optional or unnecessary in a case report. The authors herein propose sections that should be considered when drafting an entomological case report. The criteria under which entomological evidence is analyzed are discussed, as well as the limitations for each criterion. The concept of a global, standardized entomological case report is impossible to achieve due to national legislative differences, but the authors here propose a basic template which can be adapted and changed according to the needs of the practitioner. Furthermore, while the discussion is fairly detailed, capturing all differences between nations could not be accomplished, and those initiating casework for the first time are encouraged to engage other practicing forensic entomologists or professional associations within their own nation or region, to ensure a complete report is generated that meets lab or national requirements, prior to generating a finalized report.

Keywords: Calliphoridae; legislation; expert witness statement; criteria; limitations

1. Introduction

In the last ~20 years, developments in the field of forensic entomology have progressed greatly and at pace. These increased research efforts and applications have resulted in over 1100 research and review articles, and books, published since 2000 [1]. Despite the vast number of references available, one of the many challenges faced in allowing entomological evidence to be admitted into court is an understanding of what exactly forensic entomology entails, what information the arthropod evidence can provide, and the application of such information to the case at hand, in addition to meeting standards for a given legal system, such as the Daubert standard of admissibility in the USA [2] and ISO 17025 (predominantly in the European Union) [3]. Of course, such standards vary between legal systems.

Due to such standards being in place in most parts of the world, efforts have been aimed at developing specific guidelines for forensic entomology. Accreditation standards, such as those implemented by the European Association for Forensic Entomology (EAFE) for the sampling and evaluation of entomological traces, and the certification of entomological experts by the American Board of Forensic Entomology (ABFE), as well as the accreditation of laboratories (such as the Forensics Sciences Laboratory of the French Gendarmerie and the laboratory of the Belgium Institut National de Criminalistique et Criminologie) [4], have allowed for entomological evidence to be admitted into courts and analyzed as part of the legal proceedings. Similar standards are being developed in the USA by the Organization of Scientific Area Committees for Forensic Science (OSAC) [5], -Crime Scene Investigation Subcommittee, Forensic Entomology Task Group (Tomberlin, personal communication).

Forensic entomologists are often tasked with determining "how long the victim has been dead" by law enforcement and other officials [6], and may provide a written report (or expert witness statement), detailing an estimated time frame since insects first colonized the remains, and also present expert testimony in Court. The purpose of this article is to provide guidance for the preparation of an entomological case report and clarify its interpretation, with an outline of criteria, terminology, limitations/restrictions and applications of medicolegal forensic entomology, for use by both students new to the field, as well as investigators and legal counsel for clarity in a court of law. This work may also aid seasoned practitioners to improve the presentation of their findings and pave the way for a universal minimum standard in entomological case reports worldwide, adapted to suit relevant legal circumstances. However, it should be noted that such recommendations will vary in terms of their applicability depending on the location and agency involved in the investigation.

2. Criteria and Limitations

The overarching principle of forensic entomology is based on the arrival of insects to remains shortly after death, whereafter eggs or larvae are deposited. Larvae develop on the remains, and the age of the oldest developmental stages can be determined when the remains are discovered [6]. Assuming no disturbance, and ideal conditions, the age of the oldest developmental stages is determined to be close to the interval since time of death, termed the minimum post-mortem interval (minPMI). However, there are a number of factors which can cause deviation from ideal conditions, and thus affect the age determination. The role of the author of a forensic entomology report is to identify the factors present in each case, evaluate their weight and influence, and explain their impact on the minPMI estimation.

Due to the ectothermic nature of insects, much of their physiology, ecology, and behavior has been documented in relation to environmental conditions, especially temperature, but also including humidity, light intensity and wind. Insects inhabiting carrion, which include mainly flies (Diptera: e.g., the families Calliphoridae, Fanniidae, Muscidae, Phoridae, Piophilidae, Sarcophagidae, Stratiomyidae, and Syrphidae) and beetles (Coleoptera: e.g., the families Carabidae, Cleridae, Dermestidae, Histeridae, Silphidae and Staphylinidae), have been extensively studied and various criteria have been set forth to evaluate their development and succession on remains for a forensic report. The author of a forensic report should never assume (i.e., accept as being true without proof) anything; nevertheless, there are certain criteria related to insect biology which have support from scientific study and can act as guides to an analysis of insect evidence, unless there is compelling evidence to the contrary. Some of the basic criteria used by forensic practitioners, as well as their limitations, are listed below [6]:

1. Environmental weather records in the area reflect those at the body, and thus directly affect the development of the arthropods present:
   The consideration of temperature is fundamental in estimating the age of insects [7,8]. The microclimates in which insects develop at a scene can potentially vary greatly from the temperatures provided by a nearby environmental monitor (e.g., national weather station). Regardless of the debate among scientists as to whether the temperatures to which the developing insects were exposed should be taken one-to-one from that monitor or modelled site-specifically, practitioners should clearly state which method of estimating temperature was used, e.g., whether it was nearby weather station data, scene temperature logger data, or some form of regression analysis based on these two sets of data. Challenges exist with each method, and there are numerous factors that may affect the developmental patterns within the parameters of the influence of temperature [9].

2. Published developmental datasets based on laboratory studies accurately reflect developmental patterns in the nature of the insect evidence collected:
   The vast majority of published and accepted insect developmental datasets have been derived under laboratory conditions. These conditions usually applied a range of temperature profiles (constant or with daily variations), controlled humidity and specified light: dark cycles. When developing in a natural environment, none of the above-mentioned factors are controlled, and can affect development accordingly [10–12]. Temperature cycles fluctuate greatly, both daily and seasonally [13], humidity is dependent on a number of factors, including season and precipitation, and light: dark cycles are highly dependent on season and region (not to mention possible artificial lighting conditions). Although some field studies have validated laboratory data [14], the general assumption that developmental patterns observed in the laboratory are reflected in natural environments may result in an under- or overestimate of larval developmental patterns. More validation studies between laboratory and field developmental data are needed, in pursuit of increasing the accuracy and precision of entomological estimates, as well as their reliability;

3. Colonization occurred after death (i.e., no myiasis):
   In certain situations, oviposition or larviposition may occur before death, for example, when the decedent has open and possibly necrotic wounds such as decubitus ulcers (bed sores). Myiasis is the colonization of a living vertebrate host by fly larvae [15], and if the victim is not discovered until after death, it may not be known whether the colonization occurred before or after death [16,17]. While this could lead to an overestimation of time since death if not considered, it could also provide new leads for the investigation, e.g., in cases of suspected neglect, where demonstration of ante-mortem myiasis can be crucial evidence [18].

4. Specimens collected and analyzed developed on the body of the victim:
   Contamination of insect evidence can occur from other organisms that are deceased and within close proximity of the remains under scrutiny. For example, in an outdoor case, empty puparia in soil samples from a crime scene could originate from flies that had developed on a dead animal in the immediate vicinity at an earlier time.

5. Carrion-colonizing Diptera are diurnal and do not usually oviposit at night:
   Nocturnal oviposition is very rare and is thus usually excluded from analyses. Historically, it was assumed that oviposition only occurs during the day and, thus, hours of darkness were not considered when estimating the minPMI based on calculating the time of oviposition [19–21].

6. Carrion-colonizing insects (specifically Diptera and Coleoptera) have free access to the body:
   Oviposition or larviposition on the deceased occurs shortly after death without hindrance (physical and/or temporal/seasonal). However, in medico-legal cases where entomological evidence is to be obtained, a decedent may be concealed in order to prevent law enforcement from finding the body. This concealment may include

burial, wrapping or disposal in bodies of water. In such instances, carrion-colonizing arthropods are limited in their access to the remains, often only gaining access after the remains have been discovered or exposed by the elements or by scavengers. In such instances particularly, the entomological evidence obtained provides details regarding the period of environmental exposure, provided the remains have always been in the conditions of their discovery, but cannot provide more specific information regarding a time frame of the decedents' death.

7. Faunal succession patterns on, in and under the body can be used in estimating colonization intervals:

    While faunal succession patterns are somewhat predictable [22], they are seasonally and environmentally-dependent, and depend largely on the faunal species present in an area [23,24]. However, precise estimates of exact species present and their arrival patterns at remains cannot be determined without conducting field trials in many different environments, and creating a database of these findings, which is an unrealistic task and not necessarily reproducible outside of an experimental framework. Producing an entomological estimate based solely on faunal succession patterns is not likely to be robust and will have large confidence intervals. In most cases, faunal data are presented in terms of overlapping time frames, from which a minPMI can then be estimated [25–28]. In some instances, species level data can be used to interpret successional data; however, such cases are rare [29].

While the above criteria and limitations are broadly applicable to most cases, it must be noted that each case containing entomological evidence is unique and should be analyzed accordingly—there is no "one size fits all" approach.

3. Use of Terminology

There have been some disputes in recent years regarding the terminology used by forensic practitioners concerning entomological evidence [30]. Historically, entomological evidence was used to estimate the postmortem interval [31,32]. This term implies that the time of death of the decedent could be accurately estimated using arthropods present but this does not consider that there may be a delay in colonization for many different reasons, e.g., in an enclosure without insect access, such as a car trunk or locked room. All such issues would affect access to the remains by arthropods.

A myriad of alternative terms has been introduced to describe the activity of arthropods on remains. In some manner or another, each term that is used describes the time since arthropods have colonized the remains. The terms include: minPMI; post-colonization interval (PCI); and time of colonization (TOC) [33,34].

Irrespective of the terminology selected by the practitioner, it is critical that the reader understands what the term used in the report is referring to and what it means; namely, a period of time which has passed at least since the occurrence of death. The clarification of terms is important for interpretation of the report by individuals without entomological/scientific training, such as law practitioners or judges/jurors.

4. Insect Identification and Reliability of Keys

Numerous dichotomous and pictographic keys exist for the identification of arthropods based on physical characteristics. These keys are still the most frequently used means of identification for both immature and adult specimens of forensic importance. In many instances, samples may be received by the practitioner that have been damaged or are missing body parts. In such situations, the use of a dichotomous or pictographic key may not be the best avenue, as these keys reference specific body regions. Resources such as Lucidcentral [35] allow for the identification of specimens that are missing aspects critical for identification, as the data are arranged in a spreadsheet rather than a dichotomous key format. Additionally, voucher specimens from museums may also be used for comparison and identification.

With the advent of molecular identification techniques, such as DNA barcoding, arthropod identification, especially that of insect fragments, has become easier [36–40]. However, despite the vast number of researchers using databases such as GenBank, errors in gene sequences still exist, and are being continuously detected and corrected. One of the most important benefits of using techniques such as DNA for identification is the accurate differentiation of morphologically and behaviorally similar species (provided that the corresponding developmental data set are available), such as *Lucilia cuprina* (Wiedemann) and *L. sericata* (Meigen) [41], or *Hemilucilia segmentaria* (Fabricius) and *H. semidiaphana* (Rondani) (all Diptera: Calliphoridae) [37]. While these species are behaviorally and morphologically similar, they differ in their developmental patterns, so accurate identification is important to provide a reliable estimate of colonization periods [42].

Whichever method of identification is used by the practitioner to identify specimens must be mentioned in the report.

5. Recommended Sections and Explanations for an Entomological Case Report

The following proposed sections for an entomological case report have been adapted and extended from those proposed in the Standard Operating Protocol for medico-criminal case reports by the American Board of Forensic Entomology in 2009 (see Table 1 for template/summary).

1. Title indicating the contents of the report:
    *   This should include a case number or legal system reference if applicable, as well as an indication that the report is of an entomological nature.

2. Analyst/practitioner contact information (including location):
    *   This should include a working postal or email address and contact telephone number. The practitioner's title and affiliation should be included.

3. Contact information of investigating officer or law practitioner (i.e., the person requesting the report):
    *   Again, a working postal or email address and contact telephone number, plus title and affiliation of contact person included.

4. Instructions received:
    *   This section should include a brief note on when and how the practitioner was contacted and a precise description of what was being asked of them by the investigating officer or other person requesting evidentiary analysis.

5. Case information (summary based on case file):
    *   The purpose of this section is not to restate the entirety of the case file; rather, a brief summary of the biographic details of the case (date, time, location) and details pertinent to the victim(s) and entomological evidence.

6. Summary of insect evidence received, including at least a description of different developmental stages:
    *   Many practitioners relabel evidence once received, based on their own preferences or the labeling system of their laboratory. Both the original evidence details and the renamed details should be included here, to cover the bases for chain of custody.
    *   If vials containing evidence are split or repackaged for any reason, this should be indicated, with a reasonable explanation as to the reasoning behind repackaging (e.g., to change or add preservative). Vials that have been split into multiple sections must be relabeled, and new labels names indicated as well. This should follow chain of custody protocols as dictated by the regulating authority of the country.
    *   For preserved evidence, the time of collection and time of preservation should be included.

- If used, the preservation medium used by the practitioner should be indicated—often law enforcement officials do not have the necessary chemicals for preservation available at a collection scene and will use any suitable substance that is readily available (e.g., gin, vodka). Evidence is then analyzed and replaced into vials with a more standard ethanol preservative (the concentration of which must be indicated).
- If live samples were provided, a detailed timeline of collection and transportation should be provided. This includes storage conditions (e.g., in coolers), if oxygen supply was limited in a sealed container, as well as dietary medium provided during transport. If samples were further reared once reaching the practitioner, rearing details (e.g., temperature, food supplied) should also be provided.
- This would also be an appropriate section to indicate any external factors that may have affected insect colonization and development on the remains (such as concealment, found in a closed room/building with no open windows, thermostat on/off at constant temperature, as well as whether specimens had, at any point, faced refrigeration at a mortuary).

7. Environmental conditions obtained from weather stations:

- Since weather stations are not always conveniently located near crime scenes, it is advisable to use the most relevant climatic data available, from a certified meteorological organization, such as the national meteorological institution of your jurisdiction/country, and also indicate if data loggers placed at the scene after body discovery have been used to reconstruct scene data.
- The weather conditions at the time of insect collection should also be included if they were provided by the investigating officer.

8. Identification of species and biological background:

- A brief background of the species identified should be presented, including geographic range and life cycle.
- Suitable references that have been used in the analysis should also be included here. These should include references to the identification keys, voucher specimens and molecular techniques used for identification and comparison;
- If a large number of specimens were provided, and only a subset analyzed, the criteria for subset selection should be mentioned.

9. Estimation of insect age:

- This section, the bulk of the report, should be a brief summary of the estimation of the age of insect evidence based on temperature. This section should be broken down by species identified.

10. Case summary (including date range of colonization if applicable):

- This section should highlight the most important findings and/or date range of colonization if applicable.
- There are a number of ways this could be presented; it may be helpful to separate date ranges by species, with a conclusive statement encompassing the chosen range.

11. Criteria/caveats:

- There is a long list of criteria as stated above. It should not be necessary to include all of these, but definitely those most pertinent to the specific case. These can also be included wherever relevant throughout the report, rather than as a separate section.

12. Declaration:

- This should be based on the requirements of the legal system into which the report is submitted, which can vary greatly, and include a statement indicating that analyses were performed based on currently available information and, should more information become available, the findings are subject to change.

13. Signature:

    - The report should be signed in accordance with the local requirements for documents of legal value.

14. Accreditation statement (if available):

    - A list of professional qualifications of the author can be included here, including professional qualifications and the number of cases worked. This section may be omitted where national legislature does not require it, or where pre-accredited lists exist which include such information.

15. Reference list:

    - Citations identified in the report should be provided. These citations support the approaches, interpretation, and conclusions made in the report (see discussion below).

16. Supplementary documentation (if required):

    - Chain of custody documents (courier receipts etc.) if available.
    - Developmental data sets and calculations (upon request).
    - Tabulated weather data (upon request).

Table 1. Proposed template for a forensic entomological report, with summary of content.

Proposed Report Section	Summary and/or Example of Content
Title indicating the contents of the report	e.g., "Estimation of post-mortem interval based on evaluation of entomological evidence." File/case number as provided by law enforcement official, as well as location.
Analyst/practitioner contact information (including location)	Practitioner name, contact details (postal address, email address, telephone number), institutional affiliation.
Requested by	Officer/analyst/judge.
Instructions received	e.g., "Request for an estimate of post-mortem interval from entomological evidence collected from remains of (decedent/case number) on (date) at (location)".
Case information (summary based on case file)	Person contacting practitioner, date contacted, evidence received, location of discovery, scene description (e.g., indoors, sealed room, thermostat reading (if applicable)).
Summary of insect evidence received, e.g., including taxa, numbers received	e.g., "500 larval specimens preserved in (preservation medium)—rough estimates can also be provided, e.g., 3 adult specimens (preserved), 25 live larval specimens on (rearing medium)".
Environmental conditions obtained from weather stations	Local weather data for a relevant period—including an indication of where weather data were obtained from (e.g., national meteorological weather station, website, access date).
Background of species identified and analysis of evidence	Identification techniques, distribution of species, number of specimens analyzed, techniques used to age the insects based on their stage of development and environmental conditions.
Estimation of insect age	Estimation of age of insect evidence based on temperature, broken down by species.
Case summary (including date range of colonization if applicable)	If more than one species was identified, multiple date ranges should be presented with the broadest range of overlap presented.
Criteria/caveats: these can be omitted as a separate section and included where they apply	Caveats need to be stated when applicable to the case

Table 1. Cont.

Proposed Report Section	Summary and/or Example of Content
Declaration	e.g., "Practitioner reserves the right to change the document as new information becomes available; report may only be reproduced in its entirety; the contents of the report are true and accurate".
Signature	Full signature at the end of report.
Accreditation statement (if available)	e.g., "Member—American Board of Forensic Entomology", as well as professional qualifications as applicable, and membership to local certification bodies.
Reference list (*)	Developmental datasets/voucher specimens used for comparison and identification.
Supplementary documentation (if required *)	Chain of custody documentation; summary of weather data; documented analysis (e.g., for insect aging methods); summary of entomological terms; professional qualifications.

* does not necessarily need to be provided with the report, but a statement of availability on request is then necessary.

6. References and Citations Selected

References included in the case report should reflect the locality of the entomological evidence as far as possible, as well as being relevant to the species identified. Key aspects to consider when compiling/utilizing references include:

1. Locality:
   Where applicable, the datasets used should be based on insect populations as close to where the evidence was collected as possible. In cases where no local datasets are available, the practitioner should include a statement indicating that local data were not available. It must be noted, however, that developmental patterns by geographic separation may not always differ, and, in some cases, are comparable irrespective of location [34,43].
2. Species specificity:
   Where available, datasets pertaining to the actual species identified should be used. When this is not possible, it is preferable to exclude these species from analysis rather than use a dataset for a closely related species. However, the exclusion of such data is at the discretion of the practitioner, provided they maintain transparency regarding their findings, as sometimes it might be better to guide an investigation with a much more general conclusion based on data from a related species, suitably qualified with regard to accuracy, than to provide no conclusion. In such cases, the practitioner may refer to a generalized larval life cycle of the organisms concerned, or datasets for closely related species, to indicate a possible time of colonization estimate based on prevailing conditions. Irrespective of which sources a practitioner opts to utilize, these sources should be cited and the data should be accessible to any individual who is to read/analyze the submitted report.

7. Application and Conclusions

The goal of any entomological report is a reliable estimate of the TOC, which some interpret as a minPMI, of vertebrate remains by arthropods in cases where such events occurred after death. The report should be written and constructed in such a way that it is understood by individuals regardless of their level of scientific training. The report should be grounded in scientific principles and expertise, but not so saturated with scientific jargon that non-scientists struggle to understand or interpret the report. We acknowledge that an entomological report is not a scientifically peer-reviewed paper, but it should be prepared to the same high standards and, in order to meet legal standards (such as the Daubert standard in the USA, and ISO 17025), a significant element of scientific expertise is required.

Quality assurance in entomological reports is of the highest importance, and fact-based evidence will be critical. However, it is acknowledged that every report will also contain opinion-based evidence, based on the knowledge and experience of the person compiling the report, and the judge presiding over the case will need to acknowledge that each case is unique and will need to be considered, in some aspects, independently of other case reports and studies [4].

As indicated by the title of this communication, the points expressed above are recommended contents and points for consideration when compiling an entomological report for any legal purposes, including investigations of death, neglect, or stored product scenarios. As forensic practitioners, we understand that standards and legislation differ by country, and some sections may need to be revised to deviate slightly from those discussed, or the order of topics restructured to reflect the specific needs of the investigating officer or person requesting the report for its intended purpose (for example, not all entomological reports go to court). Without a global standard of legislature, the implementation of a standard entomological report may not be possible, but, at the very least, we hope this manuscript can provide a framework which entomological practitioners in any area can modify to develop a standardized report that is accepted by the respective jurisdiction in which the report is to be presented.

Author Contributions: All authors contributed equally to the development of this manuscript. All authors have read and agreed to the published version of the manuscript.

Funding: No funds were applied to this project.

Institutional Review Board Statement: Not applicable.

Data Availability Statement: Not applicable.

Acknowledgments: The authors thank Captain Melanie Pienaar (Victim Identification Center: South African Police Service) for valuable input and contributions.

Conflicts of Interest: The authors declare no conflict of interest.

References

1. Lei, G.; Liu, F.; Liu, P.; Zhou, Y.; Jiao, T.; Dang, Y.-H. A bibliometric analysis of forensic entomology trends and perspectives worldwide over the last two decades (1998–2017). *Forensic Sci. Int.* **2019**, *295*, 72–82. [CrossRef] [PubMed]
2. Solomon, S.M.; Hackett, E.J. Setting boundaries between science and law: Lessons from Daubert v. Merrell Dow Pharmaceuticals, Inc. *Sci. Technol. Hum. Values* **1996**, *21*, 131–156. [CrossRef]
3. Rodima, A.; Vilbaste, M.; Saks, O.; Jakobson, E.; Koort, E.; Pihl, V.; Sooväli, L.; Jalukse, L.; Traks, J.; Virro, K.; et al. ISO 17025 quality system in a university environment. *Accredit. Qual. Assur.* **2005**, *10*, 369–372. [CrossRef]
4. Amendt, J. Forensic entomology. *Forensic Sci. Res.* **2017**, *3*, 1. [CrossRef] [PubMed]
5. NIST. The Organization of Scientific Area Committees for Forensic Science. Available online: https://www.nist.gov/osac (accessed on 15 February 2021).
6. Tarone, A.M.; Sanford, M.R. Is PMI the hypothesis or the null hypothesis? *J. Med. Entomol.* **2017**, *54*, 1109–1115. [CrossRef] [PubMed]
7. Ratte, H.T. Temperature and insect development. In *Environmental Physiology and Biochemistry of Insects*; Hoffmann, K.H., Ed.; Springer: Berlin/Heidelberg, Germany, 1985; pp. 33–66. [CrossRef]
8. Damos, P.; Savopoulou-Soultani, M. Temperature driven models for insect development and vital thermal requirements. *Psyche* **2012**, *2012*, 123405. [CrossRef]
9. Lutz, L.; Amendt, J. Stay cool or get hot? An applied primer for using temperature in forensic entomological case work. *Sci. Justice* **2020**, *60*, 415–422. [CrossRef] [PubMed]
10. Bauer, A.M.; Bauer, A.; Tomberlin, J.K. Effects of photoperiod on the development of forensically important blow fly *Chrysomya rufifacies* (Diptera: Calliphoridae). *J. Med. Entomol.* **2020**, *57*, 1382–1389. [CrossRef] [PubMed]
11. Bauer, A.; Bauer, A.M.; Tomberlin, J.K. Impact of diet moisture on the development of the forensically important blow fly *Cochliomyia macellaria* (Fabricius) (Diptera: Calliphoridae). *Forensic Sci. Int.* **2020**, *312*, 110333. [CrossRef]
12. Chen, W.; Yang, L.; Ren, L.; Shang, Y.; Wang, S.; Guo, Y. Impact of constant versus fluctuating temperatures on the development and life history parameters of *Aldrichina grahami* (Diptera: Calliphoridae). *Insects* **2019**, *10*, 184. [CrossRef] [PubMed]

13. Bernhardt, J.; Carleton, A.M.; LaMagna, C. A comparison of daily temperature-averaging methods: Spatial variability and recent change for the CONUS. *J. Clim.* **2018**, *31*, 979–996. [CrossRef]

14. Donovan, S.E.; Hall, M.J.R.; Turner, B.D.; Moncrieff, C.B. Larval growth rates of the blowfly, *Calliphora vicina*, over a range of temperatures. *Med. Vet. Entomol.* **2006**, *20*, 106–114. [CrossRef]

15. Zumpt, F. *Myiasis in Man and Animals in The Old World: A Textbook for Physicians, Veterinarians, and Zoologists*; Butterworths: London, UK, 1965.

16. Goff, M.L.; Charbonneau, S.; Sullivan, W. Presence of Fecal Material in Diapers as a Potential Source of Error in Estimations of Postmortem Interval Using Arthropod Development Rates. *J. Forensic Sci.* **1991**, *36*, 1603–1606. [CrossRef]

17. Vanin, S.; Bonizzoli, M.; Migliaccio, M.L.; Buoninsegni, L.T.; Bugelli, V.; Pinchi, V.; Focardi, M. A case of insect colonization before the death. *J. Forensic Sci.* **2017**, *62*, 1665–1667. [CrossRef] [PubMed]

18. Hall, M.J.R.; Wall, R.L.; Stevens, J.R. Traumatic Myiasis: A Neglected Disease in a Changing World. *Annu. Rev. Entomol.* **2016**, *61*, 159–176. [CrossRef] [PubMed]

19. Williams, K.A.; Wallman, J.F.; Lessard, B.D.; Kavazos, C.R.; Mazungula, D.N.; Villet, M.H. Nocturnal oviposition behavior of blowflies (Diptera: Calliphoridae) in the southern hemisphere (South Africa and Australia) and its forensic implications. *Forensic Sci. Med. Pathol.* **2017**, *13*, 123–134. [CrossRef] [PubMed]

20. Amendt, J.; Zehner, R.; Reckel, F. The nocturnal oviposition behaviour of blowflies (Diptera: Calliphoridae) in Central Europe and its forensic implications. *Forensic Sci. Int.* **2008**, *175*, 61–64. [CrossRef] [PubMed]

21. Greenberg, B. Nocturnal oviposition behavior of blow flies (Diptera: Calliphoridae). *J. Med. Entomol.* **1990**, *27*, 807–810. [CrossRef]

22. Matuszewski, S.; Bajerlein, D.; Konwerski, S.; Szpila, K. Insect succession and carrion decomposition in selected forests of Central Europe. Part 3: Succession of carrion fauna. *Forensic Sci. Int.* **2011**, *207*, 150–163. [CrossRef]

23. Wang, J.; Li, Z.; Chen, Y.; Chen, Q.; Yin, X. The succession and development of insects on pig carcasses and their significances in estimating PMI in south China. *Forensic Sci. Int.* **2008**, *179*, 11–18. [CrossRef]

24. Schoenly, K.; Reid, W. Dynamics of heterotrophic succession in carrion revisited. *Oecologia* **1989**, *79*, 140–142. [CrossRef]

25. Anderson, G.S.; Van Laerhoven, S.L. Initial studies on insect succession on carrion in southwestern British Columbia. *J. Forensic Sci.* **1996**, *41*, 617–625. [CrossRef]

26. Amendt, J.; Krettek, R.; Niess, C.; Zehner, R.; Bratzke, H. Forensic entomology in Germany. *Forensic Sci. Int.* **2000**, *113*, 309–314. [CrossRef]

27. Wolff, M.; Uribe, A.; Ortiz, A.; Duque, P. A preliminary study of forensic entomology in Medellín, Colombia. *Forensic Sci. Int.* **2001**, *120*, 53–59. [CrossRef]

28. Archer, M.S. Annual variation in arrival and departure times of carrion insects at carcasses: Implications for succession studies in forensic entomology. *Aust. J. Zool.* **2004**, *51*, 569–576. [CrossRef]

29. Matuszewski, S. Estimating the pre-appearance interval from temperature in *Necrodes littoralis* L.(Coleoptera: Silphidae). *Forensic Sci. Int.* **2011**, *212*, 180–188. [CrossRef] [PubMed]

30. Wells, J.D. To the Editor: Misstatements Concerning Forensic Entomology Practice in Recent Publications. *J. Med. Entomol.* **2014**, *51*, 489–490. [CrossRef]

31. Nabity, P.D.; Higley, L.G.; Heng-moss, T.M. Effects of temperature on development of *Phormia regina* (Diptera: Calliphoridae) and use of developmental data in determining time intervals in forensic entomology. *J. Med. Entomol.* **2006**, *43*, 1276–1286. [CrossRef]

32. Faucherre, J.; Cherix, D.; Wyss, C. Behavior of *Calliphora vicina* (Diptera, Calliphoridae) under extreme conditions. *J. Insect Behav.* **1999**, *12*, 687–690. [CrossRef]

33. Tomberlin, J.K.; Mohr, R.; Benbow, M.E.; Tarone, A.M.; Van Laerhoven, S. A roadmap for bridging basic and applied research in forensic entomology. *Annu. Rev. Entomol.* **2011**, *56*, 401–421. [CrossRef] [PubMed]

34. Amendt, J.; Campobasso, C.P.; Gaudry, E.; Reiter, C.; LeBlanc, H.N.; Hall, M.J.R. Best practice in forensic entomology—Standards and guidelines. *Int. J. Leg. Med.* **2007**, *121*, 90–104. [CrossRef]

35. Lucid Central. Available online: https://www.lucidcentral.org/ (accessed on 15 February 2021).

36. Schroeder, H.; Klotzbach, H.; Elias, S.; Augustin, C.; Pueschel, K. Use of PCR–RFLP for differentiation of calliphorid larvae (Diptera, Calliphoridae) on human corpses. *Forensic Sci. Int.* **2003**, *132*, 76–81. [CrossRef]

37. Thyssen, P.J.; Lessinger, A.C.; Azeredo-Espin, A.M.L.; Linhares, A.X. The value of PCR-RFLP molecular markers for the differentiation of immature stages of two necrophagous flies (Diptera: Calliphoridae) of potential forensic importance. *Neotrop. Entomol.* **2005**, *34*, 777–783. [CrossRef]

38. Clarke, L.J.; Soubrier, J.; Weyrich, L.S.; Cooper, A. Environmental metabarcodes for insects: In silico PCR reveals potential for taxonomic bias. *Mol. Ecol. Resour.* **2014**, *14*, 1160–1170. [CrossRef] [PubMed]

39. Bharti, M.; Singh, B. DNA-based identification of forensically important blow flies (Diptera: Calliphoridae) from India. *J. Med. Entomol.* **2017**, *54*, 1151–1156. [CrossRef] [PubMed]

40. Yusseff-Vanegas, S.Z.; Agnarsson, I. DNA-barcoding of forensically important blow flies (Diptera: Calliphoridae) in the Caribbean Region. *PeerJ* **2017**, *5*, e3516. [CrossRef]

41. Tourle, R.; Downie, D.; Villet, M. Flies in the ointment: A morphological and molecular comparison of *Lucilia cuprina* and *Lucilia sericata* (Diptera: Calliphoridae) in South Africa. *Med. Vet. Entomol.* **2009**, *23*, 6–14. [CrossRef] [PubMed]

42. Sonet, G.; Jordaens, K.; Braet, Y.; Desmyter, S. Why is the molecular identification of the forensically important blowfly species *Lucilia caesar* and *L. illustris* (family Calliphoridae) so problematic? *Forensic Sci. Int.* **2012**, *223*, 153–159. [CrossRef]
43. Limsopatham, K.; Hall, M.J.R.; Zehner, R.; Zajac, B.K.; Verhoff, M.A.; Sontigun, N.; Sukontason, K.; Sukontason, K.L.; Amendt, J. A molecular, morphological, and physiological comparison of English and German populations of *Calliphora vicina* (Diptera: Calliphoridae). *PLoS ONE* **2018**, *13*, e0207188. [CrossRef]

Article

DNA Barcoding Identifies Unknown Females and Larvae of *Fannia* R.-D. (Diptera: Fanniidae) from Carrion Succession Experiment and Case Report

Andrzej Grzywacz [1],* , Mateusz Jarmusz [2], Kinga Walczak [1], Rafał Skowronek [3] , Nikolas P. Johnston [1] and Krzysztof Szpila [1]

[1]   Department of Ecology and Biogeography, Faculty of Biological and Veterinary Sciences, Nicolaus Copernicus University in Toruń, 87-100 Toruń, Poland; walczak.kinga00@gmail.com (K.W.); nikolaspjohoston@gmail.com (N.P.J.); szpila@umk.pl (K.S.)
[2]   Department of Animal Taxonomy and Ecology, Faculty of Biology, Adam Mickiewicz University in Poznań, 61-712 Poznań, Poland; mat.jarmusz@gmail.com
[3]   Department of Forensic Medicine and Forensic Toxicology, Faculty of Medical Sciences in Katowice, Medical University of Silesia in Katowice, 40-055 Katowice, Poland; rafal-skowronek@wp.pl
*   Correspondence: hydrotaea@gmail.com

check for updates

Citation: Grzywacz, A.; Jarmusz, M.; Walczak, K.; Skowronek, R.; Johnston, N.P.; Szpila, K. DNA Barcoding Identifies Unknown Females and Larvae of *Fannia* R.-D. (Diptera: Fanniidae) from Carrion Succession Experiment and Case Report. *Insects* **2021**, *12*, 381. https://doi.org/10.3390/insects12050381

Academic Editors: Daniel Martín-Vega and Damien Charabidze

Received: 26 February 2021
Accepted: 21 April 2021
Published: 23 April 2021

Publisher's Note: MDPI stays neutral with regard to jurisdictional claims in published maps and institutional affiliations.

Simple Summary: Insects are frequently attracted to animal and human cadavers, usually in large numbers. The practice of forensic entomology can utilize information regarding the identity and characteristics of insect assemblages on dead organisms to determine the time elapsed since death occurred. However, for insects to be used for forensic applications it is essential that species are identified correctly. Imprecise identification not only affects the forensic utility of insects but also results in an incomplete image of necrophagous entomofauna in general. The present state of knowledge on morphological diversity and taxonomy of necrophagous insects is still incomplete and identification of immature and female forms can be extremely difficult. In this study, we utilized molecular identification methods to link conspecific sexes and developmental stages of forensically important flies. We identified larvae and females of flies collected from animal and human cadavers which otherwise were morphologically unidentifiable. The present study fills a gap in taxonomy of flies and provides data facilitating application of new species as forensic indicators.

Abstract: Application of available keys to European Fanniidae did not facilitate unequivocal species identification for third instar larvae and females of *Fannia* Robineau-Desvoidy, 1830 collected during a study of arthropod succession on pig carrion. To link these samples to known species, we took the advantage of molecular identification methods and compared newly obtained cytochrome oxidase subunit I (COI) barcode sequences against sequences deposited in reference databases. As an outcome of the results obtained, we describe for the first time a third instar larva of *Fannia nigra* Malloch, 1910 and *Fannia pallitibia* (Rondani, 1866) and a female of *Fannia collini* d'Assis-Fonseca, 1966. We provide combinations of characters allowing for discrimination of described insects from other Fanniidae. We provide an update for the key by Rozkošný et al. 1997, which allows differentiation between females of *F. collini* and other species of Fanniidae. Additionally, we provide a case of a human cadaver discovered in Southern Poland and insect fauna associated with it as the first report of *F. nigra* larvae developing on a human body.

Keywords: Fanniidae; larval morphology; forensic entomology; human cadaver

1. Introduction

Fanniidae is one of the dipteran families that are attracted to and develop in decomposing animal carrion and human bodies [1–5]. Fanniids can be found at various stages of decomposition and are also known for their ability to exploit both buried remains [6,7]

and those restricted to indoor conditions [8,9]. Under certain circumstances, species of Fanniidae may be utilized as forensic indicators [4,5,8,9]. However, the broad application of Fanniidae for medico-legal purposes is inhibited by the general difficulty of species identification in this family [10] and the absence of information linking females and immature stages to the more rigorously studied adult males.

In many species the linkage of conspecific males and females is hindered by sexual dimorphism, especially differences in coloration and leg ornamentation between males and females of the same species. Morphological characters that are diagnostic in males, such as the body color, specific structures on legs and genital structures, frequently vary or are uninformative for the identification of females. Further complicating the identification of female fanniids are two issues; firstly, that some species are described only from males, and secondly females of closely related fanniids (when described) are frequently discriminated based only on a few vague characters, such as minor differences in body coloration or the number or size of fine setae. This is of particular detriment to the forensic utility of fanniids as the majority of adult specimens collected from carrion succession experiments or crime scene are females [1,2,4], and therefore in some cases the adult females are impossible to identify based on morphology alone.

The differentiation of larvae and puparia of Fanniidae from other forensically relevant dipteran families is relatively straightforward due to the characteristic dorso-ventrally flattened body, thoracic and abdominal segments equipped with fleshy projections and posterior spiracles raised on stalks. The pattern of fleshy processes on thoracic and abdominal segments is often species-specific [11,12]. However, morphological interpretations of some character states vary between various authors and require revision, without which accurate discrimination between closely related species will be challenging [13]. Furthermore, immature stages have only been described for several species attracted to carrion and dead bodies, and therefore it is likely that there is much morphological diversity among larvae yet to be discovered.

Animal carrion and dead human bodies have been observed to attract more than 50 species of fanniids worldwide [1–3,14–16], with 15 species confirmed as developing on cadavers [3,4,16–25]. However, due to taxonomic issues and difficulties in obtaining accurate species identifications, most studies refer only to a few common species of *Fannia* Robineau-Desvoidy, 1830 (e.g., *Fannia canicularis* (Linnaeus, 1761), *Fannia manicata* (Meigen, 1826), *Fannia pusio* (Wiedemann, 1830) and *Fannia scalaris* (Fabricius, 1794)) [23,26]. The overrepresentation of these species in the literature provides an incomplete representation of carrion entomofauna, which falsely implies that the most frequent and numerous elements of carrion fanniid assemblages are the commonly identified species mentioned above.

During a study on insect succession on pig carrion, we found two distinct types of third instar larvae, hereafter *Fannia* sp. 1 and *Fannia* sp. 2, and females, hereafter *Fannia* sp. 3, which were not morphologically linked to the commonly encountered carrion-breeding species of *Fannia* [11,12,27,28]. These unidentified *Fannia* spp. outnumbered many other necrophagous species [16]. Furthermore, several specimens corresponding with *Fannia* sp. 1 have also been collected from a dead human body in Southern Poland as part of a forensic investigation.

Based on these findings, we hypothesized: (1) *Fannia* sp. 1 and *Fannia* sp. 2 represent two species which are not known from third instar larvae, (2) *Fannia* sp. 3 represents a species unknown from the female sex and (3) either *Fannia* sp. 1 or *Fannia* sp. 2 is conspecific with *Fannia* sp. 3. To validate our hypotheses and assist in identifying unknown *Fannia* spp. we utilized DNA barcoding. First, we obtained cytochrome oxidase subunit I (COI) barcode sequences, and secondly we compared those against sequences deposited in reference databases.

2. Materials and Methods

2.1. Sampling

Materials for this study were collected during a field study of arthropod succession on pig carrion, conducted in Central Poland and from a human cadaver discovered in Southern Poland. The carrion succession experiment was carried out in spring, summer and autumn of 2012 and 2013. Four carcasses were used in every study season (24 pig carcasses total). Comprehensive description of the experimental design and field protocol of the carrion succession experiment can be found in Jarmusz & Bajerlein 2019 and Jarmusz et al. 2020 [16,29]. Insects have been identified using the keys of Rozkošný et al. [27] and Chillcott [28] for immature stages. For morphological examination, larvae were cleaned with a fine brush and anterior body parts were slide mounted in Hoyer's medium, using concave slides. Images were taken with a Leica DFC450 C digital camera mounted on a Leica DM2500 LED microscope (Leica Camera AG, Germany). Terminology follows Courtney et al. [30] for the general morphology. For family-specific structures, particularly for the processes covering body segments, we are following Lyneborg [12] and Grzywacz et al. [20]. Voucher specimens have been deposited in the collection of the Department of Ecology and Biogeography, Nicolaus Copernicus University in Toruń.

2.2. DNA Extraction, Amplification and Sequencing

Total genomic DNA was isolated from thoracic muscles in females and thoracic and abdominal tissues in larvae using a DNeasy Blood & Tissue Kit (Qiagen, Valencia, CA, USA) following the manufacturer's protocol. Isolated DNA was quantified with a Qubit 3.0 fluorometer using dsDNA High Sensitivity Assay Kit (Life Technologies, Inc., Carlsbad, CA, USA) following the manufacturer's instructions. To obtain the COI barcode region, we used the primers TY-J-1460 and C1-N-2191 [31,32]. We performed a standard 25-µl PCR for each sample using 1 × PCR buffer, 0.2 mM dNTPs, 0.2 µM of each primer, 2 mM MgCl$_2$, 1 U of Taq DNA polymerase (Qiagen) and 1–2 µL of the DNA template. The PCR cycles were as follows: 94 °C for 2 min, 30 cycles 94 °C for 30 s, 50 °C for 30 s and 70 °C for 45 s, followed by a final extension at 70 °C for 10 min.

The PCR products were electrophoresed in a 1% agarose gel, stained with GelRed (Biotium, Darmstadt, Germany) and photographed with a gel documentation system. For sequencing, we only used samples without obvious polymorphisms (multiple bands from a single PCR product). The PCR products were purified with AMPure XP (Beckman Coulter, Carlsbad, CA, USA) (1 × ratio of beads to sample volume). Purified products were re-suspended in TE buffer and the DNA yield was measured using a Qubit 3.0 fluorometer and 2100 Bioanalyzer with the High Sensitivity DNA Analysis kit. Cycle sequencing reactions were carried out using the PCR product (5–20 ng/µL of template DNA) and fluorescent Big Dye terminators (Applied Biosystems, San Francisco, CA, USA). Final products of sequencing were resolved using an automated DNA sequencer at the Laboratory of Molecular Biology Techniques, UAM (Poznań, Poland). Both forward and reverse strands were edited and then assembled using SeqMan II ver. 4.0 (DNASTAR, Lasergene, Madison, WI, USA).

2.3. Sequence Alignment and Data Analysis

Obtained sequences were identified by comparison to sequences available in the NCBI database (National Center for Biotechnology Information, Bethesda, MD, USA) using the Basic Local Alignment Search Tool (BLAST) and BOLD v4 (Barcode of Life Data Systems) using BOLD Identification System [33]. All COI barcode sequences available for *Fannia*, including library of COI reference sequences available for forensically relevant Fanniidae [10], have been downloaded from BOLD database and supplemented with newly obtained data. DNA sequences were aligned using MAFFT v7 [34] and visually inspected and trimmed to a 658-bp long barcode fragment in Seaview 4.4.0 [35]. For the graphical presentation of our data, we performed Neighbor Joining (NJ) phylogenetic analysis using the pairwise distance with 1000 bootstrap replications. We pruned from the NJ analysis sequences not assigned to species and when multiple sequences were available for a species, we used up to 20 randomly selected sequences.

3. Case Report

On 14th October 2016, at noon, the corpse of a woman was discovered in a forest near the DK86 road in Southern Poland. The body was in a state of active decomposition except the head, where signs of advanced decomposition were observed. The body was naked, and clothes were scattered around it. No signs of unnatural death were observed. The deceased was schizophrenic, in the last period of her life she stopped taking medications and showed psychotic symptoms. According to the findings of the investigation, the woman was last seen alive on 10th September. Entomological material was collected during the inspection of the crime scene. Larvae and pupae of Diptera were immediately preserved in 96% ethanol. Entomological material was identified according to keys of Szpila [36], Fremdt et al. [37], Martín-Vega et al. [38] and Rozkošný et al. [27]. Preimaginal stages of the following species were recorded: third instar larvae of *Calliphora vomitoria* (Linnaeus, 1758), third instar larvae and pupae of *Chrysomya albiceps* (Wiedemann, 1819), third instar larvae of *Fannia* sp. 1 and third instar larvae of *Stearibia nigriceps* (Meigen, 1826). The oldest developmental stages noticed on the corpses were pupae of the *Ch. albiceps*. However, the temperatures in the area of the crime scene between when the deceased was last seen alive and the discovery of the corpse, were low (average 11.1 °C) and close to or below the lower developmental threshold reported for this species [39,40]. Using thermal data provided by a prosecutor, we estimated that the minimum time necessary for pupariation of *Ch. albiceps* was considerably longer than the time the deceased person was missing. Analysis of the *C. vomitoria* larvae gave minimum PMI of 17–18 days before the body was found, before 28th September 2016. The investigation was legally discontinued, as no evidence was found that indicated criminal activity had contributed to the woman's death.

4. Results

4.1. Molecular Data and Identification

We successfully amplified the full 658bp COI barcode sequence for two specimens of *Fannia* sp. 1, and two specimens of *Fannia* sp. 3. Due to poor quality material, it was only possible to amplify a short 311 bp fragment for *Fannia* sp. 2. All sequences can be accessed in GenBank through the following accession numbers: MW670438 (Fannia_sp_1_1), MW670439 (Fannia_sp_1_2), MW670440 (Fannia_sp_2), MW670441 (Fannia_sp_3_1) and MW670442 (Fannia_sp_3_2). The newly sequenced COI barcodes were then supplemented with additional 4421 COI barcode sequences, representing 67 *Fannia* species, retrieved from the online databases BOLD. Among those sequences 114 have not been assigned to species or recognized as uncertain identifications.

Comparison of COI barcodes for *Fannia* sp. 1, sp. 2 and sp. 3 with COI data deposited in BOLD and GenBank, enabled identification of these unknown specimens. Sequences obtained for *Fannia* sp. 1 (Fannia_sp_1_1 and Fannia_sp_1_2) were identified as *Fannia nigra* Malloch, 1910, based on 100% similarity with sequences provided from a male (KY511202) and a female (KY511203) of *F. nigra* by Grzywacz et al. [10].

Sequences obtained for *Fannia* sp. 3 (Fannia_sp_3_1 and Fannia_sp_3_2) were identified as *Fannia collini* d'Assis-Fonseca, 1966, based on 99.85% similarity with sequences obtained from male specimens (KY511166, KY511167, KY511168) of *F. collini* by Grzywacz et al. [10].

The sequence obtained for *Fannia* sp. 2 (Fannia_sp_2) was identified as *Fannia pallitibia* (Rondani, 1866) based on 100% similarity with an unpublished sequence assigned to *F. pallitibia* and available in BOLD.

To validate obtained identifications, the COI dataset retrieved from BOLD was supplemented with two additional sequences of *F. pallitibia* (voucher specimens jka09-04927 and KWi-1777 accessible through the Finnish Museum of Natural History (FMNH)) and a newly obtained sequence of *Fannia pruinosa* (Meigen, 1826), a species closely related with *F. pallitibia*. COI barcode sequences, referring to unidentified *Fannia* sp. 1, 2 and 3, were compared against the combined dataset using SpeciesIdentifier v1.8 [41]. This approach confirmed identifications obtained from the BOLD Identification System. Fannia_sp_2 sequence was found 100% similar with *F. pallitibia* (jka09-04927). Additionally, phylogenetic analysis confirmed those identifications (Figure 1), and sequences Fannia_sp_1_1 and Fannia_sp_1_2 clustered with those referring to *F. nigra* (BS = 87%), Fannia_sp_2 with *F. pallitibia* (BS = 100%) and sequences Fannia_sp_3_1 and Fannia_sp_3_2 with *F. collini* (BS = 99%).

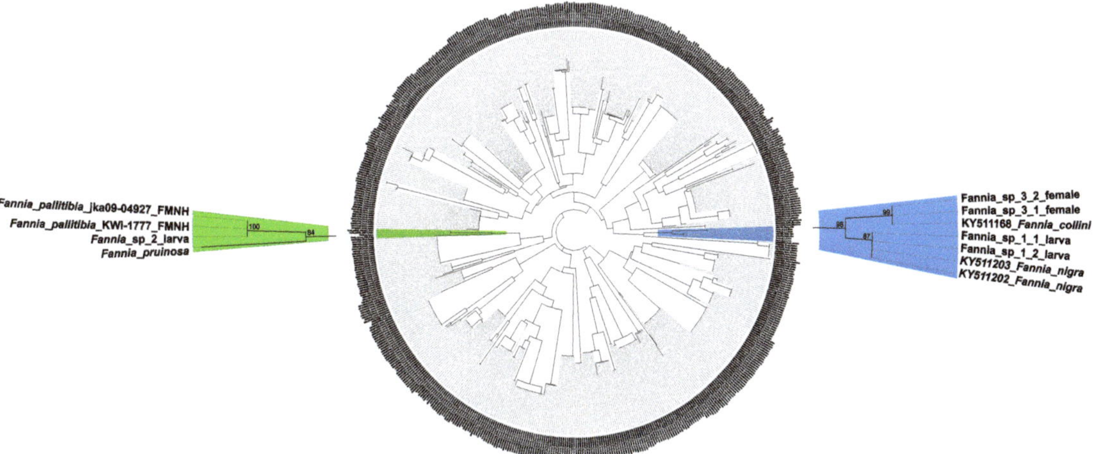

Figure 1. Neighbor joining phylogenetic analysis of 596 COI barcode sequences representing 67 species of *Fannia* retrieved from BOLD and GenBank supplemented with six newly obtained sequences, including three query morpho-species and *Fannia pruinosa*. Values above nodes indicate support for sequence clusters obtained from 1000 bootstrap replications.

4.2. Fannia nigra Malloch, 1910

In Figure 2A,B and Figure 3. Material examined. Third instar larvae from pig carrion: 2, 23 V 2012; 2, 2 VI 2012; 2, 5 VI 2012; 6, 7 VI 2012; 1, 11 VI 2012; 1, 13 VI 2012; 2, 15 VI 2012; 2, 22 VI 2012; 1, 25 VI 2012; 3, 2 VII 2012; 2, 6 VII 2012; 2, 19 VII 2012; 2, 2 VIII 2012; 1, 4 VIII 2012; 1, 7 VIII 2012; 4, 10 VIII 2012; 2, 12 VIII 2012; 2, 23 VIII 2012; 1, 27 VIII 2012; 2, 3 IX 2012; 4, 11 IX 2012; 1, 5 X 2012; 1, 11 X 2012; 2, 14 X 2012; 2, 15 X 2012; 2, 19 X 2012; 10, 23 X 2012; 20, 25 X 2012; 18, 27 X 2012; 19, 3 XI 2012; 36, 7 XI 2012; 6, 11 XI 2012; 34, 15 XI 2012; 7, 26 XI 2012; 17, 3 XII 2012; 1, 26 V 2013; 1, 29 V 2013; 1, 30 V 2013; 2, 1 VI 2013; 1, 2 VI 2013; 10, 3 VI 2013; 2, 4 VI 2013; 2, 5 VI 2013; 13, 6 VI 2013; 6, 7 VI 2013; 7, 9 VI 2013; 11, 11 VI 2013; 4, 13 VI 2013; 5, 15 VI 2013; 1, 17 VI 2013; 25, 21 VI 2013; 157, 26 VI 2013; 42, 30 VI 2013; 24, 4 VII 2013; 2, 11 VII 2013; 1, 18 VII 2013;1, 6 VIII 2013; 6, 9 VIII 2013; 3, 10 VIII 2013; 1, 12 VIII 2013; 1, 14 VIII 2013; 1, 16 VIII 2013; 2, 18 VIII 2013; 4, 20 VIII 2013; 4, 22 VIII 2013; 1, 30 VIII 2013; 14, 4 IX 2013; 3, 11 IX 2013; 5, 18 IX 2013; 2, 22 X 2013; 1, 23 X 2013; 4, 25 X 2013; 4, 28 X 2013; 9, 30 X 2013; 3, 1 XI 2013; 14, 3 XI 2013; 10, 5 XI 2013; 20, 7 XI 2013; 19, 9 XI 2013; 13, 12 XI 2013; 8, 15 XI 2013; 2, 18 XI 2013; 23, 25 XI 2013; 12, 2 XII 2013; 14, 16 XII 2013; 3, 12 II 2014; hornbeam-oak forest, Biedrusko, Poland, M. Jarmusz leg. Third instar larvae from a human cadaver: 8, 14 X 2016; Sarnów near Będzin, Poland.

Third instar larva dorso-ventrally flattened; young larvae creamy-white, mature larvae yellowish to brownish.

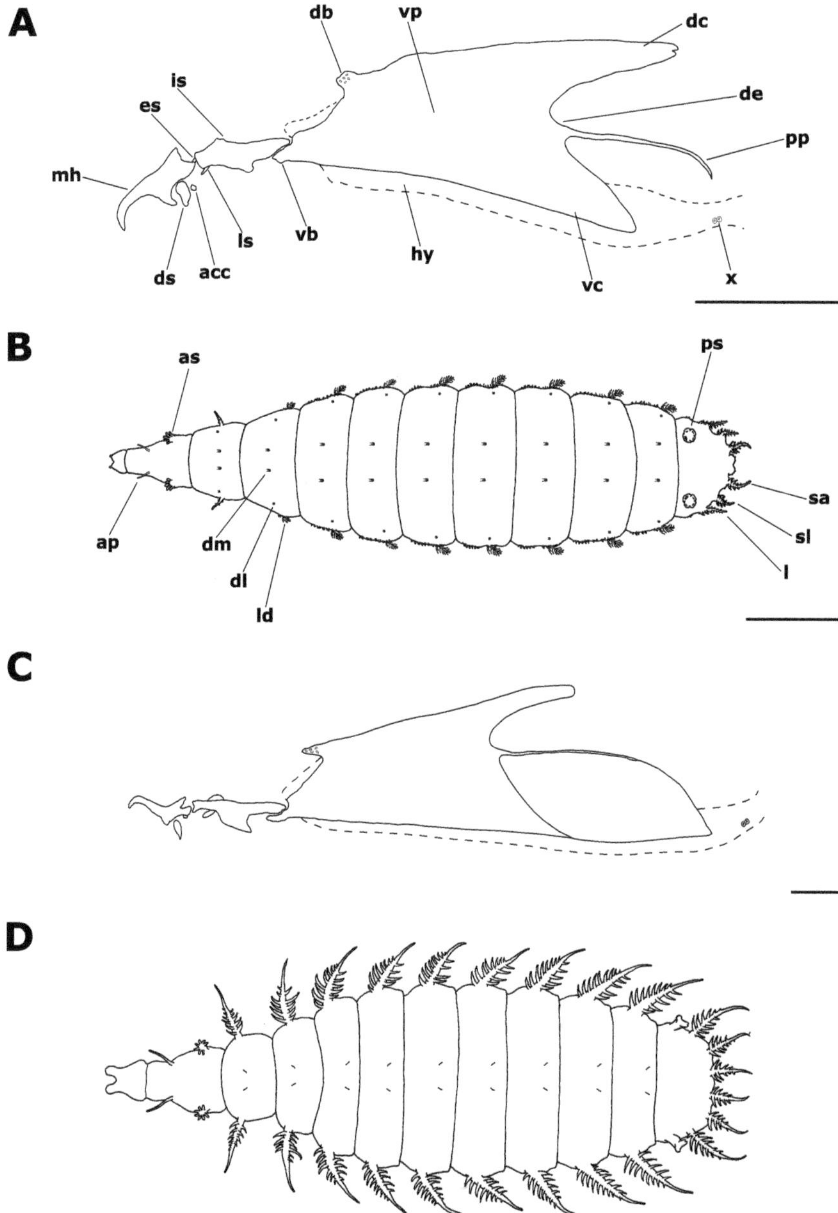

Figure 2. Third instar larvae of *Fannia*: (**A**) cephaloskeleton, *Fannia nigra*; (**B**) larva in dorsal view, *Fannia nigra*; (**C**) cephaloskeleton, *Fannia pallitibia*; (**D**) larva in dorsal view, *Fannia pallitibia*. Scale bare 100 μm (**A**,**C**) and 1 mm (**B**,**D**). Abbreviations: acc, accessory stomal sclerite; ap, anterior process; as, anterior spiracle; db, dorsal bridge; dc, dorsal cornu, de, dorsal extension; dl, dorsolateral process; dm, dorsomedian process; ds, dental sclerite; es, epistomal sclerite; hy, hypopharynx; is, intermediate sclerite; l, lateral process; ld, laterodorsal process; ls, labial sclerite; mh, mouthhook; pp, posterior projection; ps, posterior spiracle; sa, subapical process; sl, sublateral process; vb, ventral bridge; vc, ventral cornu; vp, vertical plate; x, sensory organ X.

Figure 3. Third instar larva of *Fannia nigra*: (**A**) anterior body end with cephaloskeleton; (**B**) seventh abdominal segment, dorsal view; (**C**) posterior body end, dorsal view; (**D**) seventh abdominal segment and anal division, dorsal view; (**E**) larva in dorsal view. Scale bar 100 μm (**A**), 250 μm (**B**–**D**) and 1 mm (**E**). Abbreviations: dl, dorsolateral process; dm, dorsomedian process; l, lateral process; ld, laterodorsal process; sa, subapical process; sl, sublateral process.

Cephaloskeleton. Cephaloskeleton distinctly chitinized (Figure 3A). Mouthhook (*mh*) with apical part downcurved and basal part of *mh* anteriorly equipped with a distinct, ventrolateral extension (Figure 2A). Posterodorsal and posteroventral angles of *mh* drawn out into pointed processes. Paired dental sclerites (*ds*) and accessory stomal sclerites (*acc*) below the basal part of *mh*. Intermediate sclerite (*is*) elongated and H-shaped, with a broad crossbeam. In the lateral view crossbeam visible as a distinct process directed postero-ventrally. Epistomal sclerite (*es*) lies freely between anterior arms of *is*, above the crossbeam; *es* not visible in lateral view. A pair of labial sclerites (*ls*) present below the crossbeam. Basal sclerite (*bs*) consists of very broad vertical plate (*vp*), dorsal cornu (*dc*) and ventral cornu (*vc*). Parastomal bar not developed. Both *vp* connected antero-dorsally by a perforated dorsal bridge (*db*) and antero-ventrally by a ventral bridge (*vb*). Anterior margin of *vp*, between *vb* and *db*, equipped with lighter portion of sclerotisation (possibly additional patch of sclerotisation increasing with larval age). The length of *dc* is about the same or slightly longer than *vc*. Dorsal extension (*de*) of *vc* equipped with well sclerotized and down-curved posterior projection (*pp*). Vertical plate (*vp*) and *vc* are connected below for their entire length by a weakly chitinized hypopharynx, bearing longitudinal ridges. In lower posterior part of *vc*, a sensory organ X (*x*), equipped with paired sensilla is present.

Anterior and posterior spiracles. Anterior spiracles (*as*) fan-shaped, with about 6–9 relatively long lobes, each. Posterior spiracles (*ps*) well separated from lateral margin of anal division (*ad*) (Figure 3C). Posterior spiracles (*ps*) slightly raised on stalks above the surface of *ad* (Figure 3D). Each of the three respiratory slits (*rs*) placed on a short undistinguished finger-like lobe. The plate of each of *ps* without distinct spiracular tufts (*st*), which are probably reduced to a form of single sensillum or complex of trichoid sensilla.

Pattern of processes. Thoracic (T1–3) and abdominal (A1–7) segments equipped with relatively short or inconspicuous projections (Figures 2B and 3E). On T1, two pairs of lateral processes (*lp*), each in form of a cluster of short prominences, present below *as*. Anterior margin of T1 dorsal surface with a pair of forwardly directed, short anterior processes (*ap*). Dorsomedian (*dm*), dorsolateral (*dl*), laterodorsal (*ld*), lateroventral (*lv*), ventrolateral (*vl*) and ventromedian (*vm*) processes present from the second thoracic (T2) to the seventh abdominal (A7) segment. Dorsomedians (*dm*) weakly developed, and present in form of a basal ring equipped with a cluster of minute projections (Figure 3D). Dorsolateral (*dl*) present as a circle of minute projections. Both *dm* and *dl* of similar size, but *dl* devoid of the basal ring. On the thoracic segments *dl* are situated in the anterior part of each segment, between *dm* and *ld*, but when on abdominal segments *dl* are placed on the posterior part of each segment, close to the base of *ld*. The strongest processes on thoracic and abdominal segments are *ld* (Figures 2B and 3B). On T2 *ld* form a process with minute lateral projections. On remaining segments, T3–A7, *ld* in form a short process equipped with lateral projections. On T2 *ld* directed anteriorly, on remaining segments *ld* directed posteriorly. Rows of minute, simple projections precede abdominal *ld* (Figure 3B). Thoracic *lv* form a circle of minute projections placed either on the anterior part of the T2 or in the middle part of the T3. On T3 *lv* followed by a short row of minute projections. Abdominal *lv* present as very short stalks with lateral projection, placed on the posterior part of each segment, and preceded with a row of minute projections. Thoracic *vm* present on the anterior part of segments, and abdominal *vm* on the posterior margin A1–7, all have a form of a pair of circles of minute projections. Second pair of *vm* indistinct. Ventrolaterals (*vl*) in form of small, yet distinct tufts of projections and are well marked on T2–A7 and *ad*.

Sublateral processes (*sl*) and lateral (*l*) processes on anal division of similar length (Figure 3D). Subapical (*sa*) processes appear somewhat, yet not distinctly longer. All three pairs of processes equipped with distinct lateral projections, mostly simple, not bifurcated. Lateral projections are longest in basal half of each process and shorten towards apical part of process.

Integumental sculpture. Thoracic and abdominal segments with a pattern of small, discrete elements, some of which grouped in irregular polygons (Figure 3B). Elements grouped in polygons separated by fine lines, while others widely spaced. Dorsally, integumental pattern is more or less obscured by uniformly smooth, dark areas, arranged in transverse lines. Anterior margin of T3–A7 dorsally covered with smooth polygonal plates arranged in a single row. Posterior margin of T1–A7 dorsally and anterior margin of T1–A1 ventrally, covered with fine, dense, convex elements devoid of projections. Anterior margin of A2–7 with transverse line of scale-like elements ventrally, arranged in groups of at least three elements.

4.3. Fannia pallitibia (Rondani, 1866)

In Figure 2C,D and Figure 4. Material examined. Third instar larvae from pig carrion: 1, 19 VII 2012; 3, 7 XI 2013; hornbeam-oak forest, Biedrusko, Poland, M. Jarmusz leg.

Figure 4. Third instar larva of *Fannia pallitibia*: (**A**) anterior body end with cephaloskeleton; (**B**) sixth abdominal segment, dorsal view; (**C**) posterior body end, dorsal view; (**D**) seventh abdominal segment and anal division, dorsal view; (**E**) larva in dorsal view. Scale bar 100 µm (**A**), 250 µm (**B–D**) and 1 mm (**E**). Abbreviations: ap, anterior process; l, lateral process; ld, laterodorsal process; sa, subapical process; sl, sublateral process.

Third instar larva dorso-ventrally flattened, young larvae creamy-white, mature larvae yellowish to brownish.

Cephaloskeleton. Cephaloskeleton well chitinised, with postero-ventral part of basal sclerite transparent (Figure 4A). Apical part of *mh* directed dorsally, with hook-like distal part (Figure 2C). Basal part of *mh* distinctly bifurcated posteriorly, anteriorly equipped with an indistinct ventrolateral extension. Paired scale-shaped *ds* and minute *acc* present below basal part of *mh*. Intermediate sclerite (*is*) long, yet in general appearance not enlarged, arms relatively narrow. In the lateral view crossbeam of *is* distinctly elongated postero-ventrally. Epistomal sclerite (*es*) well visible in lateral view; a pair of labial sclerites (*ls*) present below the crossbeam. Basal sclerite (*bs*) consists of very broad, vertical plate (*vp*), dorsal cornu (*dc*) and ventral cornu (*vc*). Parastomal bar not developed. Both *vp* connected antero-dorsally by a perforated dorsal bridge (*db*) and antero-ventrally by a ventral bridge (*vb*). The anterior margin of *vp*, between *vb* and *db*, equipped with lighter portion of sclerotisation (possibly additional patch of sclerotisation increasing with larval age). The length of *dc* is the same as *vc*. Dorsal extension (*de*) of *vc* equipped with indistinctly sclerotized, yet well visible, down-curved posterior projection (*pp*). Ventral cornu (*vc*) generally smooth, extends posteriorly with weakly pigmented scale-like element. Upper margin of this element is delimited by *pp* and lower margin by a hypopharynx. Through the entire length pairs of *vp* and *vc* are connected below by a weakly chitinized hypopharynx, bearing longitudinal ridges. In lower posterior part of *vc*, a sensory organ X (*x*), is present and equipped with paired sensilla.

Anterior and posterior spiracles. Anterior spiracles (*as*) small, rounded, equipped with about 6 indistinct lobes. Posterior spiracles (*ps*) well separated from each other, placed on the lateral margin of anal division (*ad*) (Figure 4C). Posterior spiracles raised on distinct, apically bifurcated stalks (Figure 4D). Two respiratory slits (*rs*) placed on single, distinct finger-like lobe, the third *rs*, the outer one, placed on a second finger-like lobe. Spiracular tufts (*st*) indistinct, most likely reduced to a form of single sensillum or complex of trichoid sensilla.

Pattern of processes. Thoracic (T1–3) and abdominal (A1–7) segments with long laterodorsals (*ld*), remaining processes inconspicuous or absent (Figures 2D and 3E). Processes close to anterior spiracles (*as*) not present. Anterior margin of T1 with a pair of forwardly directed, long anterior processes (*ap*). Dorsomedians (*dm*) weakly developed and in form of minute filiform processes. Thoracic *dm* located in the mid part of each segment and directed anteriorly. Abdominal *dm* present on the posterior part of each segment and directed posteriorly and. Dorsolaterals (*dl*) absent or indistinguishable under the stereomicroscope (Figure 4D). Laterodorsals (*ld*) strong. Thoracic *ld* equipped with long, bifurcated lateral projections in the basal half, while abdominal *ld* equipped with long projections in the basal two thirds. Thoracic *ld* not directed posteriorly, abdominal *ld* directed posteriorly. Lateroventrals (*lv*) in a form of a cone, built from a circle of minute projections. First pair of *lv* (on T2) minute and placed close to the base of *ld*. Thoracic *lv* smaller than abdominal *lv*. One pair of ventromedians (*vm*), in form of a circle of minute projections, present on each thoracic segment. Second pair of *vm* indistinct. Each abdominal segment carries two or three pairs of *vm* in form of a sensillum on low callus. Thoracic *vm* present on the anterior part of segments, and abdominal *vm* on the posterior margin of each segment. Thoracic ventrolaterals (*vl*) very small, in form of circles of minute projections. Abdominal *vl* small, yet well visible, present as cones composed from minute projections. Sublateral processes (*sl*), lateral (*l*) and subapical (*sa*) processes long and of similar length (Figures 2D and 4C). Three pairs of processes on anal division equipped with distinct, bifurcated lateral projections in basal half.

Integumental sculpture. Thoracic and abdominal segments with a pattern of small, discrete elements (Figure 4B). Elements on outer margins of thoracic and abdominal segments and on the basal part of laterodorsals (*ld*) wart-like in shape. In the middle part of segment, towards lateral margins, elements are grouped in irregular polygons. Polygons separated by fine lines, while other elements are widely spaced. Dorsally, integumental pattern more or less obscured by uniformly smooth, darker areas (Figure 4C,E). Anterior margin of T3–A7 dorsally covered with smooth polygonal plates arranged in a single row. Posterior margin of T1–A7 dorsally, and anterior margin of T1–A1 ventrally, covered with fine, dense, convex elements devoid of projections.

4.4. Fannia collini d'Assis-Fonseca, 1966

In Figure 5. Material examined. Adult insects: 2♀♀, 29 IV 2012; 1♀, 1 V 2012; 1♀, 6 V 2012; 1♀, 7 V 2012; 1♀, 8 V 2012; 3♀♀, 9 V 2012; 1♀, 13 V 2012; 2♀♀, 14 V 2012; 2♀♀, 15 V 2012; 1♀, 16 V 2012; 2♀♀, 17 V 2012; 5♀♀, 18 V 2012; 9♀♀, 19 V 2012; 5♀♀, 20 V 2012; 5♀♀, 21 V 2012;

1♀, 22 V 2012; 2♀♀, 23 V 2012; 1♀, 24 V 2012; 4♀♀, 26 V 2012; 1♀, 29 V 2012; 1♀, 30 V 2012; 3♀♀, 5 VI 2012; 6♀♀, 7 VI 2012; 6♀♀, 9 VI 2012; 1♀, 13 VI 2012; 3♀♀, 13 VI 2012; 3♀♀, 15 VI 2012; 3♀♀, 28 VI 2012; 1♀, 2 VII 2012; 3♀♀, 6 VII 2012; 2♀♀, 10 VII 2012; 1♀, 22 VII 2012; 1♀, 23 VII 2012; 2♀♀, 24 VII 2012; 3♀♀, 25 VII 2012; 3♀♀, 26 VII 2012; 2♀♀, 27 VII 2012; 1♀, 29 VII 2012; 3♀♀, 31 VII 2012; 3♀♀, 1 VIII 2012; 5♀♀, 2 VIII 2012; 3♀♀, 3 VIII 2012; 3♀♀, 4 VIII 2012; 1♀, 8 VIII 2012; 2♀♀, 10 VIII 2012; 7♀♀, 17 VIII 2012; 1♀, 19 IX 2012; 5♀♀, 20 IX 2012; 2♀♀, 21 IX 2012; 3♀♀, 22 IX 2012; 1♀, 25 IX 2012; 1♀, 27 IX 2012; 1♀, 28 IX 2012; 1♀, 29 IX 2012; 3♀♀, 1 X 2012; 1♀, 2 X 2012; 3♀♀, 8 X 2012; 1♀, 9 X 2012; 1♀, 10 X 2012; 1♀, 14 X 2012; 1♀, 8 V 2013; 1♀, 9 V 2013; 1♀, 15 V 2013; 1♀, 16 V 2013; 1♀, 17 V 2013; 5♀♀, 18 V 2013; 12♀♀, 19 V 2013; 10♀♀, 20 V 2013; 6♀♀, 21 V 2013; 4♀♀, 22 V 2013; 1♀, 23 V 2013; 6♀♀, 24 V 2013; 7♀♀, 25 V 2013; 1♀, 26 V 2013; 12♀♀, 27 V 2013; 6♀♀, 28 V 2013; 1♀, 29 V 2013; 27♀♀, 30 V 2013; 2♀♀, 31 V 2013; 2♀♀, 1 VI 2013; 3♀♀, 2 VI 2013; 37♀♀, 3 VI 2013; 3♀♀, 4 VI 2013; 1♀, 5 VI 2013; 3♀♀, 6 VI 2013; 11♀♀, 7 VI 2013; 2♀♀, 9 VI 2013; 1♀, 11 VI 2013; 2♀♀, 13 VI 2013; 2♀♀, 31 VII 2013; 1♀, 2 VIII 2013; 1♀, 3 VIII 2013; 3♀♀, 6 VIII 2013; 1♀, 7 VIII 2013; 3♀♀, 8 VIII 2013; 4♀♀, 9 VIII 2013; 1♀, 10 VIII 2013; 1♀, 12 VIII 2013; 1♀, 16 VIII 2013; 1♀, 18 VIII 2013; 1♀, 27 VIII 2013; 1♀, 30 VIII 2013; 1♀, 11 IX 2013; 1♀, 23 IX 2013; 1♀, 25 IX 2013; 2♀♀, 1 X 2013; 2♀♀, 5 X 2013; 10♀♀, 6 X 2013; 1♀, 7 X 2013; 2♀♀, 8 X 2013; 1♀, 10 X 2013; 3♀♀, 16 X 2013; 1♀, 17 X 2013; 3♀♀, 21 X 2013; 2♀♀, 22 X 2013; 3♀♀, 24 X 2013; 1♀, 9 XI 2013; hornbeam-oak forest, Biedrusko, Poland, M. Jarmusz leg.

Figure 5. Female of *Fannia collini*: (**A**) lateral view; (**B**) head in anterior view; (**C**) legs in dorsal aspect. Scale bar 1 mm (**A**) and 500 μm (**B,C**).

Description of the female.

Body length 4.8 mm, wing length 4.2 mm.

Head. Eyes sparsely haired. Palpi uniform, not flattened, dark, slightly longer than proboscis (Figure 5A,B). Genae, parafacials and fronto-orbital plates grey, frontal vitta brownish in frontal view (Figure 5B) and black in dorsal aspect. Parafacials narrow and bare. Frontal vitta about as broad as fronto-orbital plates. Proboscis dusted. Three pairs of strong and long frontal setae with at least three pairs of shorter setae between them; two pairs of orbital setae. Upper postocular setulae uniserial. Lower orbital setae inserted close to outer margin of fronto-orbital plates. Arista dark, sparsely haired.

Thorax. Ground-color black, greyish dusted without distinct stripes (Figure 5A). Proepisternum bare. Presutural acrostichals biserial, postsutural acrostichals triserial, dorsocentrals 2 + 3. Notopleuron without additional short hairs. Two prealar setae, anterior seta stronger and about half as long as posterior notopleural seta. Proepisternum, meron bare. Calypters yellowish, lower one broad, projecting beyond upper one. Scutellum with two pairs of long setae. Spiracles yellowish.

Wings. Basicosta and veins yellowish-brown. Halters whitish. Wings membrane hyaline.

Legs. Generally dark, except fore tibiae which are narrowly pale at base (Figure 5C). Fore femora with a row of short posterodorsal setae. Fore tibia with only preapical setae. Mid femora apically with row of anterior setae and posterior setae. Mid tibia with 1 anterodorsal and 1 posterodorsal setae. Hind femora with 3 long anteroventral setae in apical half. Inner posterior side of hid coxa with seta. Hind tibia with 1 anterodorsal, 2–3 anteroventral, 1 median and 1 preapical dorsal setae.

Abdomen. Uniformly dark, with gray pruinosity (Figure 5A), without distinct spots or median vitta. Segments 1–3 with indistinct brownish coloration in dorsal view.

5. Discussion

5.1. Taxonomy

Using DNA barcoding we were able to assign unknown *Fannia* morpho-types to species and reveal their potential forensic utility. However, accuracy of identification by means of DNA barcoding depends on many factors, such as availability of extensive reference library of gene sequences [42]. Taxonomic representation of Fanniidae in reference databases is uneven, with sampling biased towards common species while many other species are underrepresented or represented only by single sequences [10]. Even though undersampling may increase error rate of DNA barcoding [43], in our previous study we observed very high accuracy in species identification using COI barcode sequences in Fanniidae [10]. A distinct barcoding gap was revealed even for species in which females are indistinguishable, e.g., *F. aequilineata* Ringdahl, 1945 and *F. latipalpis* (Stein, 1892) [10]. The misidentification of voucher specimens may be detrimental for end users of reference libraries [42]. However, the identifications we obtained in this study are considered to be valid, as all query sequences had the highest similarity with sequences obtained from what we consider to be accurately identified voucher specimens.

The present state of knowledge on Fanniidae taxonomy is still incomplete. Even though recent studies have improved our taxonomic understanding of this family [44,45], conspecific specimens belonging to different sexes or developmental stages are yet to be described for many species [27]. This gap in knowledge could have severe implications, for example, in Europe where females of many species remain unknown (i.e., *F. alpina* Pont, 1970, *F. brinae* Albuquerque, 1951, *F. conspecta* Rudzinski, 2003, *F. fasciculata* (Loew, 1873), *F. limbata* (Tiensuu, 1938), *F. pseudonorvegica* d'Assis-Fonseca, 1966, *F. rabdionata* Karl, 1940 and *F. ringdahlana* Collin, 1939). This is of particular concern when conspecific males are known to be attracted to carrion and thus by extension are forensically important (i.e., *F. conspecta*) [2,46]. Because of the incompleteness of Fanniidae taxonomy, researchers and practitioners should utilize molecular techniques such as DNA barcoding to verify the identity of specimens which cannot be identified with current keys and/or are discordant with morphological descriptions. The utility of this approach is exemplified in this study, in which molecular barcoding allowed the identification and description of the female of *F. collini*, a species previously known only from males.

Immature stages are still unknown for the great majority of Fanniidae. More than 50 species of fanniids have been reported from decomposing cadavers worldwide [1,2,10]. However, many of these records are based on only a few adult specimens collected from a single cadaver and as such may represent an accidental occurrence rather than a true association. The presence of immature stages can help to confirm a species association with carrion, but this is reliant on the ability to accurately identify these larval stages. The utility of larvae to confirm species association with carrion was exemplified during our two-year carrion succession experiment. Throughout this experiment only two males and two females of *F. nigra* were collected (Jarmusz unpubl.). However, DNA barcoding facilitated association of many unidentified larvae with adult conspecifics, and thus *F. nigra* has been recognized as regular element of necrophagous fauna [16] and as discussed below, a valuable forensic indicator.

The two distinct larval morpho-types which we have examined can easily be placed in the identification key provided by Rozkošný et al. [27], and thus differentiated from remaining European larvae of Fanniidae. First and foremost, the pattern and structure of abdominal dorsolaterals (*dl*) in *F. nigra* and *F. pallitibia* are unique among larvae of Fanniidae. In *F. nigra* thoracic and abdominal *dl* consist of a minutely projected ring, and in *F. pallitibia dl* are completely absent. It should be noted that it is possible that *dl* are present in *F. pallitibia* but in an extremely reduced form and as such we were unable to distinguished them using a stereomicroscope. Nevertheless, in the remaining *Fannia*, even when thoracic *dl* are weakly developed, abdominal *dl* are present as at least minute stalks with a few projections. Only *Euryomma peregrinum* (Meigen, 1826) lacks abdominal *dl*, yet even this species displays *dl* on T2. In addition to the *dl* pattern, *F. nigra* can be easily differentiated from other Fanniidae by the following combination of characters: uniquely small thoracic and abdominal processes, specifically, minute *dm* and *dl*, very short *ap* and *ld* and the latter with only minute lateral projections. Similarly, *F. pallitibia* can be distinguished from other known larvae by the following combination of characters: posterior spiracles placed on lateral margin of the anal division, *dm* minute and filiform, *dl* absent and all processes on the anal division of the same length and equipped basally with long, bifurcated projections.

Fannia collini was known only from Great Britain [27,47], until recent collections of this species were reported from Czech Republic (a single male specimen collected in 2012 in Bohemia) [46] and Central Poland (more than 20 males collected during carrion succession experiments; first records originating from 2006) [1]. Even though *F. collini* has been considered a rare European fanniid, we currently recognize it as a regular and relatively frequent element of carrion arthropod assemblages in Central Europe. Females of *Fannia collini* run to the couplet *no 49* in the key for European Fanniidae [27] and key out as *Fannia immutica* Collin, 1939. However, females of *Fannia collini* are morphologically discordant with females of *F. immutica*. A single COI barcode sequence was available for *F. immutica* in the online GenBank database (MF874564), and our molecular analysis showed interspecific distance between *F. immutica* and *F. collini* was higher than 12%. Based on literature data [27,28,47] and examination of specimens deposited in the collection of the Natural History Museum (London, UK), we propose the following combination of characters for the separation of these two species in the key of Rozkošný et al. [27]:

49.   Two prealar setae, anterior prealar seta longer than half length of posterior notopleural seta and inserted nearer to suture than to supraalar seta ... 49a

49a.   Hind tibia with 8–10 short anterodorsal setae in addition with 1 strong seta; lower orbital setae in the middle of fronto-orbital plates ... *F. immutica* Collin

-   Hind tibia with 1 strong anterodorsal seta, without additional setae above; lower orbital setae inserted nearer to outer margin of fronto-orbital plates ... *F. collini* d'Assis-Fonseca

-   Only one weak prealar seta inserted nearer to supra-alar seta ... 50

5.2. Forensic Importance

Imprecise identification of material collected during carrion succession experiments gives an incomplete image of necrophagous entomofauna. Comparison of COI barcode sequences against those in depository databases enabled unequivocal identification of fanniids and thus better understanding of the composition of carrion insect assemblages in Central Europe. For this reason, the results we obtained in this study, specifically identifications of *Fannia* sp. 1 as *Fannia nigra*, have already been used in practice and detailed analysis of *F. nigra* association with carrion has been published in Jarmusz et al. [16]. A significant correlation between the appearance time of *F. nigra* third instar larvae and the onset of active decay of carrion decomposition process was revealed by Jarmusz et al. [16]. We confirm this association with data obtained from a case study, that is, third instars of *F. nigra* have been collected from a human body which was in the active stage of decomposition. Recent re-analysis of data provided by Sonet et al. [48] revealed *F. nigra* adult specimen has already been collected from a human cadaver [10], despite the authors not being able to identify it to species level.

Analysis of entomological evidence obtained from the aforementioned casework did not provide unequivocal PMI estimations. According to laboratory observations of Grassberger et al. [39], *Ch. albiceps* larvae in central Europe do not develop below 15 °C. However, Richards et al. [40], after reanalysis of data obtained by Grassberger et al., found that lower developmental threshold for pupariation is $T_0 = 11.65$ °C and for eclosion is $T_0 = 10.10$ °C. Additional data from other studies confirm *Ch. albiceps* is able to develop in temperatures below 15 °C [49,50]. We used the lower development threshold of *Ch. albiceps* identified by Richards et al. [40] to identify the development time, however, our results were unsatisfactory for the prosecutor in the context of last time the deceased person was seen alive. Estimations from *Ch. albiceps* significantly overestimated, and those from *C. vomitoria* underestimated mPMI.

Fannia nigra, was utilized to help refine the calculation of mPMI in this case. According to Jarmusz et al. [16], the appearance time on carrion, (the time until when first specimen of a given taxon was recorded on a carcass) for third instar larvae identified herein as *F. nigra* is 34.8 days in autumn [51]. After analysis of entomological evidence, we conclude that (1) thermal requirements for the development of immature stages of Central European population of *Ch. albiceps* require reinterpretation; (2) the time of cadaver exposure was longer than mPMI estimated from *C. vomitoria* larvae; and (3) PMI estimation obtained from information about *F. nigra* association with carrion were the most congruent with the time the deceased person was last seen alive. In this study we provide, for the first time, morphological data which enable prompt and easy identification of third instars and puparia of *F. nigra* and thus facilitate its broad application as forensic indicator.

Four larvae of *F. pallitibia* have been collected during our carrion succession experiment. As such, *F. pallitibia* was either a rare element of necrophagous fauna or the four larvae were randomly present in the surrounding environment, e.g., in soil beneath the corpse, and their presence was solely an artefact of application of pitfall traps used as part of the collecting protocol. Thus, until future studies investigate this relationship, we refrain from considering *F. pallitibia* as a forensic indicator.

This study revealed, adults of *F. collini* are a regular element of carrion insect assemblages in Central Europe, and in some cases the most abundant fanniid species (data unpublished). As such, we preliminarily assumed immature stages representing *Fannia* sp. 1 or *Fannia* sp. 2 were conspecific with the adult *Fannia* sp. 3 (identified as *F. collini)*, however this was not confirmed. It is possible that, *F. collini* repeatedly visit carrion in adult stage, likely to obtain protein meal, yet its larvae do not develop in decomposing carrion or cadavers. This behavior is commonly observed in other insects [52], e.g., hundreds of adults of *Thricops simplex* (Wiedemann, 1817) and *Pollenia* Robineau-Desvoidy, 1830 have been reported from pig carrion [51,53], however, not a single report of larvae is available for either species. Another, somewhat unlikely, possibility is that larvae of *F. collini* are present on dead bodies, yet despite intensive sampling, they have not been collected. Nevertheless, this study emphasizes the importance of taxonomic skills among forensic entomologists. In particular, a wide knowledge of morphological diversity of adults and immature stages of insects.

6. Conclusions

The present study fills a gap in taxonomy of Fanniidae and our knowledge of morphological diversity of the preimaginal instars of *Fannia*. DNA barcoding enabled the assignment of unidentified larvae and females to species and revealed their potential forensic utility. *Fannia nigra* and *F. collini* appeared to be repeatable elements of arthropod carrion assemblages in Central Europe.

Author Contributions: Conceptualization, A.G.; methodology, A.G.; software, A.G.; formal analysis, A.G. and K.S.; investigation, A.G., M.J., K.W., R.S., N.P.J. and K.S.; resources, A.G., M.J., R.S. and K.S.; data curation, A.G. and M.J.; writing—original draft preparation, A.G.; writing—review and editing, A.G., M.J., K.W., R.S., N.P.J. and K.S.; visualization, A.G.; funding acquisition, A.G. All authors have read and agreed to the published version of the manuscript.

Funding: This project has received funding from the European Union's Horizon 2020 research and innovation programme under grant agreement No 823827 SYNTHESYS+ and the Polish Ministry of Science and Higher Education grant IUVENTUS PLUS (grant no. 0146/IP1/2015/73).

Institutional Review Board Statement: Because only carcasses were used for arthropod succession experiment, this study did not require any approval from the Ethics Committee for Animal Experimentation.

Informed Consent Statement: The prosecutor's office agreed to a scientific study of the presented case and publication of its results.

Data Availability Statement: Newly obtained sequences were deposited in GenBank.

Acknowledgments: We would like to express our appreciation to Jere Kahanpää, who granted us access to unpublished COI barcode sequences obtained at the Finnish Museum of Natural History.

Conflicts of Interest: The authors declare no conflict of interest.

References

1. Grzywacz, A.; Jarmusz, M.; Szpila, K. New and noteworthy records of carrion-visiting *Fannia* Robineau-Desvoidy (Diptera: Fanniidae) of Poland. *Entomol. Fenn.* **2018**, *29*, 169–174. [CrossRef]
2. Grzywacz, A.; Castro, C.P. New records of *Fannia* Robineau-Desvoidy (Diptera: Fanniidae) collected on pig carrion in Portugal with additional data on the distribution of *F. conspecta* Rudzinski. *Entomol. Fenn.* **2012**, *23*, 169–176. [CrossRef]
3. Mądra, A.; Frątczak, K.; Grzywacz, A.; Matuszewski, S. Long-term study of pig carrion entomofauna. *Forensic Sci. Int.* **2015**, *252*, 1–10. [CrossRef]
4. Velásquez, Y.; Martínez-Sánchez, A.; Rojo, S. First record of *Fannia leucosticta* (Meigen) (Diptera: Fanniidae) breeding in human corpses. *Forensic Sci. Int.* **2013**, *229*, 13–15. [CrossRef] [PubMed]
5. Vasconcelos, S.D.; Costa, D.L.; Oliveira, D.L. Entomological evidence in a case of a suicide victim by hanging: First collaboration between entomologists and forensic police in north-eastern Brazil. *Aust. J. Forensic Sci.* **2019**, *51*, 231–239. [CrossRef]
6. Bourel, B.; Tournel, G.; Hédouin, V.; Gosset, D. Entomofauna of buried bodies in northern France. *Int. J. Legal Med.* **2004**, *118*, 215–220. [CrossRef] [PubMed]
7. Mariani, R.; García-Mancuso, R.; Varela, G.L.; Inda, A.M. Entomofauna of a buried body: Study of the exhumation of a human cadaver in Buenos Aires, Argentina. *Forensic Sci. Int.* **2014**, *237*, 19–26. [CrossRef] [PubMed]
8. Benecke, M.; Lessig, R. Child neglect and forensic entomology. *Forensic Sci. Int.* **2001**, *120*, 155–159. [CrossRef]
9. Benecke, M.; Josephi, E.; Zweihoff, R. Neglect of the elderly: Forensic entomology cases and considerations. *Forensic Sci. Int.* **2004**, *146S*, S195–S199. [CrossRef]
10. Grzywacz, A.; Wyborska, D.; Piwczyński, M. DNA barcoding allows identification of European Fanniidae (Diptera) of forensic interest. *Forensic Sci. Int.* **2017**, *278*, 106–114. [CrossRef]
11. Holloway, B.A. Larvae of New Zealand Fanniidae (Diptera: Calyptrata). *N. Z. J. Entomol.* **1984**, *11*, 239–258. [CrossRef]
12. Lyneborg, L. Taxonomy of European *Fannia* larvae (Diptera, Fanniidae). *Staatl. Mus. Nat.* **1970**, *215*, 1–28.
13. Szpila, K.; Grzywacz, A. Larvae of the North American Calyptratae flies of forensic importance. In *Forensic Entomology: The Utility of Arthropods in Legal Investigations*; Byrd, J.H., Tomberlin, J.K., Eds.; CRC Press: Boca Raton, FL, USA, 2020; pp. 531–545. [CrossRef]
14. Grisales, D.; Lecheta, M.C.; Aballay, F.H.; de Carvalho, C.J.B. A key and checklist to the Neotropical forensically important "Little House Flies" (Diptera: Fanniidae). *Zoologia* **2016**, *33*, 1–16. [CrossRef]

15. Fiedler, A.; Halbach, M.; Sinclair, B.; Benecke, M. What is the edge of a forest? A diversity analysis of adult Diptera found on decomposing piglets inside and on the edge of a western German woodland inspired by a courtroom question. *Entomol. Heute* **2008**, *20*, 173–191.

16. Jarmusz, M.; Grzywacz, A.; Bajerlein, D. A comparative study of the entomofauna (Coleoptera, Diptera) associated with hanging and ground pig carcasses in a forest habitat of Poland. *Forensic Sci. Int.* **2020**, *309*, 110212. [CrossRef] [PubMed]

17. Quiroga, N.I.; Domínguez, M.C. A new species of the genus *Fannia* Robineau-Desvoidy (Diptera: Fanniidae) belonging to the canicularis species group, collected on pig carrion in the Yungas of the province of Jujuy, Argentina. *Stud. Neotrop. Fauna Environ.* **2010**, *45*, 95–100. [CrossRef]

18. Faria, L.S.; Paseto, M.L.; Franco, F.T.; Perdigão, V.C.; Capel, G.; Mendes, J. Insects breeding in pig carrion in two environments of a rural area of the state of minas gerais, Brazil. *Neotrop. Entomol.* **2013**, *42*, 216–222. [CrossRef]

19. Bohart, G.E.; Gressitt, J.L. Filth-inhabiting flies of Guam. *Bernice Pauahi Bish. Mus. Bull.* **1951**, *204*, 1–151.

20. Grzywacz, A.; Pape, T.; Szpila, K. Larval morphology of the lesser housefly, *Fannia canicularis*. *Med. Vet. Entomol.* **2012**, *26*, 70–82. [CrossRef]

21. Carmo, R.F.R.; Oliveira, D.L.; Barbosa, T.M.; Soares, T.F.; Souza, J.R.B.; Vasconcelos, S.D. Visitors versus colonizers: An empirical study on the use of vertebrate carcasses by necrophagous Diptera in a rainforest fragment. *Ann. Entomol. Soc. Am.* **2017**, *110*, 492–500. [CrossRef]

22. Pont, A.C. A Revision of Australian Fanniidae (Diptera: Calyptrata). *Aust. J. Zool. Suppl. Ser.* **1977**, *51*, 1–60. [CrossRef]

23. Smith, K.G.V. *A Manual of Forensic Entomology*; British Museum (Natural History): London, UK; Cornell University Press: Ithaca, NY, USA, 1986.

24. Domínguez, M.C.; Pont, A.C. Fanniidae (Insecta: Diptera). *Fauna N. Z.* **2014**, *71*, 1–91.

25. Huckett, H.C. The Muscidae of California: Exclusive of subfamilies Muscinae and Stomoxyinae. *Bull. Calif. Insect Surv.* **1975**, *18*, 1–148.

26. de Carvalho, C.J.B.; Mello-Patiu, C.A. de Key to the adults of the most common forensic species of Diptera in South America. *Rev. Bras. Entomol.* **2008**, *52*, 390–406. [CrossRef]

27. Rozkošný, R.; Gregor, F.; Pont, A.C. The European Fanniidae (Diptera). *Acta Sci. Nat. Brno* **1997**, *31*, 1–80.

28. Chillcott, J.G. A revision of the Nearctic species of Fanniinae (Diptera: Muscidae). *Can. Entomol.* **1960**, *92*, 5–295. [CrossRef]

29. Jarmusz, M.; Bajerlein, D. Decomposition of hanging pig carcasses in a forest habitat of Poland. *Forensic Sci. Int.* **2019**, *300*, 32–42. [CrossRef]

30. Courtney, G.W.; Sinclair, B.J.; Meier, R. Morphology and terminology of Diptera larvae. In *Contributions to a Manual of Palaearctic Diptera (with Special Reference to Flies of Economic Importance)*; Papp, L., Darvas, B., Eds.; Science Herald Press: Budapest, Hungary, 2000; pp. 85–161.

31. Grzywacz, A.; Wallman, J.F.; Piwczyński, M. To be or not to be a valid genus: The systematic position of *Ophyra* R.-D. revised (Diptera: Muscidae). *Syst. Entomol.* **2017**, *42*, 714–723. [CrossRef]

32. Bernasconi, M.V.; Valsangiacomo, C.; Piffaretti, J.C.; Ward, P.I. Phylogenetic relationships among Muscoidea (Diptera: Calyptratae) based on mitochondrial DNA sequences. *Insect Mol. Biol.* **2000**, *9*, 67–74. [CrossRef]

33. Ratnasingham, S.; Hebert, P.D.N. BOLD: The Barcode of Life Data System (http://www.barcodinglife.org). *Mol. Ecol. Notes* **2007**, *7*, 355–364. [CrossRef] [PubMed]

34. Katoh, K.; Standley, D.M. MAFFT multiple sequence alignment software version 7: Improvements in performance and usability. *Mol. Biol. Evol.* **2013**, *30*, 772–780. [CrossRef]

35. Gouy, M.; Guindon, S.; Gascuel, O. SeaView version 4: A multiplatform graphical user interface for sequence alignment and phylogenetic tree building. *Mol. Biol. Evol.* **2010**, *27*, 221–224. [CrossRef]

36. Szpila, K. Key for the Identification of Third Instars of European Blowflies (Diptera: Calliphoridae) of Forensic Importance. In *Current Concepts in Forensic Entomology*; Amendt, J., Campobasso, C., Pietro Goff, M.L., Grassberger, M., Eds.; Springer Netherlands: Dordrecht, The Netherlands, 2010; pp. 43–56. [CrossRef]

37. Fremdt, H.; Szpila, K.; Huijbregts, J.; Lindström, A.; Zehner, R.; Amendt, J. *Lucilia silvarum* Meigen, 1826 (Diptera: Calliphoridae)— A new species of interest for forensic entomology in Europe. *Forensic Sci. Int.* **2012**, *222*, 335–339. [CrossRef]

38. Martín-Vega, D.; Díaz-Aranda, L.M.; Baz, A. The immature stages of the necrophagous fly *Liopiophila varipes* and considerations on the genus Liopiophila (Diptera: Piophilidae). *Dtsch. Entomol. Z.* **2014**, *61*, 37–42. [CrossRef]

39. Grassberger, M.; Friedrich, E.; Reiter, C. The blowfly *Chrysomya albiceps* (Wiedemann) (Diptera: Calliphoridae) as a new forensic indicator in Central Europe. *Int. J. Legal Med.* **2003**, *117*, 75–81. [CrossRef]

40. Richards, C.S.; Paterson, I.D.; Villet, M.H. Estimating the age of immature *Chrysomya albiceps* (Diptera: Calliphoridae), correcting for temperature and geographical latitude. *Int. J. Legal Med.* **2008**, *122*, 271–279. [CrossRef]

41. Meier, R.; Shiyang, K.; Vaidya, G.; Ng, P.K.L. DNA barcoding and taxonomy in Diptera: A tale of high intraspecific variability and low identification success. *Syst. Biol.* **2006**, *55*, 715–728. [CrossRef]

42. Collins, R.A.; Cruickshank, R.H. The seven deadly sins of DNA barcoding. *Mol. Ecol. Resour.* **2013**, *13*, 969–975. [CrossRef]

43. Meyer, C.P.; Paulay, G. DNA Barcoding: Error rates based on comprehensive sampling. *PLoS Biol.* **2005**, *3*, e422. [CrossRef]

44. Kahanpää, J.; Haarto, A. Notes on Fanniidae (Diptera) of Finland, with a description of the female of *Fannia stigi* Rognes, 1982. *Entomol. Fenn.* **2013**, *24*, 179–185. [CrossRef]

45. Zhang, D.; Wang, Q.; Wang, M. Description of females of *Fannia imperatoria* Nishida and *Phaonia vagata* Xue & Wang (Diptera: Muscoidea). *Entomol. Fenn.* **2019**, *22*, 274–278. [CrossRef]
46. Barták, M.; Preisler, J.; Kubík, Š.; Šuláková, H.; Sloup, V. Fanniidae (Diptera): New synonym, new records and an updated key to males of European species of *Fannia*. *Zookeys* **2016**, *593*, 91–115. [CrossRef] [PubMed]
47. d'Assis Fonseca, E.C.M. Diptera Cyclorrhapha Calyptrata Section (b) Muscidae. *Handb. Identif. Br. Insects* **1968**, *10*, 1–118.
48. Sonet, G.; Jordaens, K.; Braet, Y.; Bourguignon, L.; Dupont, E.; Backeljau, T.; De Meyer, M.; Desmyter, S. Utility of GenBank and the Barcode of Life Data Systems (BOLD) for the identification of forensically important Diptera from Belgium and France. *Zookeys* **2013**, *365*, 307–328. [CrossRef] [PubMed]
49. Marchenko, M.I. Medicolegal relevance of cadaver entomofauna for the determination of the time of death. *Forensic Sci. Int.* **2001**, *120*, 89–109. [CrossRef]
50. Shabani Kordshouli, R.; Grzywacz, A.; Akbarzadeh, K.; Azam, K.; AliMohammadi, A.; Ghadi Pasha, M.; Ali Oshaghi, M. Thermal requirements of immature stages of *Chrysomya albiceps* (Diptera: Calliphoridae) as a common forensically important fly. *Sci. Justice* **2021**. [CrossRef]
51. Matuszewski, S.; Bajerlein, D.; Konwerski, S.; Szpila, K. Insect succession and carrion decomposition in selected forests of Central Europe. Part 2: Composition and residency patterns of carrion fauna. *Forensic Sci. Int.* **2010**, *195*, 42–51. [CrossRef] [PubMed]
52. Grzywacz, A.; Hall, M.J.R.; Pape, T.; Szpila, K. Muscidae (Diptera) of forensic importance—An identification key to third instar larvae of the western Palaearctic region and a catalogue of the muscid carrion community. *Int. J. Legal Med.* **2017**, *131*, 855–866. [CrossRef] [PubMed]
53. Szpila, K. Annotated list of blowflies (Diptera: Calliphoridae) recorded during studies of insect succession on large carrion in Poland. *Dipteron* **2017**, *33*, 85–93.

MDPI

Technical Note

Technical Note: Effects of Makeshift Storage in Different Liquors on Larvae of the Blowflies *Calliphora vicina* and *Lucilia sericata* (Diptera: Calliphoridae)

Senta Niederegger 

Department of Forensic Entomology, Institute of Legal Medicine at the University Hospital Jena, 07740 Jena, Germany; senta.niederegger@med.uni-jena.de

Simple Summary: Sometimes, police need to collect fly maggots as evidence. If the proper equipment is not at hand, alternatives might need to be found. This evidence can later be given to a forensic entomologist for further examination. The alternative methods, however, can have unknown effects on the samples. We placed maggots in different alcoholic beverages and measured size changes happening over time to provide experts with such information. Our results show that low alcohol beverages can cause samples to shrink. With knowledge about these specific effects, the samples can still be used for casework in forensic entomology.

Abstract: Unexpected findings of forensically important insects might prompt makeshift storage in alternative liquids if the proper equipment is lacking. The assessment of whether such evidence can still be used and correctly interpreted can be difficult. In this study, the effects of using alcoholic beverages as storing agents for post-feeding larvae of *Calliphora vicina* and *Lucilia sericata* were analyzed. Larvae were killed with boiling water (HWK) or placed alive into four alcoholic liquids: two spirits, vodka and brandy, and two liquors, Jägermeister and peppermint schnapps. Storage effects were documented after one day, nine days, and one month and compared to larvae treated according to guidelines for forensic entomology. Results show that the method of killing larvae is more important than the storing medium. Storage of HWK larvae in high-alcohol/low-sugar spirits had almost negligible effects on both species, while all fresh larvae shrank significantly. High sugar contents of the beverages might additionally lead to shrinkage of larvae.

Keywords: forensic entomology; biological variation; death time estimation; alternative storage; casework

Citation: Niederegger, S. Technical Note: Effects of Makeshift Storage in Different Liquors on Larvae of the Blowflies *Calliphora vicina* and *Lucilia sericata* (Diptera: Calliphoridae). *Insects* **2021**, *12*, 312. https://doi.org/10.3390/insects12040312

Academic Editors: Damien Charabidze and Daniel Martín-Vega

Received: 26 February 2021
Accepted: 29 March 2021
Published: 1 April 2021

1. Introduction

Forensically relevant flies deposit their eggs on cadavers and carrion where their offspring hatch, feed, and develop [1,2]. This process was described and even denominated as a "biological clock" for the estimation of the minimal postmortem interval (mPMI) [3,4].

Oftentimes, human bodies colonized by fly larvae are discovered within a domestic setting. In many instances, investigators might presume natural death. Police personnel, as a precaution, might still want to collect and preserve a number of larvae. Circumstances indicating a crime might become known only after the body was cremated and the apartment cleaned. It can take an unpredictable amount of time for a previously unsuspicious death to change into the result of a crime.

The guidelines of forensic entomology [5], which recommend using boiling water for killing and fixation and ethanol for storage of fly larvae, should always be followed when collecting and storing insect evidence. The best medium for storage in forensic entomology casework is 70–80% ethanol [6], while some even suggest 70–95% ethanol [7]. The effect of numerous other media of known fly larvae killing and preservation methods have been investigated, from formaldehyde [6,8] to Kahle's solution [8] and San Veino [9]. Investigators, however, might not always be able to provide ethanol, other chemicals, or hot water when larvae are unexpectedly found at a scene. In such cases, imaginative caseworkers might turn to materials at hand for emergency storage of larvae. Oftentimes, a range of alcoholic spirits or liquors are present right at the scene or a liquor store is nearby.

The aim of this study was to evaluate whether such evidence can still be interpreted and the results presented in court. To achieve this, post-feeding larvae of two species were treated as recommended in the guidelines, but stored in easily achievable alcoholic liquids, as well as 70% ethanol. Furthermore, living larvae were killed by transferring them directly into the tested liquors. Effects of storage up to one month were investigated.

2. Materials and Methods

Larvae of two forensically important fly species were used in this study: *Calliphora vicina* Robineau-Desvoidy, 1830 and *Lucilia sericata* (Meigen, 1826) from laboratory colonies kept in a climatic chamber at 21 °C with 60% humidity. Larval age was synchronized by 24 h periods, during which flies were offered minced meat for oviposition. Emerging larvae were left undisturbed in a rearing container for six days until most had reached peak length and started migration.

The larvae were extracted from the rearing container and randomly assigned to two treatments and five storing liquids. Fresh larvae were transferred directly into their assigned liquids. HWK (hot-water-killed) larvae were killed by placing not more than 20 individuals at a time in a tea strainer and doused with boiling water for at least 20 s; after dabbing with a paper towel, they were transferred to their assigned liquids. The abbreviation HWK (hot-water-killed) was established for this method of killing before [6] and maintained even though boiling and not just hot water was used.

To investigate the effect of alcoholic beverages as storage agents, popular spirits like vodka (Kaliskaya, 37.5% vol.) and brandy (Chantré, 36% vol.), as well as liquor specialties like Jägermeister (35% vol.), a well-known brand of herbal liquor as representative for locally diverse brands, and peppermint schnapps (18% vol.) due to its low alcohol content, were used. Fresh larvae were also stored in 70% ethanol. A second trial was performed five days later, for a total of 350 larvae (Table 1).

Table 1. Average sizes (±SD [1]) and size ranges of larvae in first measurement (1 day) in mm, controls are HWK larvae stored in 70% ethanol for 1 day.

Sizes	*Calliphora vicina*			*Lucilia sericata*		
	Fresh $n = 100$	**HWK** [2] $n = 80$	**Control** $n = 20$	**Fresh** $n = 75$	**HWK** [2] $n = 60$	**Control** $n = 15$
average	14.7 (±1.2)	15.9 (±1.2)	16.1 (±0.9)	12.7 (±0.9)	14.0 (±1.5)	15.2 (±0.7)
range	12.0–17.01	13.3–18.4	14.5–17.9	10.3–15.1	10.8–16.3	13.6–16.3

[1] SD = standard deviation, [2] HWK = hot-water-killed.

HWK post-feeding larvae of *C. vicina* and *L. sericata* stored in 70% ethanol were used as control and their lengths at the respective time served as reference points for both HWK and fresh larvae.

For each measurement, larvae were extracted from their vials, placed into a petri dish, photographed with a camera and a Zeiss Stereomicroscope Stemi 2000, and put back into their original vials without changing the storing medium. Length of the larvae was measured using Zeiss software in the images yielded from the microscope and documented.

In the first trial, maggots were measured after 2 h, but because at least 25% of the fresh larvae were still alive after this time, the data were disregarded and the measurement was not repeated. Subsequent length measurements were performed after 1 day, 9 days, and 1 month. These periods were deemed reasonable for domestic deaths to potentially change from unsuspicious to ominous.

A nonparametric Mann–Whitney U test was performed to compare samples of different sample sizes to their control group. Statistical analyses were conducted using IBM SPSS Statistics 26.

Storage temperature was not altered from laboratory average temperature of 20 °C, as a study of Richards et al. [10] showed no statistically significant effect of storage temperature between −25 °C and +24 °C in larval length and weight.

3. Results

When living fly larvae were placed directly into their respective storage liquids, all animals squirmed and contracted (Figure 1a). Larvae were still alive for up to three hours in their respective media. Killing with boiling water, on the other hand led to death and instant straightening of the whole body (Figure 1b). The first length measurements were performed after one day (Table 1).

(a)   (b)

Figure 1. Larvae of *Calliphora vicina* after one day in 70% ethanol, (**a**) fresh: larvae placed into liquid while alive, (**b**) HWK: larvae hot-water-killed before placement in liquid.

The first experiment compared lengths of HWK larvae of both species stored in alternative alcoholic liquids to their counterparts treated according to the guidelines for forensic entomology [5] in 70% ethanol (Figure 2 and Table 2).

After one day, sizes of all *C. vicina* larvae in all spirits and liquors were within two percent points compared to controls, and thus not significantly different. Larvae in peppermint schnapps over time remained shorter than the control, but in a statistically nonrelevant range. Samples in Jägermeister shrank most and resulted in significantly shorter larvae than the controls. Lengths of larvae in the two spirits, vodka and brandy, in contrast, slightly increased to more than 100% of the size of control larvae over time, but differences were not statistically significant (Table 2).

Larvae of *L. sericata* in alternative liquids were 6–18% smaller than their controls on day one, and thus significantly different. Over time, larvae stored in brandy expanded the most and their lengths reached 99% of the size of controls. Larvae in vodka reached 97% of the size of controls by the end of the storage time of one month. These differences were no longer statistically significant (Table 2). Sizes of larvae in Jägermeister and peppermint schnapps differed more from controls over time. Lengths of larvae in Jägermeister shrank from 91% on day 1 to 83% of the size of controls after one month. Larvae in peppermint schnapps shrank from 82% to 79% on day nine, but reversed to 81% by one month storage time (Figure 2).

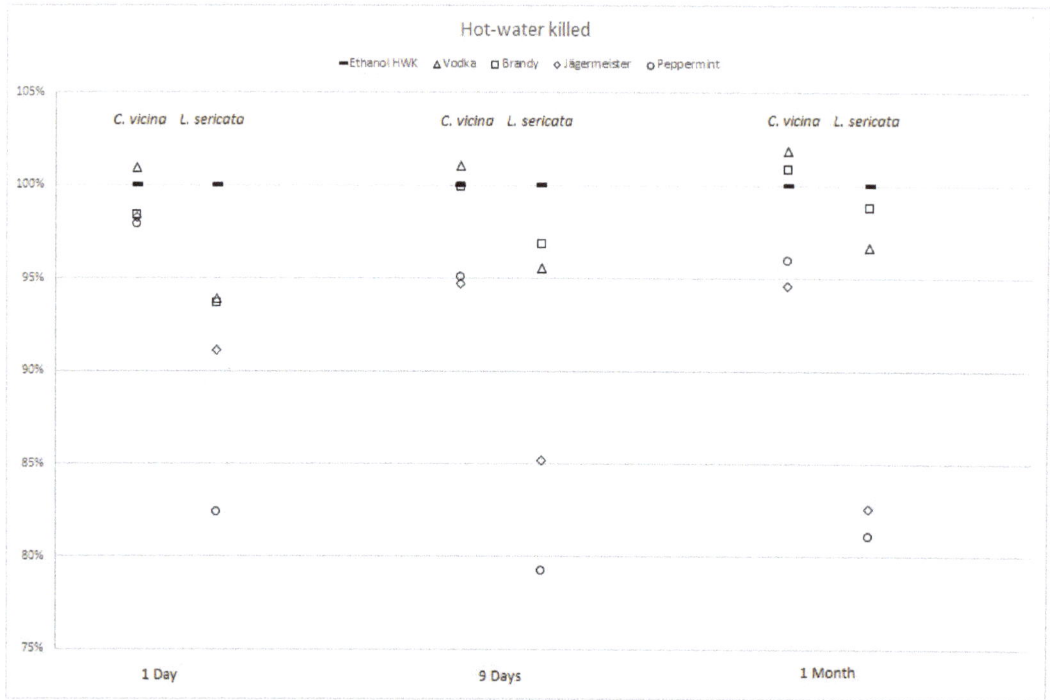

Figure 2. Percentage changes in lengths for *Calliphora vicina* (left column) and *Lucilia sericata* (right column) post-feeding larvae after hot-water-killing (HWK) and storage in four alcoholic liquids for 1 day, 9 days, and 1 month compared to HWK larvae in 70% ethanol (=100%).

Table 2. Average sizes of *Calliphora vicina* and *Lucilia sericata* larvae in mm, standard deviations (±SD), and number of larvae (n) measured after 1 day (1 d) and after 1 month (1 m) in alcoholic liquids, alcoholic content given in % vol., grey fields indicate controls.

Alcoholic Liquid	*Calliphora vicina*				*Lucilia sericata*			
	Fresh		HWK [1]		Fresh		HWK [1]	
	1 d	1 m	1 d	1 m	1 d	1 m	1 d	1 m
Ethanol 70%	12.8 * (±1.2) n = 20	13.8 * (±1.2)	16.3 (±0.9) n = 20	16.1(±0.9)	12.2 * (±0.9) n = 15	12.8 * (±0.8)	15.5 (±0.6) n = 15	15.2 (±0.7)
Vodka 37.5%	13.1 * (±1.0) n = 20	15.2 * (±0.8)	16.5 (±1.2) n = 20	16.4 (±1.1)	11.5 * (±0.6) n = 15	12.8 * (±0.9)	14.6 * (±0.7) n = 15	14.7 (±0.7)
Brandy 36%	14.2 * (±1.8) n = 20	15.3 * (±1.0)	16.0 (±1.2) n = 20	16.3 (±1.2)	12.1 * (±0.8) n = 15	13.2 * (±1.0)	14.5 * (±0.8) n = 15	15.1 (±1.0)
Jägermeister 35%	13.7 * (±1.7) n = 20	14.6 * (±1.1)	16.0 (±1.1) n = 20	15.2 * (±1.1)	11.7 * (±1.0) n = 15	12.3 * (±0.6)	14.1 * (±0.7) n = 15	12.6 * (±1.1)
Peppermint 18%	13.9 * (±1.8) n = 20	14.6 * (±1.3)	16.0 (±1.3) n = 20	15.5 (±1.3)	12.1 * (±1.0) n = 15	12.3 * (±0.9)	12.8 * (±1.0) n = 15	12.4 * (±0.9)

[1] HWK = hot-water-killed, * statistically significant differences compared to controls (Mann–Whitney U test).

The second experiment compared fresh larvae stored in 70% ethanol and alternative liquids without previous hot-water-killing to the same control group of HWK larvae in 70% ethanol (Figure 3).

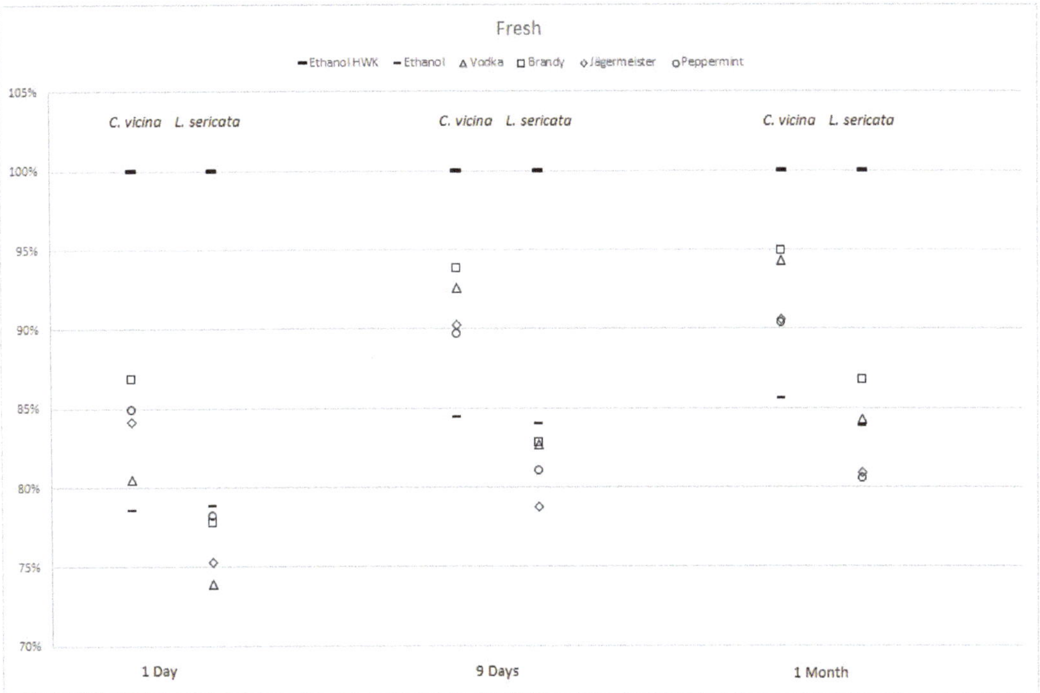

Figure 3. Percentage changes in lengths for *Calliphora vicina* (left columns) and *Lucilia sericata* (right columns) post-feeding larvae stored alive in five alcoholic liquids for 1 day, 9 days, and 1 month compared to hot-water-killed (HWK) larvae in 70% ethanol.

All freshly stored larvae were smaller than their HWK controls and all differences were statistically significant (Table 2). On day one, *C. vicina* larvae freshly stored in ethanol were 22% smaller, and fresh *L. sericata* larvae in ethanol were 21% smaller than their controls. These values represented the lowest discrepancy for *L. sericata*, while it was the highest recorded for *C. vicina*.

While the differences in sizes between fresh larvae of *L. sericata* and their control group were larger than in *C. vicina*, the range of variation between liquids tested was larger in *C. vicina*. Sizes of fresh larvae in all liquids over time approximated the sizes of control larvae for both species, but differences remained significantly different (Table 2).

4. Discussion

The effect of numerous methods of scientific killing and preservation for fly larvae have been investigated [1,6–10] to date. No investigations, however, have been conducted on the effects of improvised methods, such as the use of alcoholic beverages for storage of fly larvae, which might become necessary due to a lack of equipment.

For this study, post-feeding *C. vicina* and *L. sericata* aged six days were used. Numerous aspects concerning the larvae of these two forensically important species are well investigated [11–44]. At constant temperatures of about 20 °C, they develop at very similar rates. Days to maximum length, and thus the beginning of the nonfeeding stage, is given for *C. vicina* as 5.12 days and for *L. sericata* as 5.81 days [45]. This temperature regime, however, is marginally suboptimal for *L. sericata* [46,47], which resulted in slightly diminished oviposition, hatching activity, and sample sizes for this species (Tables 1 and 2).

Adams and Hall [6] noted that the rate of expansion in larvae was highest during the first 3 h in 80% ethanol. In our study, at least 25% of the fresh larvae were still alive after two hours. Placement of living larvae into liquids led to contraction in all samples, which is due to the long drowning and suffocation process. Some larvae were still alive after three hours. The early measurement data could therefore not be incorporated. This resulted in a lack of data on the initial length of larvae tested. There is, to our knowledge, only one method to measure living larvae using a geometrical micrometer [48]. This method, however, is not suited for measurement of 350 mobile larvae in a reasonable amount of time.

The first measurements were performed the next day to ensure the death of all larvae. Subsequent measurements after nine days and one month were conducted to account for potential revelations of new information in casework. It takes at least a few days for a previously unsuspicious domestic death to change into a potential crime due to newly discovered evidence. It can also take up to a few weeks until the samples are sent to a forensic entomologist. Handling of samples was restricted to avoid damages.

Even though the age of larvae was synchronized by limiting oviposition for 24 h, the range in sizes was considerable (Table 1), with maximal sizes consistent with literature [10,34,46]. Biological variation in length within larvae of the same species was larger than the largest experimental difference in average lengths (Table 2). Size ranges are seldom specified in publications on development of forensically relevant insects. The provision of such data should be encouraged in publications on forensic entomology. It could contribute to the awareness of biological variation in entomological samples. Furthermore, analyses should not be based on larval sizes alone, but should incorporate the developmental stage in order to differentiate between an unusually large feeding larva and an unusually small post-feeding individual. Such tasks are especially difficult for nonexpert police personnel. It is therefore important to develop illustrated guidelines designed specifically for nonexperts in order to capacitate police personnel to perform useful insect collections.

The effects of alternative storage over all alcoholic beverages and both treatments were smaller in *C. vicina* than in *L. sericata* (Figure 2 and Table 2). This indicates that *C. vicina* might be a more robust fly species, which is also reflected in a wider temperature tolerance [47].

Comparison of HWK larvae showed that sizes of *C. vicina* larvae stored in the spirits vodka and brandy, as well as the liquor peppermint schnapps, were statistically similar to their controls in all measurements (Figure 2, Table 2). The effects of vodka and brandy were also statistically nonsignificant in HWK larvae of *L. sericata* after nine days and after one month. The European spirits regulation (EU Spirituosenverordnung [49]) defines vodka as a spirit gained from distillation of potatoes or grain. The maximal content of sugar allowed is 8 g per liter, alcohol content must be at least 37.5% vol. Brandy is a liquor produced by distilling wine and the sugar content allowed is no more than 35 g per liter, alcohol content must be at least 36% vol. Herbal liquors like Jägermeister and peppermint schnapps are required to contain at least 80 g sugar per liter and at least 15% vol. of alcohol.

This indicates that sugar content needs as much attention as alcohol content when trying to determine possible effects of storage media on samples of forensically important fly larvae.

The largest effect detected in this study did not arise from the choice of storing media, but treatment of larvae before storage. While HWK larvae of both species stored in vodka and brandy were similar to their HWK controls in 70% ethanol, all fresh larvae were significantly different. Size differences of freshly stored samples to controls diminished over time for both species, where the effect was more pronounced in *C. vicina* larvae (Figure 3). Longer storage times might partly balance initial shrinkage effects caused by missing fixation via HWK. Interestingly, of all storage media, 70% ethanol had the largest effect on fresh larvae of *C. vicina* over all measurements, while it had the least effect of fresh *L. sericata* larvae. This result must be investigated further.

The lengths of most larvae collected and stored in alternative liquids were smaller than those of their HWK controls in 70% ethanol. This indicates a need for cautious estimations of minimal PMIs when interpreting specimen stored in alternative liquids with lower alcohol concentrations. Calculations based on size might underestimate the real development time of inspected larvae stored in media similar to those investigated here.

5. Conclusions

This study shows that storage of larvae in an alcoholic liquid is preferable to omitting collection at all due to lack of proper equipment. Subpar storage methods in weaker liquors than 70% ethanol did not have fatal effects on fly larvae or make them unusable for further examinations. The effect on fresh larvae was significant shrinkage compared to control samples. This can lead to an underestimation of developmental times. For larvae previously HWK on the other hand, spirits with high alcohol and low sugar contents were found to have almost negligible effects. HWK, and thus fixation of the larvae, should therefore always be aimed for when collecting fly larvae. It might help counteract effects of subpar storage media and increase comparability to existing data. It furthermore sterilizes the larvae and reduces degradation during longer storage times.

Exact documentation of treatments and storage liquids is imperative, especially in cases with makeshift methods. Casework in forensic entomology can never incorporate all influencing factors. Even less when working with external samples and/or a small amount of specimen and a lack of supplemental data. Such caveats must always be pointed out when asked specific questions in order to help an investigation. The denomination of insects developing on a human body as a "biological clock" increased expectations for the science of forensic entomology. In reality, the method cannot be as accurate as a clock, as determination of minimal PMI based on insect development must always be an estimation.

Funding: We acknowledge support by the German Research Foundation and the Open Access Publication Fund of the Thueringer Universitaets-und Landesbibliothek Jena Projekt-Nr. 433052568.

Institutional Review Board Statement: Not applicable.

Data Availability Statement: The data presented in this study are available on request from the corresponding author.

Acknowledgments: Thanks go to Roland Spieß for valuable editorial contributions and to Gita Mall for administrative support. Two anonymous reviewers provided constructive comments, which improved the manuscript.

Conflicts of Interest: The author declares no conflict of interest.

References

1. Byrd, J.H.; Tomberlin, J. *Forensic Entomology: The Utility of Arthropods in Legal Investigations*, 3rd ed.; CRC Press: Boca Raton, FL, USA; Taylor & Francis Group: Boca Raton, FL, USA, 2019; p. 620.
2. Gennard, D.E. *Forensic Entomology. An Introduction*; Wiley Chichester: Hoboken, NJ, USA, 2007; p. 223.
3. Catts, E.P.; Goff, M.L. Forensic entomology in criminal investigations. *Annu. Rev. Entomol.* **1992**, *37*, 253–272. [CrossRef] [PubMed]
4. Greenberg, B.; Kunich, J.C. *Entomology and the Law. Flies as Forensic Indicators*; Cambridge University Press: Cambridge, UK, 2002; Volume 1, p. 306.
5. Amendt, J. Best practice in forensic entomology—Standards and guidelines. *Int. J. Leg. Med.* **2007**, *121*, 90–104. [CrossRef] [PubMed]
6. Adams, Z.J.O.; Hall, M.J.R. Methods used for the killing and preservation of blowfly larvae, and their effect on post-mortem larval length. *Forensic Sci. Int.* **2003**, *138*, 50–61. [CrossRef] [PubMed]
7. Bugelli, V.; Campobasso, C.P.; Verhoff, M.A.; Amendt, J. Effects of different storage and measuring methods on larval length values for the blow flies (Diptera: Calliphoridae) *Lucilia sericata* and *Calliphora vicina*. *Sci. Justice* **2017**, *57*, 159–164. [CrossRef] [PubMed]
8. Day, D.M.; Wallman, J.F. Effect of preservative solutions on preservation of *Calliphora augur* and *Lucilia cuprina* larvae (Diptera: Calliphoridae) with implications for post-mortem interval estimates. *Forensic Sci. Int.* **2008**, *179*, 1–10. [CrossRef] [PubMed]
9. Tantawi, T.I.; Greenberg, B. The Effect of Killing and Preservative Solutions on Estimates of Maggot Age in Forensic Cases. *J. Forensic Sci.* **1993**, *38*, 702–707. [CrossRef] [PubMed]
10. Richards, C.S.; Rowlinson, C.C.; Hall, M.J.R. Effects of storage temperature on the change in size of *Calliphora vicina* larvae during preservation in 80% ethanol. *Int. J. Leg. Med.* **2013**, *127*, 231–241. [CrossRef] [PubMed]
11. Smith, K.E.; Wall, R. The use of carrion as breeding sites by the blowfly *Lucilia sericata* and other Calliphoridae. *Med. Vet. Entomol.* **1997**, *11*, 38–44. [CrossRef] [PubMed]
12. Hayes, E.J.; Wall, R.; E Smith, K. Measurement of age and population age structure in the blowfly, *Lucilia sericata* (Meigen) (Diptera: Calliphoridae). *J. Insect Physiol.* **1998**, *44*, 895–901. [CrossRef]

13. Clark, K.; Evans, L.; Wall, R. Growth rates of the blowfly, *Lucilia sericata*, on different body tissues. *Forensic Sci. Int.* **2006**, *156*, 145–149. [CrossRef] [PubMed]

14. Davies, L. Lifetime reproductive output of *Calliphora vicina* and *Lucilia sericata* in outdoor caged and field populations; flight vs. egg production? *Med. Vet. Entomol.* **2006**, *20*, 453–458. [CrossRef] [PubMed]

15. Wooldridge, J.; Scrase, L.; Wall, R. Flight activity of the blowflies, *Calliphora vomitoria* and *Lucilia sericata*, in the dark. *Forensic Sci. Int.* **2007**, *172*, 94–97. [CrossRef] [PubMed]

16. Tarone, A.M.; Foran, D.R. Generalized Additive Models and *Lucilia sericata* Growth: Assessing Confidence Intervals and Error Rates in Forensic Entomology. *J. Forensic Sci.* **2008**, *53*, 942–948. [CrossRef] [PubMed]

17. Gallagher, M.B.; Sandhu, S.; Kimsey, R. Variation in Developmental Time for Geographically Distinct Populations of the Common Green Bottle Fly, *Lucilia sericata* (Meigen). *J. Forensic Sci.* **2010**, *55*, 438–442. [CrossRef] [PubMed]

18. Gunn, A.; Bird, J. The ability of the blowflies *Calliphora vomitoria* (Linnaeus), *Calliphora vicina* (Rob-Desvoidy) and *Lucilia sericata* (Meigen) (Diptera: Calliphoridae) and the muscid flies *Muscina stabulans* (Fallén) and *Muscina prolapsa* (Harris) (Diptera: Muscidae) to colonise buried remains. *Forensic Sci. Int.* **2011**, *207*, 198–204. [CrossRef]

19. Tarone, A.M.; Picard, C.J.; Spiegelman, C.; Foran, D.R. Population and Temperature Effects on *Lucilia sericata* (Diptera: Calliphoridae) Body Size and Minimum Development Time. *J. Med. Entomol.* **2011**, *48*, 1062–1068. [CrossRef] [PubMed]

20. Mai, M.; Amendt, J. Effect of different post-feeding intervals on the total time of development of the blowfly *Lucilia sericata* (Diptera: Calliphoridae). *Forensic Sci. Int.* **2012**, *221*, 65–69. [CrossRef] [PubMed]

21. Roe, A. Development Modeling of *Lucilia sericata* and *Phormia regina* (Diptera: Calliphoridae). Ph.D. Thesis, Nebraska-Lincoln, Lincoln, NE, USA, 2014.

22. Aubernon, C.; Charabidze, D.; Devigne, C.; Delannoy, Y.; Gosset, D. Experimental study of *Lucilia sericata* (Diptera Calliphoridae) larval development on rat cadavers: Effects of climate and chemical contamination. *Forensic Sci. Int.* **2015**, *253*, 125–130. [CrossRef]

23. Roe, A.; Higley, L.G. Development modeling of *Lucilia sericata* (Diptera: Calliphoridae). *Peer J.* **2015**, *3*, e803. [CrossRef] [PubMed]

24. Bernhardt, V.; Pogoda, W.; Verhoff, M.A.; Toennes, S.W.; Amendt, J. Estimating the age of the adult stages of the blow flies *Lucilia sericata* and *Calliphora vicina* (Diptera: Calliphoridae) by means of the cuticular hydrocarbon *n*-pentacosane. *Sci. Justice* **2017**, *57*, 361–365. [CrossRef]

25. Robinson, L.; Bryson, D.; Bulling, M.; Sparks, N.; Wellard, K. Post-feeding activity of *Lucilia sericata* (Diptera: Calliphoridae) on common domestic indoor surfaces and its effect on development. *Forensic Sci. Int.* **2018**, *286*, 177–184. [CrossRef]

26. Dimitrov, R.E.; Amendt, J.; Rothweiler, F.; Zehner, R. Age determination of the adult blow fly *Lucilia sericata* (Diptera: Calliphoridae) through quantitative pteridine fluorescence analysis. *Forensic Sci. Med. Pathol.* **2020**, *16*, 641–648. [CrossRef] [PubMed]

27. Kaib, M. Die Fleisch-und Blumenduftrezeptroen auf der Antenne der Schmeißfliege *Calliphora vicina*. *J. Comp. Physiol.* **1974**, *95*, 105–121. [CrossRef]

28. Reiter, C. Zum Wachstumsverhalten der Maden der blauen Schmeißfliege *Calliphora vicina*. *Int. J. Leg. Med.* **1984**, *91*, 295–308. [CrossRef]

29. Nunes, M.V.; Kenny, N.; Saunders, D. The photoperiodic clock in the blowfly *Calliphora vicina*. *J. Insect Physiol.* **1990**, *36*, 61–67. [CrossRef]

30. Sadler, D.; Fuke, C.; Court, F.; Pounder, D. Drug accumulation and elimination in *Calliphora vicina* larvae. *Forensic Sci. Int.* **1995**, *71*, 191–197. [CrossRef]

31. Saunders, D.S.; Bee, A. Effects of larval crowding on size and fecundity of the blow fly, *Calliphora vicina* (Diptera: Calliphoridae). *Eur. J. Entomol.* **1995**, *92*, 615–622.

32. Faucherre, J.; Cherix, D.; Wyss, C. Behavior of *Calliphora vicina* (Diptera, Calliphoridae) Under Extreme Conditions. *J. Insect Behav.* **1999**, *12*, 687–690. [CrossRef]

33. Kaneshrajah, G.; Turner, B. *Calliphora vicina* larvae grow at different rates on different body tissues. *Int. J. Leg. Med.* **2004**, *118*, 242–244. [CrossRef] [PubMed]

34. Donovan, S.E.; Hall, M.J.R.; Turner, B.D.; Moncrieff, C.B. Larval growth rates of the blowfly, *Calliphora vicina*, over a range of temperatures. *Med. Vet. Entomol.* **2006**, *20*, 106–114. [CrossRef]

35. Spieß, R.; Schoofs, A.; Heinzel, H.-G. Anatomy of the stomatogastric nervous system associated with the foregut in *Drosophila melanogaster* and *Calliphora vicina* third instar larvae. *J. Morphol.* **2008**, *269*, 272–282. [CrossRef]

36. Szpila, K.; Pape, T.; Rusinek, A. Morphology of the first instar of *Calliphora vicina*, *Phormia regina* and *Lucilia illustris* (Diptera, Calliphoridae). *Med. Vet. Entomol.* **2008**, *22*, 16–25. [CrossRef] [PubMed]

37. Martín-Vega, D.; Hall, M.J.R. Estimating the age of *Calliphora vicina* eggs (Diptera: Calliphoridae): Determination of embryonic morphological landmarks and preservation of egg samples. *Int. J. Leg. Med.* **2016**, *130*, 845–854. [CrossRef] [PubMed]

38. Zajac, B.K.; Amendt, J.; Verhoff, M.A.; Zehner, R. Dating Pupae of the Blow Fly *Calliphora vicina* Robineau–Desvoidy 1830 (Diptera: Calliphoridae) for Post Mortem Interval—Estimation: Validation of Molecular Age Markers. *Genes* **2018**, *9*, 153. [CrossRef] [PubMed]

39. Niederegger, S.; Gorb, S.; Jiao, Y. Contact behaviour of tenent setae in attachment pads of the blowfly *Calliphora vicina* (Diptera, Calliphoridae). *J. Comp. Physiol. A* **2002**, *187*, 961–970. [CrossRef] [PubMed]

40. Niederegger, S.; Spieß, R. Cuticular muscle attachment sites as a tool for species determination in blowfly larvae. *Parasitol. Res.* **2012**, *110*, 1903–1909. [CrossRef] [PubMed]

41. Niederegger, S.; Wartenberg, N.; Spiess, R.; Mall, G. Influence of food substrates on the development of the blowflies *Calliphora vicina* and *Calliphora vomitoria* (Diptera, Calliphoridae). *Parasitol. Res.* **2013**, *112*, 2847–2853. [CrossRef]
42. Marchenko, M.I. Medicolegal relevance of cadaver entomofauna for the determination of the time of death. *Forensic Sci. Int.* **2001**, *120*, 89–109. [CrossRef]
43. Smith, K.G.V. *A Manual of Forensic Entomology*; The Trustees of the British Museum (Natural History): London, UK, 1986; Volume 1.
44. Smith, K.E.; Wall, R. Asymmetric competition between larvae of the blowflies *Calliphora vicina* and *Lucilia sericata* in carrion. *Ecol. Entomol.* **1997**, *22*, 468–474. [CrossRef]
45. Davies, L.; Ratcliffe, G.G. Development rates of some pre-adult stages in blowflies with reference to low temperatures. *Med. Vet. Entomol.* **1994**, *8*, 245–254. [CrossRef]
46. Grassberger, M.; Reiter, C. Effect of temperature on *Lucilia sericata* (Diptera: Calliphoridae) development with special reference to the isomegalen- and isomorphen-diagram. *Forensic Sci. Int.* **2001**, *120*, 32–36. [CrossRef]
47. Niederegger, S.; Pastuschek, J.; Mall, G. Preliminary studies of the influence of fluctuating temperatures on the development of various forensically relevant flies. *Forensic Sci. Int.* **2010**, *199*, 72–78. [CrossRef] [PubMed]
48. Villet, M.H. An inexpensive geometrical micrometer for measuring small, live insects quickly without harming them. *Entomol. Exp. Appl.* **2007**, *122*, 279–280. [CrossRef]
49. EUR-Lex, Access to Europen Union Law, Regulation (EU) 2019/787 on the Definition, Description, Presentation and Labelling of Spirit Drinks (EU-Spirituosenverordnung). Available online: https://eur-lex.europa.eu/legal-content/DE/TXT/PDF/?uri=CELEX:32019R0787&from=EN (accessed on 18 March 2021).

Article

Puparia Cleaning Techniques for Forensic and Archaeo-Funerary Studies

Jennifer Pradelli [1], Fabiola Tuccia [1], Giorgia Giordani [2] and Stefano Vanin [3,4,*]

1   School of Applied Sciences, University of Huddersfield, Queensgate, Huddersfield HD1 3DH, UK; jennifer.pradelli@gmail.com (J.P.); tuccia.fabiola@gmail.com (F.T.)

2   Dipartimento di Farmacia e Biotecnologie (FABIT), Alma Mater Studiorum Università di Bologna, 40126 Bologna, Italy; giorgia.giordani.gg@gmail.com

3   Dipartimento di Scienze della Terra dell'Ambiente e della Vita (DISTAV), Università di Genova, Corso Europa 26, 16132 Genova, Italy

4   National Research Council, Institute for the Study of Anthropic Impact and Sustainability in the Marine Environment (CNR-IAS), Via de Marini 6, 16149 Genova, Italy

*   Correspondence: stefano.vanin@unige.it or stefano.vanin@gmail.com; Tel.: +39-3481360682

Simple Summary: In forensic entomology, the correct identification of the species colonizing a body is fundamental. In old cases, puparia of Diptera represent the only entomological evidence available. Their identification is made particularly difficult not only because the lack of identification keys, but also because the presence on their surface of elements (dust, soil, dry decomposition fluids, bacteria, etc.) that can cover the diagnostic characters. Because of their fragility and the low amount of DNA, six cleaning techniques based on chemical and physical treatments have been tested. The results of this study indicate that cleaning via warm water/soap, the sonication and treatment with a sodium hydroxide solution are the best methods to achieve a good quality of the samples.

Abstract: Diptera puparia may represent both in forensic and archaeo-funerary contexts the majority of the entomological evidence useful to reconstruct the peri and post-mortem events. Puparia identification is quite difficult due to the lack of identification keys and descriptions. In addition, external substances accumulated during the puparia permanence in the environment make the visualization of the few diagnostic characters difficult, resulting in a wrong identification. Six different techniques based on physical and chemical treatments have been tested for the removal of external substances from puparia to make identification at species level feasible. Furthermore, the effects of these methods on successful molecular analyses have also been tested as molecular identification is becoming an important tool to complement morphological identifications. The results of this study indicate that cleaning via warm water/soap, the sonication and treatment with a sodium hydroxide solution are the best methods to achieve a good quality of the samples.

Keywords: Diptera; identification; forensic entomology; funerary archaeoentomology

Citation: Pradelli, J.; Tuccia, F.; Giordani, G.; Vanin, S. Puparia Cleaning Techniques for Forensic and Archaeo-Funerary Studies. *Insects* **2021**, *12*, 104. https://doi.org/10.3390/insects12020104

Received: 31 December 2020
Accepted: 21 January 2021
Published: 26 January 2021

Publisher's Note: MDPI stays neutral with regard to jurisdictional claims in published maps and institutional affiliations.

1. Introduction

One of the most important taxa involved in the decomposition processes of animal organic matter is Diptera [1]. Flies, belonging to Calliphoridae, Sarcophagidae, and Muscidae families, are particularly important in legal investigation being the first colonisers of a body after death [2]. In medico-legal forensic entomology, the estimation of the minimum post-mortem interval (minPMI) and other evaluations about the relocation and/or concealing of a body, are possible only after an accurate identification at species level [3]. In fact, Diptera development, and more generally insect development, is temperature dependent but species specific, very often population related [4]. Moreover, the habitat preferendum, phenology, digging attitude, chronobiology, and distribution that represent the knowledge used to answer the investigative questions depend on the species [5–8].

Being associated with human decomposition Diptera are also commonly recovered from archaeological excavations. The study of insects associated with ancient human remains, such as natural and anthropogenic mummies or graves has been defined as Funerary Archaeoentomology [9]. This discipline provides information not only about thanatology (the scientific study of death and all the body modifications that happen after it), but also about funerary practices [10], season of death, social habits, and hygiene and health condition of past populations [11]. Also, in this case, any evaluation and interpretation of the archeological hypothesis needs a correct identification of the species [12].

Flies are holometabolous insect. Generally, the life cycle of flies includes egg, larval, pupal and adult stages. During the pupariation process, the larval cuticle goes through a series of chemical and physical changes, with the final formation of a hard case known as "puparium" [13]. It acts as a protective case in which metamorphosis takes place. After the adult emergence, the puparium is left empty on the site. Due to its high resistance to decay, puparia can be found in crime scenes [14,15], and they are particularly important in old cases when other developmental stages are no longer present. Moreover, these structures can be found even in archaeological contexts, where they might be the only traces of insect activity left after centuries or millennia [11,16–18].

The morphological identification of puparia is challenging due to the presence of a few diagnostic features on their outer surface. Most of the distinctive features are found in the posterior region, such as posterior spiracles and anal plate, and, on the ventral side of abdominal segment number 7, such as the size, shape, and distribution of spiculae [19]. It is worth mentioning that oral sclerites can be analysed from a puparium but often, especially in archaeological contexts, they are no longer present [2]. Puparia, depending on the context and on their conservation, are very often coated by external substances, like dust, decomposed fluids, dirt, fibres and soil debris which might cover and hide the above-mentioned diagnostic characters making difficult, if not impossible, a correct identification of the specimens [20].

During the past decades, several methods and techniques of insect cleaning, designed especially for adult beetles belonging to museum collections or for immature stages prepared for scanning electron microscopy (SEM) observation, have been described [21–26]. In literature, cleaning techniques are categorised into two main groups: methods based on a mechanical removal of the dirt particles, and methods based on a soaking system using different solvents [25]. The selection of which method is the most suitable for a specific specimen is strictly related to the state of insects' conservation (how fragile the specimen is, the developmental stage considered, how old the sample is, etc.) and to the chemical and physical nature of the substance deemed to cover it. In principle, in order to be correctly identified, specimens have to preserve all the distinctive features after the cleaning treatment. Therefore, avoiding any damage to the sample is a priority. In practice, all methods and techniques affect the state of preservation of specimens, both molecularly and morphologically, although the extent of these effects can vary significantly based on the amount of time each sample is processed. Thus, it is important to balance the efficiency in processing entomological samples. In addition, because of the more and more common application of DNA techniques for species identification [27], also used for puparia identification [28], cleaning processes should not interfere with the DNA extractability and integrity.

In order to better understand which cleaning technique was the most suitable to be used for puparia, six chemical-physical methods have been tested. Two different experiments have been performed. The first one aimed to evaluate the efficiency of each method in removing external substances, improving the visual assessment of diagnostic features. The second one was designed to investigate the compatibility of each cleaning technique with potential molecular identification. Procedure guidelines are presented and tooltips for each method are listed.

2. Materials and Methods

Six methods were selected from the literature according to their ability to dissolve or remove specific substances (Table 1). Costs and availability of solutions were considered to select methods affordable by a standard laboratory. Diptera puparia in the families Calliphoridae, Sarcophagidae, Muscidae, and Sphaeroceridae, from forensic and several archaeological contexts were selected from the FLEA collection (Forensic Lab for Entomology and Archaeology based till 2019 at the University of Huddersfield (UK) and now at the University of Genoa (Italy)), and used for this study. All the specimens (ranging from 0.3 to 1 cm depending on the family) were visibly covered by substances of an unknown chemical composition deriving from the context of origin and therefore likely including non-insect organic material, botanical and soil residuals and other debris. As a result, the original appearance of the specimens was concealed. After a preliminary qualitative evaluation under a stereomicroscope (Leica M60, Leica, Wetzlar, Germany), the most adequate method according to the substances present was applied to the puparia (5–8 puparia tested for each cleaning method).

Table 1. Cleaning methods tested with reference and target substances.

Method	Suitable for	References
Warm Water and Soap solution (WH$_2$O)	Fibres Dust Sludge	[16]
Sonication (SON)	Dross Soil debris Sand Botanical residues	[29]
Glacial Acetic Acid (GAA)	Inorganic crystals	[30]
Sodium Hydroxide solution (NaOH)	Putrefactive liquids Any organic matter	[31,32]
Hydrochloric Acid/Sodium Bicarbonate (ZAN)	Oily substances Grease	[30]
Sodium Hypochlorite (BL)	Organic matter Bacteria Mould/Fungi	[33,34]

A pictorial archive of specimens before and after treatments was created using a Keyence VHX-S90BE digital microscope, equipped with Keyence VH-Z250R and VH-Z20R lens and VHX-2000 Ver.2.2.3.2 software (Keyence, Osaka, Japan).

(a) Warm water and soap solution: puparia were soaked in a solution of warm water (~60 °C) and commercial dish soap (depending on the brand of dish soap, component percentages might vary: sodium linear dodecyl benzene sulfonate, sodium lauryl alcohol triethoxy sulfate, lauric/myristic monoethanolamide, hydrotrope mixture, magnesium sulfate, colorant, petrolatum, perfume, ethyl alcohol 95%, deionized water) for approximately 10–30 min depending on the substances attached on their surface, and then they were wiped with paintbrushes. The processed samples were then rinsed with deionized water and air dried.

(b) Sonication: puparia were placed separately inside vials filled with deionized water and then individually sonicated between 5 and 15 s, depending on the preservation status, using a sonicator bath (QH Kerry Ultrasonic Limited, f = 50 Hz). The samples were rinsed with clean deionized water and air dried.

(c) Glacial acetic acid: puparia were gently wiped with a paintbrush soaked in glacial acetic acid (CH$_3$COOH) or totally immersed in the acid for 5 min. They were then rinsed several times with deionized water, in order to stop the chemical reactions, and air dried afterwards.

(d) Sodium hydroxide solution: puparia were immersed in sodium hydroxide (NaOH) 10% solution either for 5 or for 10 min. The solution was prepared by adding sodium hydroxide solid crystal to water. The samples were then washed gently in running deionized water to stop the chemical reaction, and air dried.

(e) Hydrochloric acid/sodium bicarbonate: this method was described by Zangheri [30] in order to clean Coleoptera from museum collections. The method combines several different solutions in a pre-set order. Puparia were first immersed in distilled water for 24 h, and then placed in a clean vial with hydrochloric acid (HCl) for 10 min and soaked in a saturated solution of sodium bicarbonate (NaHCO$_3$) for 15 min immediately afterwards. Finally, puparia were wiped with paintbrushes. The samples were washed with deionized water and air dried, prior to being microscopically observed.

(f) Sodium hypochlorite: puparia were soaked for 5 and 10 min in a 5% solution of sodium hypochlorite (NaOCl). The specimens were washed under deionized running water and air dried before identification.

All the molecular analyses were performed on modern puparia of *Lucilia sericata* (Meigen, 1826) obtained from a breeding colony at the University of Huddersfield (UK). Puparia were subjected to two different treatments prior to DNA extraction. The first batch of puparia underwent previously described cleaning procedures immediately after the adults' emergence; straight after DNA extractions had been performed (as a control, three puparia without any cleaning treatment were selected). The second batch, after adults' emergence, was placed in small pierced plastic boxes containing a mixture of decontaminated horse blood, cat food, and ground soil, mimicking the conditions of thanatocoenosis and taphocoenosis. The containers were closed and stored inside the laboratory at room temperature. After seven days of incubation inside the mixture, the six cleaning techniques were applied to the puparia (as a control, six puparia were selected, three not placed in the mixture and not cleaned with any methods and three placed in the mixture but not cleaned with any techniques). In addition, further sodium hydroxide concentrations (saturated and 1%) were tested.

All DNA extractions were performed in triplicate on a single empty puparium using the QIAamp DNA Investigator Kit (QIAGEN, Redwood City, CA, USA). The manufacturer's protocol was followed, and slightly modified by additional use of Proteinase K (100 µg/mL) from PROMEGA (Madison, WI, USA). Elution was performed with 50 µL of Buffer ATE. Quantification was performed using a Qubit® 3.0 Fluorometer (Thermo Scientific, Waltham, MA, USA). Universal LCO-1490 Forward primer (5′ GGTCAACAAATCATAAAGATATTGG-3′) and HCO-2198 Reverse primer (5′-TAAACTTCAGGG TGACCAAAAAATCA-3′) were used [35,36] to amplify the mitochondrial COI gene (658 bp long) using Polymerase Chain Reaction (PCR). Master-mix reactions of 20 µL final volume were prepared following the PROMEGA GoTaq® Flexi Polymerase protocol, which included Colourless GoTaq Flexi Buffer (5×), MgCl2 (25 mM), primers (IDT) (10 pmol/µL), Nucleotide Mix (10 mM), GoTaq DNA Polymerase (5 u/µL) and 2–4 µL of DNA template. The amplification programme (initial heat activation step at 95 °C for 10 min, 35 cycles of 95 °C for 1 min, 49.8 °C for 1 min, 72 °C for 1 min, and a final extension step at 72 °C for 10 min) was set up on a BioRad C1000 Thermal Cycler (Bio-Rad Laboratories, Inc., Hercules, CA, USA). A standard gel electrophoresis, in 1.5% agarose gel stained with Midori Green Advanced DNA Stain (Geneflow, Elmhurst, UK), was used to check PCR products. In case of positive results, 15 µL of PCR products were purified using QIAquick PCR Purification Kit® (QIAGEN, Hilden, Germany) following the manufacturer's instructions. Purified amplicons were sequenced by Eurofins (Eurofins Operon MWG, Ebersberg, Germany) following the standard Sanger method. For species identification purposes, DNA sequences were searched on GenBank database through BLASTn® tool (NCBI, Bethesda, MD, USA).

3. Results

All the methods selected impacted the external appearances differently and led to various amplification results. Most of them removed the bulk of external substances after the first cleaning attempt. Examples of before and after-treatments are shown in Figure 1.

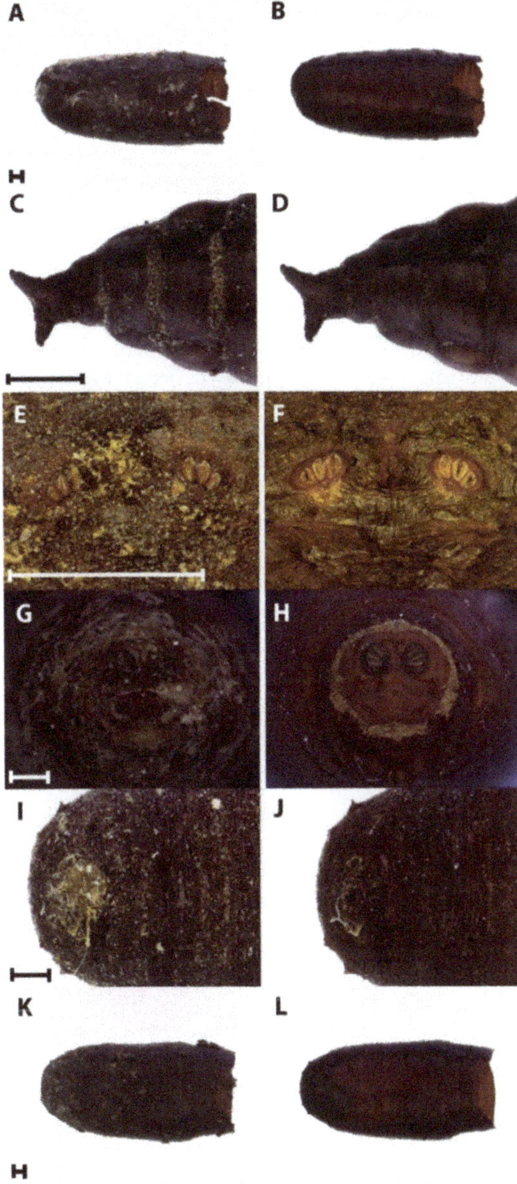

Figure 1. Puparia before and after treatment: (**A,B**) Water and soap solution, (**C,D**) Glacial Acetic Acid, (**E,F**) Sonication; (**G,H**) Sodium hydroxide 10% solution, (**I,J**) Hydrochloric acid/sodium bicarbonates solutions, (**K,L**) Sodium hypochlorite 1–5% solution. Scale bars: 500 μm.

Despite the excellent visual results, it is worth mentioning that it is not always possible to achieve a totally cleaned puparium surface, due to the nature of the substances covering the puparia, which are a heterogeneous mixture. However, in the majority of the cases, specimens result cleaned enough to make visible the diagnostic features and allowing their identification. Different types of substances can simultaneously cover the external surface of a puparium. Hence, according to the composition of each substance, it may be necessary to use more than one method or to perform the same cleaning method several times to obtain a perfectly clean surface. Even though it may seem reasonable and fair to proceed until reaching the highest level of cleanliness, those multiple and/or combined methods affect the structure of the puparium. A brief qualitative analysis and tool tips for each method are listed below.

(a) Warm water and soap solution: This is the most affordable and the most effective method. The permanence of the puparia in warm water and soap can be prolonged as long as the operator is aware of the positive correlation between time and softness. This means that, during the final brushing, the operator needs to pay attention not to crush the puparium, which becomes more fragile.

(b) Sonication: This method is particularly effective on encrusted debris. It works also on desiccated muddy or sludgy material, but in those cases, it is a time-consuming process. In fact, desiccated material, once rehydrated, usually stains the water inside the vial, not allowing a precise check on the status of the specimens treated. A multiple and/or prolonged sonication can widen the cracks present naturally on the puparia after the eclosion of the adults. In worst cases, posterior spiracles and anal plate can be ripped out from the puparium by the vibration, with the consequent loss of identification features. It is suggested, especially on archaeological samples, to check carefully the conservation status (presence of cracks on the surface) of each specimen prior to treat them with sonication. The more cracks are present, the less time the specimens should be kept in the sonication bath.

(c) Glacial acetic acid: This method is effective at dissolving inorganic crystals. Commonly used by coleopterists, it was not previously tested on dipterous puparia. Due to its corrosive nature, low quantities and several rinsing steps are suggested. Some archaeological samples have to be evaluated closely before using acetic acid. In specific cases, due to the process of per-mineralisation (fossilisation process, during which mineral deposit creates a cast of the organism), a total or partial substitution of the organic matter can happen to the pupae. In these cases, acetic acid can destroy the sample totally.

(d) Sodium hydroxide solution: The solution is very effective on samples covered by organic substances such as putrefactive liquids. This method is also commonly used to diaphanise larvae for slide microscopy [37].

(e) Hydrochloric acid/sodium bicarbonate: It is the most time-consuming method as it involves an initial 24-h immersion in water. It is also the least effective of all methods, usually leaving a thin residue layer behind. Hence, additional cleaning with one of the other methods is also required.

(f) Sodium hypochlorite: It is a common chemical and easily present in an entomological laboratory. It is known to disinfect and to react with many natural pigments. However, the solution is not particularly effective as a cleaning solution and, as a minor result, it decolours the specimens.

In term of DNA extractability, the results of the first group of puparia treated with cleaning procedures immediately after the adults' emergence are presented in Table 2. DNA was positively extracted from the controls, from all the samples that were immersed in sodium hydroxide 10% solution for 5 and 10 min, from sonicated samples, from samples washed with water and soap, and from samples brushed with glacial acetic acid. However, DNA extraction failed when samples were immersed in glacial acetic acid, in bleach for 5 and 10 min, and with samples treated with the combination of hydrochloric acid and sodium bicarbonate solutions.

Table 2. Quantifications of DNA (average ± stdev) extracted from the first group of puparia cleaned immediately after adults' emergence (CNTRL = control; NaOH = sodium hydroxide; SON = sonication; GAA = glacial acetic acid; WH$_2$O = warm water and soap; ZAN = hydrochloric acid/sodium bicarbonate solutions; BL = bleach; ✓ positive results for the expected fragment, ✗ negative results for the expected fragment).

Samples	DNA (ng/μL)	PCR
CNTRL	0.427 ± 0.26	✓
NaOH 10%, 5′	<0.001	✓
NaOH 10%, 10′	<0.001	✓
SON	0.045 ± 0.034	✓
GAA immersed	0.408 ± 0.109	✗
GAA paintbrush	<0.001	✓
WH$_2$O	<0.001	✓
ZAN	<0.001	✗
BL 5′	<0.001	✗
BL 10′	<0.001	✗

Results of the second group of puparia, which were placed in a mixture of decontaminated horse blood, cat food, and ground soil for a week and then cleaned, are presented in Table 3. DNA was extracted from all the controls: the three control puparia who were not placed in the mixture and not cleaned with any methods (CNTRL1) gave good quality PCR amplification; in contrast, the three control puparia placed in the mixture but not cleaned with any techniques (CTRLN 2) did not show any amplification. PCR was successful from samples immersed for 5 and 10 min in sodium hydroxide solutions (1%, 10% and saturated), from sonication, and from the samples brushed with glacial acetic acid. One sample washed in warm water/soap solution, and one sample immersed in glacial acetic acid, also showed positive results. The rest of the puparia cleaned with warm water/soap solution, immersed in glacial acetic acid and bleach, and treated with hydrochloric acid/sodium bicarbonate solutions did not show any positive results.

Table 3. Quantifications of DNA (average ± stdev) extracted from the second group of puparia placed in the mixture for a week and then cleaned (CNTRL1 = puparia not placed in the mixture and not cleaned; CNTRL2 = puparia placed in the mixture and not cleaned; NaOH = sodium hydroxide; SON = sonication; GAA = glacial acetic acid; WH$_2$O = water/soap; ZAN = hydrochloric acid/sodium bicarbonate solutions; BL = bleach; ✓ positive results for the expected fragment, ✗ negative results for the expected fragment).

Samples	DNA (ng/μL)	PCR
CNTRL_1	0.871 ± 0.219	✓
CNTRL_2	4.251 ± 1.988	✗
NaOH 10%, 5′	<0.001	✓
NaOH 10%, 10′	<0.001	✓
NaOH 1%, 5′	<0.001	✓
NaOH 1%, 10′	<0.001	✓
NaOH sat, 5′	<0.001	✓
NaOH sat, 10′	<0.001	✓
SON	0.053 ± 0.081	✓
GAA immersed	0.345 ± 0.373	✗
GAA paintbrush	<0.001	✓
WH$_2$O	<0.001	✗
ZAN	<0.001	✗
BL 5′	<0.001	✗
BL 10′	<0.001	✗

The mean quantification of both puparia groups was quite low, in agreement with the scarce availability of tissue in a single puparium suitable to extract nucleic acids from. In fact, most of the samples showed concentrations below the detectable threshold of the Qubit 3.0 fluorometer (0.001 ng/µL). However, DNA was successfully amplified, and the fragments successfully sequenced allowing to confirm the identification of all the specimens as *Lucilia sericata*.

4. Conclusions

All six methods selected successfully cleaned the puparia. However, if morphological and molecular analyses are taken into account together, the best methods, with positive results in both analyses, were the warm water/soap, sonication and sodium hydroxide solutions. Hydrochloric acid/sodium bicarbonate solutions, bleach, and glacial acetic acid immersion are, therefore, not recommended to clean entomological samples if molecular analysis is intended to be carried out. Furthermore, prolonged and multiple treatments with any of the cleaning methods might result in damage of insect remains and negatively affect DNA analysis.

Author Contributions: Conceptualization, S.V.; methodology, S.V. and G.G.; investigation, J.P., F.T., and G.G.; data curation, J.P., F.T., and G.G.; writing—original draft preparation, S.V. and J.P.; writing—review and editing, S.V., J.P., F.T., and G.G.; supervision, S.V.; funding acquisition, S.V. All authors have read and agreed to the published version of the manuscript.

Funding: The work of J.P., G.G., F.T. was funded by the Leverhulme Trust Doctoral Scholarship program.

Conflicts of Interest: The authors declare no conflict of interest. The funders had no role in the design of the study; in the collection, analyses, or interpretation of data; in the writing of the manuscript, or in the decision to publish the results.

References

1. Smith, K.G. *A Manual of Forensic Entomology*; Trustees of the British Museum: London, UK, 1986.
2. Goff, M.L. Early Postmortem Changes and Stages of Decomposition. In *Current Concepts in Forensic Entomology*; Amendt, J., Goff, M.L., Campobasso, C.P., Grassberger, M., Eds.; Springer: Dordrecht, The Netherlands, 2010; pp. 1–24.
3. Amendt, J.; Campobasso, C.P.; Gaudry, E.; Reiter, C.; LeBlanc, H.N.; Hall, M.J. Best practice in Forensic Entomology—Standards and guidelines. *Int. J. Legal. Med.* **2007**, *121*, 90–104. [CrossRef] [PubMed]
4. Richards, C.S.; Paterson, I.D.; Villet, M.H. Estimating the age of immature *Chrysomya albiceps* (Diptera: Calliphoridae), correcting for temperature and geographical latitude. *Int. J. Legal. Med.* **2008**, *122*, 271–279. [CrossRef]
5. Bostock, E.; Green, E.W.; Kyriacou, C.P.; Vanin, S. Chronobiological studies on body search, oviposition and emergence of *Megaselia scalaris* (Diptera, Phoridae) in controlled conditions. *Forensic. Sci. Int.* **2017**, *275*, 155–159. [CrossRef] [PubMed]
6. Reibe, S.; Madea, B. Use of *Megaselia scalaris* (Diptera: Phoridae) for post-mortem interval estimation indoors. *Parasitol. Res.* **2010**, *106*, 637–640. [CrossRef] [PubMed]
7. Vanin, S.; Caenazzo, L.; Arseni, A.; Cecchetto, G.; Cattaneo, C.; Turchetto, M. Records of *Chrysomya albiceps* in Northern Italy: An ecological and forensic perspective. *Mem. Inst. Oswaldo Cruz* **2009**, *104*, 555–557. [CrossRef]
8. Vanin, S.; Tasinato, P.; Ducolin, G.; Terranova, C.; Zancaner, S.; Montisci, M.; Ferrara, P.; Turchetto, M. Use of *Lucilia* species (Diptera: Calliphoridae) for forensic investigations in Southern Europe. *Forensic. Sci. Int.* **2008**, *177*, 37–41. [CrossRef] [PubMed]
9. Huchet, J.B. L'archeoentomologie Funeraire: Une approche originale dans l'interprétation des sépultures. *Bull. Mem. Soc. Anthropol. Paris* **1996**, *8*, 299–311. [CrossRef]
10. Huchet, J.B.; Greenberg, B. Flies, Mochicas and burial practices: A case study from Huaca de la Luna, Peru. *J. Archaeol. Sci.* **2010**, *37*, 2846–2856. [CrossRef]
11. Vanin, S.; Huchet, J.B. Forensic Entomology and Funerary Archaeoentomology. In *Taphonomy of Human Remains: Forensic Analysis of the Dead and the Depositional Environment*; Schotsmans, E.M., Márquez-Grant, N., Forbes, S.L., Eds.; Wiley: Manhattan, CA, USA, 2017; pp. 167–186.
12. Giordani, G.; Erauw, C.; Eeckhout, P.A.; Owens, L.S.; Vanin, S. Patterns of camelid sacrifice at the site of Pachacamac, Peruvian Central Coast, during the Late Intermediate Period (AD1000–1470): Perspectives from funerary archaeoentomology. *J. Archaeol. Sci.* **2020**, *114*, 105065. [CrossRef]
13. Martín-Vega, D.; Hall, M.J.R.; Simonsen, T.J. Resolving confusion in the use of concepts and terminology in intrapuparial development studies of cyclorrhaphous Diptera. *J. Med. Entomol.* **2016**, *53*, 1249–1251. [CrossRef]
14. Ulf Grote, Mark Benecke Möglichkeiten der Zusammenarbeit von Forensischer Entomologie und Archäologie am Beispiel eines frühmittelalterlichen Gräberfeldes. *Archäologisches Zellwerk. Festschrift für Helmut Roth zum 60 Geb*; Dobiat, C., Leidorf, K., Eds.; Internationale Archäologie/Studia Honoria; Marie Leidorf Verlag, Rahden/Westf: Rahden, Germany, 2001; Volume 16, pp. 47–59.

15. Baumjohann, K.; Benecke, M. Insect Traces and the Mummies of Palermo—A Status Report. *Entomol. Heute* **2019**, *31*, 73–93.
16. Giordani, G.; Tuccia, F.; Floris, I.; Vanin, S. First record of *Phormia regina* (Meigen, 1826) (Diptera: Calliphoridae) from mummies at the Sant'Antonio Abate Cathedral of Castelsardo, Sardinia, Italy. *PeerJ* **2018**, *6*, e4176. [CrossRef] [PubMed]
17. Pradelli, J.; Rossetti, C.; Tuccia, F.; Giordani, G.; Licata, M.; Birkhoff, J.M.; Vanin, S. Environmental necrophagous fauna selection in a funerary hypogeal context: The putridarium of the Franciscan monastery of Azzio (northern Italy). *J. Archaeol. Sci. Rep.* **2019**, *24*, 683–692. [CrossRef]
18. Giordani, G.; Vanin, S. Description of the puparium and other notes on the morphological and molecular identification of *Phthitia empirica* (Diptera, Sphaeroceridae) collected from animal carcasses. *Egypt J. Forensic. Sci.* **2020**, *10*, 1–8. [CrossRef]
19. Giordani, G.; Grzywacz, A.; Vanin, S. Characterization and identification of puparia of *Hydrotaea* Robineau-Desvoidy, 1830 (Diptera: Muscidae) from forensic and archaeological contexts. *J. Med. Entomol.* **2019**, *56*, 45–54. [CrossRef]
20. Samerjai, C.; Sanit, S.; Sukontason, K.; Klong-Klaew, T.; Kurahashi, H.; Tomberlin, J.K.; Sukontason, K.L. Morphology of puparia of flesh flies in Thailand. *Trop. Biomed.* **2014**, *31*, 351–361.
21. Nelson, H.G. A method of cleaning insects for study. *Coleopt Bull.* **1949**, *3*, 89–92.
22. Keirans, J.; Clifford, C.; Corwin, D. *Ixodes sigelos*, n. sp. (Acarina: Ixodidae), a parasite of rodents in Chile, with a method for preparing ticks for examination by scanning electron microscopy. *Acarologia* **1976**, *18*, 217–225.
23. Corwin, D.; Clifford, C.M.; Keirans, J.E. An improved method for cleaning and preparing ticks for examination with the scanning electron microscope. *J. Med. Entomol.* **1979**, *16*, 352–353. [CrossRef]
24. Jenkins, J. A simple device to clean insect specimens for museums and scanning electron microscopy. *Proc. Entomol. Soc. Wash* **1991**, *93*, 204–205.
25. Harrison, J.d.G. Cleaning and preparing adult beetles (Coleoptera) for light and scanning electron microscopy. *Afr. Entomol.* **2012**, *20*, 395–401. [CrossRef]
26. Schneeberg, K.; Bauernfeind, R.; Pohl, H. Comparison of cleaning methods for delicate insect specimens for scanning electron microscopy. *Microsc Res. Tech.* **2017**, *80*, 1199–1204. [CrossRef] [PubMed]
27. Amendt, J.; Zehner, R.; Johnson, D.G.; Wells, J. Future Trends in Forensic Entomology. In *Current Concepts in Forensic Entomology*; Amendt, J., Goff, M.L., Campobasso, C.P., Grassberger, M., Eds.; Springer: Dordrecht, The Netherlands, 2010; pp. 353–368.
28. Mazzanti, M.; Alessandrini, F.; Tagliabracci, A.; Wells, J.D.; Campobasso, C.P. DNA degradation and genetic analysis of empty puparia: Genetic identification limits in forensic entomology. *Forensic. Sci. Int.* **2010**, *195*, 99–102. [CrossRef] [PubMed]
29. Ronderos, M.M.; Spinelli, G.R.; Sarmiento, P. Preparation and mounting of biting midges of the genus *Culicoides* Latreille (Diptera: Ceratopogonidae) to be observed with a scanning electron microscope. *Trans. Am. Entomol. Soc.* **2000**, *126*, 125–132.
30. Zangheri, P. *Il Naturalista Esploratore, Raccoglitore, Preparatore, Imbalsamatore*; Hoepli Editore: Milano, Italy, 1981; 506p.
31. Gurney, A.B.; Kramer, J.P.; Steyskal, G.C. Some techniques for the preparation, study, and storage in microvials of insect genitalia. *Ann. Entomol. Soc. Am.* **1964**, *57*, 240–242. [CrossRef]
32. Sukontason, K.L.; Ngern-Klun, R.; Sripakdee, D.; Sukontason, K. Identifying fly puparia by clearing technique: Application to forensic entomology. *Parasitol. Res.* **2007**, *101*, 1407–1416. [CrossRef]
33. Gifawesen, C.; Funke, B.; Proshold, F. Control of antifungal-resistant strains of *Aspergillus niger* mold contaminants in insect rearing media. *J. Econ. Entomol.* **1975**, *68*, 441–444. [CrossRef]
34. Stueben, M.; Linsenmair, K.E. Advances in insect preparation: Bleaching, clearing and relaxing ants (Hymenoptera: Formicidae). *Myrmecol. News* **2008**, *12*, 15–21.
35. Folmer, O.; Hoeh, W.; Black, M.; Vrijenhoek, R. Conserved primers for PCR amplification of mitochondrial DNA from different invertebrate phyla. *Mol. Mar. Biol. Biotechnol.* **1994**, *3*, 294–299.
36. Tuccia, F.; Giordani, G.; Vanin, S. A general review of the most common COI primers for Calliphoridae identification in Forensic Entomology. *Forensic. Sci. Int. Genet.* **2016**, *24*, e9–e11. [CrossRef]
37. Tuccia, F.; Giordani, G.; Vanin, S. A combined protocol for identification of maggots of forensic interest. *Sci. Justice* **2016**, *56*, 264–268. [CrossRef] [PubMed]

Article

The Growth Model of Forensically Important *Lucilia sericata* (Meigen) (Diptera: Calliphoridae) in South Korea

Sang Eon Shin, Ji Hye Park, Su Jin Jeong and Seong Hwan Park *

Department of Legal Medicine, Korea University College of Medicine, Seoul 02841, Korea; shinfbr@nate.com (S.E.S.); jhp09@korea.kr (J.H.P.); rainofsujin@naver.com (S.J.J.)
* Correspondence: pelvis@korea.ac.kr

Simple Summary: This study provides a detailed growth data for Lucilia sericata (Meigen) collected in South Korea. With the growth data, authors compared different minimum ADH models and found little differences. However, the logalithmic model was the best fit among differenct models.

Abstract: Development of forensically important *Lucilia sericata* (Meigen) was analyzed in South Korea. Rearing was replicated five times at seven constant temperatures between 20–35 °C to elucidate changes in accumulated degree hours, based on developmental stage and body length, and 2673 individuals were statistically analyzed. The results indicated that the optimum temperature, the base temperature, and the overall thermal constant were 22.31 °C ($\pm$1.21 °C, 95% CI), 9.07 °C, and 232.81 $\pm$ 23 (mean $\pm$ SD) accumulated degree days, respectively. In the minimum ADH models of each development stage, nonlinear regression graphs were parallel at the immature stages. Based on the scatter plot (n = 973) of immature stages using ADH values and body length, the logarithmic model using $Log_{10}ADH$ as the dependent variable was identified as the best fitting regression model. Additionally, the adjusted R^2 value and mean square of error were 0.911 and 0.007, respectively. This is the first forensically focused study on the development of *L. sericata* for the estimation of minimum postmortem interval in South Korea. In future studies, we intend to study the development of other necrophagous fly species and to identify parameters for the determination of age at post-feeding and pupal stages.

Keywords: development; minimum postmortem interval (PMI-min); rearing; calliphoridae; *Lucilia sericata*

Citation: Shin, S.E.; Park, J.H.; Jeong, S.J.; Park, S.H. The Growth Model of Forensically Important *Lucilia sericata* (Meigen) (Diptera: Calliphoridae) in South Korea. *Insects* **2021**, *12*, 323. https://doi.org/10.3390/insects12040323

Academic Editors: Damien Charabidze and Daniel Martín-Vega

Received: 22 February 2021
Accepted: 1 April 2021
Published: 6 April 2021

Publisher's Note: MDPI stays neutral with regard to jurisdictional claims in published maps and institutional affiliations.

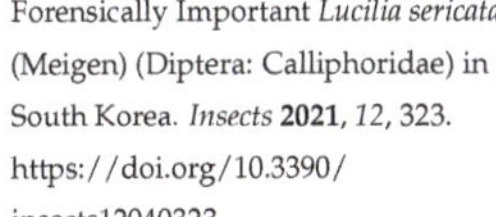

1. Introduction

In medico-legal entomology, insects are used as scientific evidence to solve cases related to the time since death (TSD), entomotoxicology, abuse, and neglect, etc. This field focuses primarily on the time at which insect eggs (or larva in Sarcophagidae) are laid on the body after death to estimate the minimum postmortem interval (PMI-min) [1]. PMI-min is assumed to be most accurately predicted by calculating the age of immature insects [2], even when the body is badly decayed. As a real case, fly pupae in the soil and maggots found in the freezer for body preservation were collected 45 days after the discovery of a putrefied male cadaver in fallow ground. The PMI-min was estimated at 10 days before the discovery of the body, based on egg laying time from the growth rate of insects, the distribution of pupae toward pupariation sites, weather information, and so on [3].

The growth rate of insects is strongly influenced by temperature and can be presented as an S-shaped velocity curve at constant temperatures [4]. Further, the growth rate of immature insects is considered to have a linear relationship with developmental temperatures [5]. In these linear models, energy budgets designated for physiological development are considered to remain constant through the various developmental stages [5]. However, base temperature varies from species to species and can also vary with geographic location [5]. Considering these dependent relationships, the values of accumulated degree hours or days (ADH or ADD, respectively) for specific developmental stages (while estimating the age of forensically important insects) and the base temperature should be predetermined by setting constant temperatures in rearing experiments [5–7].

Lucilia sericata (Meigen), the earliest arriving necrophagous fly species on corpses, is known as one of the most dominant forensically important species in the temperate zone of the Northern Hemisphere as well as in both urban and suburban areas [2,7,8] and has been found in such places as apartments in Germany [9], some stagnant water in a city of Spain [10], the Iwate prefecture in Japan [11], indoors in Italy [12], and indoors in South Korea [13]. In addition, this species is causing myiasis in South Korea, and the importance of myiasis with this species as an indicator of a poor hygienic condition and a lack of due care is ever growing in an aging society such as is South Korea [14]. Consequently, numerous studies on the growth of *L. sericata* have been performed in several countries [15–20]. Anderson [18] documented the minimum and maximum time taken to reach each developmental stage as a way to estimate the time since death. Shortly afterward, Grassberger and Reiter [19] illustrated morphological length and stage changes using isomegalen- and isomorphen-diagrams. Nevertheless, Roe and Higley [21] emphasized that methodological inconsistencies in the previous studies made it difficult to apply error rates or confidence intervals to cases within a given region. These inconsistencies stimulated the launch of studies on blind validation [22] and field validation [23] of development datasets.

The goal of the present study was to generate practical development data for *L. sericata*, the most common indoor insect species in South Korea [13]. The rearing experiments were replicated five times at seven different temperatures, held constant throughout the investigation, to analyze the changes in ADH or ADD according to developmental stage and body length.

2. Materials and Methods

2.1. Identification and Rearing of Adult Flies

Maggots of *L. sericata* were collected from autopsies in northeastern Seoul, Korea. After their emergence in incubators, the adults were identified by the following morphological characters: 6 to 8 occipital setae behind the vertical bristle, acrostichal bristles 2 + 3, and dorsocentral bristles 3 + 3 [24]. The adult flies were provided with a damp paper towel as a water source, along with a mixture of powdered milk (50%) and dry granular sugar (50%). Newly identified adult flies were occasionally added to acryl cages (the dimensions of $40 \times 40 \times 40$ cm^3), which were constructed to prevent odor and the trapping of flies in the folds. A mesh cloth (20×20 cm) was used for the lateral sides and was attached using Velcro tape. This design facilitated internal cleaning and also provided ventilation. Moreover, the size of the mesh was small enough to prevent the intrusion of coffin flies (Phoridae) (Figure 1).

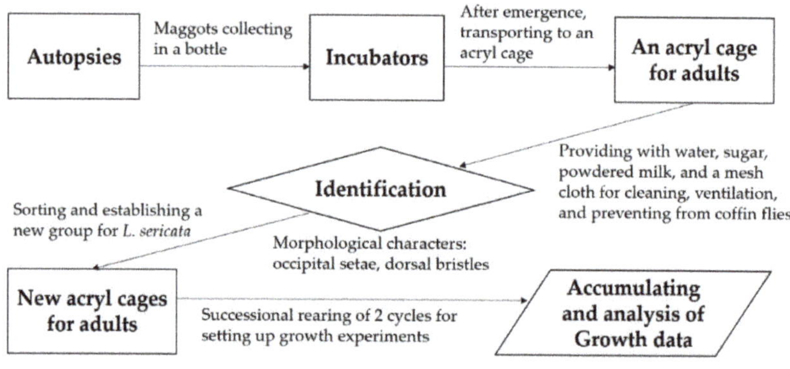

Figure 1. A diagram showing the preparation process for the growth experiment of *Lucilia sericata*.

2.2. Sampling and Rearing of Maggots

Fresh pork liver was sliced into pieces (approximately 50 g in weight) and the pieces were frozen at −20 °C until use. They were thawed slowly at 25 °C for 24 h in order to maintain freshness and minimize blood leakage. Eggs were collected from a piece of fresh pork liver within 40 min of the beginning of egg laying. Eggs were then separated from each other by soaking in sodium sulfate solution (2%) and rinsing with distilled water [19]. Twenty-five of these moist eggs were then deposited onto a new piece of fresh pork liver (50 g) using a small moist brush to prevent them from drying. Ten bottles (diameter 10 cm, height 9 cm) containing the eggs and liver were placed at the center of a growth chamber (50 × 50 × 50 cm) to reduce the effects of location. The process of rearing—from eggs to adult stages—was duplicated five times at 70% relative humidity with a photoperiod (h) of 16:8 (L:D) at seven constant temperatures—namely, 2 °C, 22 °C, 24 °C, 26 °C, 28 °C, 32 °C, and 35 °C (for a total of 35 experiments). This was done considering the possibility of diapause [18,25] and the upper temperature threshold [19,26]. When the movement trace of post-feeding larvae could be observed, dry wood sod (depth 6 cm) was added to the 10 bottles for pupation. Once the first adult fly emerged, the bottles were transferred to acryl cages to continue the rearing of adult flies. Notably, the selected temperature of the growth chambers was not assigned to a single temperature. This was done to distinguish between the effect of the selected temperature and the mechanical error of the growth chamber [27]. Moreover, the center temperature of the growth chambers was measured for temperature correction. Regardless of body length or developmental stage, four individuals were removed from one bottle every 12 h. Afterward, the same bottle was replaced, and the other bottles in the chamber were shuffled. Specimens (four individuals) were killed by submersion in boiling water for 30 s to prevent shrinkage. Specimens were then preserved in an 80% ethanol solution [28] and placed in a freezer (−20 °C).

2.3. Body Length, Larval Stages, and the Optimum Development Temperature

Body length was measured using micrographs (Olympus, SZX10) and calculated using Microsoft Office Excel 2007 (Microsoft Corp., Redmond, WA, USA) (Table S1). Larval stages were determined based on the condition of the crop and the number of posterior spiracle slits [5]. Additionally, the minimum time taken to reach each developmental stage was based on the time at which the first observed individual was discovered (Table S2). The optimum development temperature was statistically estimated from the inflection in the sigmoid model of growth rate [27]. Additionally, ADH (or ADD) was calculated using the following equations [5] (Table S3):

$$\text{Time (h)} \times (\text{temperature} - \text{base temperature}) = \text{ADH (°H)} \tag{1}$$

$$\text{Time (days)} \times (\text{temperature} - \text{base temperature}) = \text{ADD (°D)} \tag{2}$$

2.4. Data Fit and Statistical Analysis

SigmaPlot (version 10.0) and Microsoft Office Excel 2007 were used for plotting all graphs and for performing basic statistical analyses. Two-way ANOVA, without replication, was conducted using the SAS program (ProcMIXED, SAS9.4) [29] to determine the differences among minimum mean hours spent in each development stage and temperature ($p \leq 0.05$) [23]. In addition, the growth data from egg to adult were fitted with a four-parameter sigmoid model to determine the minimum growth rate (y^0) as well as the optimum development temperature (x^0):

$$F(x) = y^0 + a/\left(1 + \exp\left(-\left(x - x^0\right)/b\right)\right) \tag{3}$$

where y^0 is the minimum developmental rate, x^0 is the inflection or the optimum development temperature in the sigmoid curve, "a" is the difference between the maximum and minimum developmental rates, and "1/b" is the steepness of the sigmoid curve [27,30].

The fitted curve for growth rate was compared with rearing results from previous studies using 95% confidence and 95% prediction intervals. In addition, a scatter plot was produced to illustrate the correlations among the following variables: body length, ADH, and growth stage, including the transition stages in the first and second instar. Using the scatter plot, linear regression and nonlinear regression analyses were performed to conform to the growth model of the immature stages using ADH and Log_{10}ADH values.

3. Results

3.1. Body Length and Minimum Development Time

Among 8750 eggs (25 eggs $\times$ 10 bottles $\times$ 7 temperatures $\times$ 5 replicates), 2673 individuals were sampled and statistically analyzed (sample coverage, 32.7%), including 200 outliers. Body length values (mean $\pm$ SD) were 1.17 $\pm$ 0.13 mm (egg), 2.45 $\pm$ 0.65 mm (first instar), 6.29 $\pm$ 1.54 mm (second instar), 13.16 $\pm$ 2.40 mm (third instar), 12.12 $\pm$ 1.99 mm (post-feeding larva), and 7.73 $\pm$ 0.63 mm (pupa) (Figure 2). Additionally, the minimum development time (mean $\pm$ SD, n = sample size) from egg to adult stages at each of the seven temperatures was 20.60 $\pm$ 1.53 days (20 °C, n = 499); 16.42 $\pm$ 1.54 days (22 °C, n = 357); 14.78 $\pm$ 0.61 days (24 °C, n = 361); 12.75 $\pm$ 0.96 days (26 °C, n = 332); 11.70 $\pm$ 0.84 days (28 °C, n = 360); 10.90 $\pm$ 0.55 days (32 °C, n = 390); and 10.70 $\pm$ 0.45 days (35 °C, n = 374). Values for the minimum mean development time were significantly different among the developmental stages (F = 53.8; df = 5; $p \leq 0.05$) and temperatures (F = 3.6; df = 6; $p \leq 0.05$).

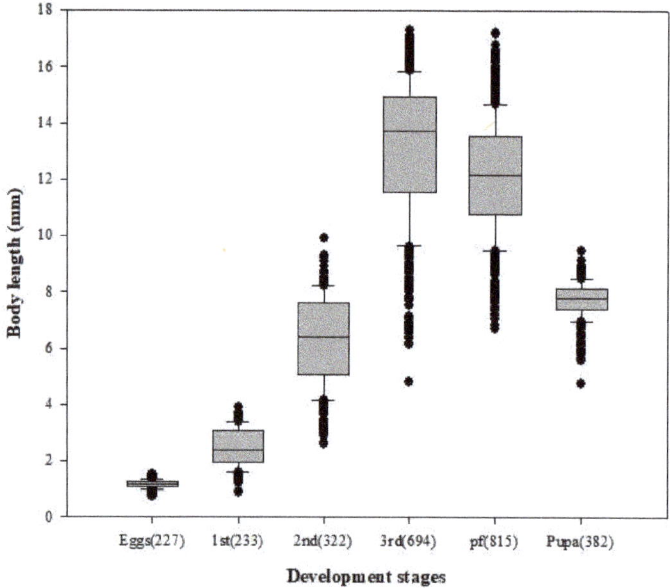

Figure 2. Boxplot of body length according to developmental stages in *L. sericata* (n = sample size). Body length was greatest at the feeding third instar stage but decreased during pupation.

3.2. Base Temperature, Optimum Temperature, and Comparisons with Previous Studies

The base temperature was calculated as 9.07 °C (Table 1), and the growth data were fitted with the four-parameter sigmoid model. The statistically adjusted R^2 value was 0.93, and the mean square error (MSE) was 3.10; coefficient values were calculated as 28.56, 3.28, 22.31, and 11.14 for a, b, x^0, and y^0, respectively. The optimum temperature (or the inflection (x^0)) was estimated as 22.31 °C ($\pm$1.21 °C, 95% CI) (Figure 3). Additionally, the growth rate in the present study corresponded to that reported in most previous studies in the 95% prediction interval (Figure 3).

Table 1. Base temperatures and *p* values for *Lucilia sericata* according to development stage.

Stage	Regression Equation	R^2	Base Temperature (°C)	*p*-Value
Egg	Y = 39.9305x − 479.1665	0.8701	12.0000	0.0207 *
First instar	Y = 25.4630x − 164.0213	0.5673	6.4416	0.1416
Second instar	Y = 21.3656x − 147.9693	0.6338	6.9256	0.1071
Third instar	Y = 12.5778x − 92.7091	0.6265	7.3709	0.1106
Post-feeding	Y = 9.2416x − 65.3248	0.3773	7.0686	0.2703
Pupa	Y = 4.4917x − 46.8452	0.9152	10.4293	0.0108 *
Egg to adult	Y = 1.9040x − 17.2758	0.9907	9.0734	<0.0004 *

* *p*-value < 0.05.

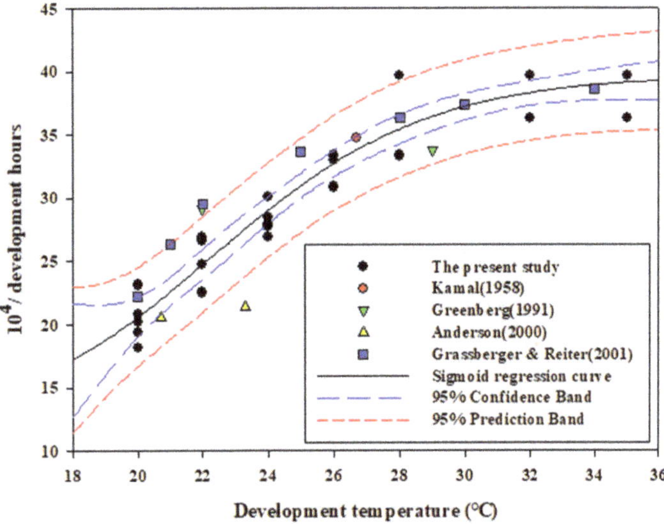

Figure 3. Four-parameter sigmoid model for the growth rate of *Lucilia sericata* from eggs to adults at five temperature regimes with 95% confidence and 95% prediction intervals. Although rearing was duplicated five times at seven temperatures, the dots of the present study overlapped due to similar results. Data from most previous studies fell within the 95% prediction interval.

3.3. Minimum ADH Models and Scatter Plots

When plotted, the minimum ADH models based on the same development stages ran parallel at feeding larval stages. However, the plots curved upward at the post-feeding and pupal stages (Figure 4). In addition, the scatter plots (*n* = 2566) developed from ADH values and body length presented a constant relationship during the feeding larval stage (≤1551.60 ADH). Moreover, minimum ADH values at each developmental stage were estimated as follows: first instar: 203.16 ADH; second instar: 524.64 ADH; third instar: 812.64 ADH; post-feeding third instar: 1551.60 ADH; and pupa: 2492.04 ADH (Figure 5). The first and second instar larvae transitioning to the next developmental stage and characterized by one additional slit under the posterior spiracle slits [4] presented relatively narrow ADH ranges—from 454.3 ADH to 622.3 ADH and from 812.6 ADH to 1612.4 ADH (Figure 5).

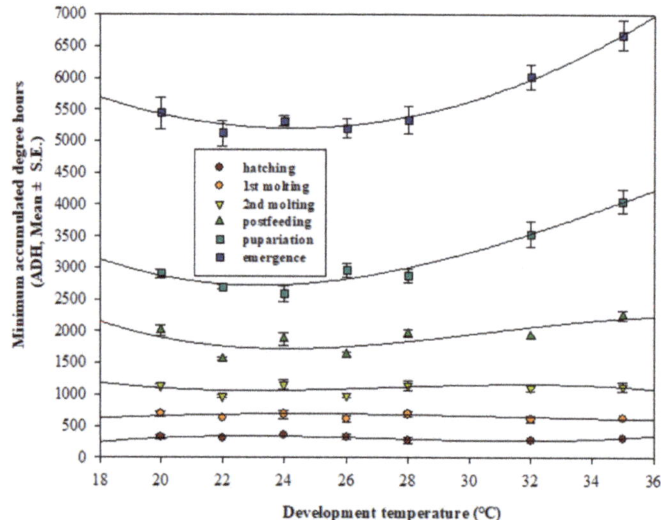

Figure 4. Minimum accumulated degree hours (ADH) model for *Lucilia sericata* developed by accumulated minimum development hours at each developmental stage, at seven temperature regimes, with a base temperature of 9.07 °C. Regression curves were fitted with polynomial cubic equations.

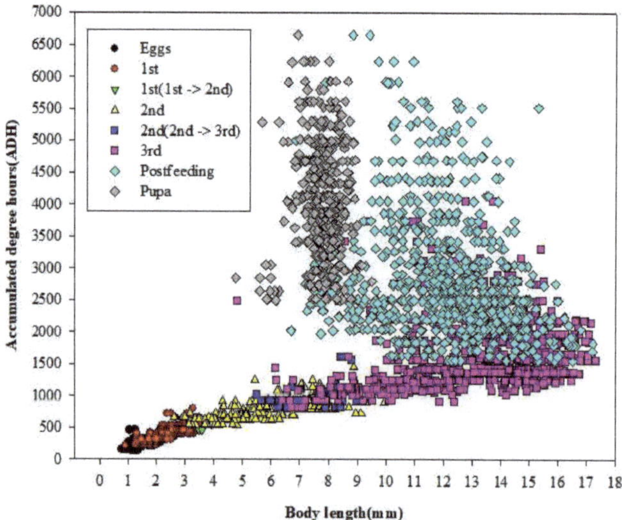

Figure 5. Scatter plot (*n* = 2566) of *Lucilia sericata* developed based on ADH values and body lengths at each developmental stage. It shows a constant relationship during feeding larval stages (≤1551.60 ADH), minimum ADH values for each developmental stage, and the possibility of transition forms as a forensic indicator.

3.4. Linear and Nonlinear Regressions during Immature Stages

Both linear regression and nonlinear regression were performed using ADH values ($f(x)$) and body length (x) during the feeding larval stage (≤1551.60 ADH) in the scatter plot (n = 973). It is important to note that the adjusted R^2 value of the secondary model was higher than that of the linear model, whereas the MSE using $\mathrm{Log_{10}ADH}$ was lower (Table 2). Therefore, the logarithmic model (2 Parameter I) using $\mathrm{Log_{10}ADH}$ was estimated as the best fitting regression model (Figure 6), considering the R^2_{adj} value (0.911) and MSE (0.007) (Table 2).

Table 2. Linear and nonlinear regression models during immature stages using ADH and Log$_{10}$ADH values ($n = 973$).

Model	Y	Regression Equation	R^2_{adj}	SE	MSE
Linear	ADH	$Y = 238.508 + 83.076x$	0.872	142.800	204×10^2
Quadratic	ADH	$Y = 41.570 + 158.911x - 4.814x^2$	0.920	112.729	127×10^2
Logarithm	ADH	$Y = 38.392 + 466.960 \, Log_{10}(x)$	0.899	126.993	161×10^2
Linear	Log$_{10}$ADH	$Log_{10}(Y) = 2.469 + 0.053x$	0.744	0.141	0.020
Quadratic	Log$_{10}$ADH	$Log_{10}(Y) = 2.228 + 0.146x - 0.006x^2$	0.892	0.091	0.008
Logarithm	Log$_{10}$ADH	$Log_{10}(Y) = 2.295 + 0.327 \, Log_{10}(x)$	0.911	0.083	0.007

SE: standard error of estimation, MSE: mean square error.

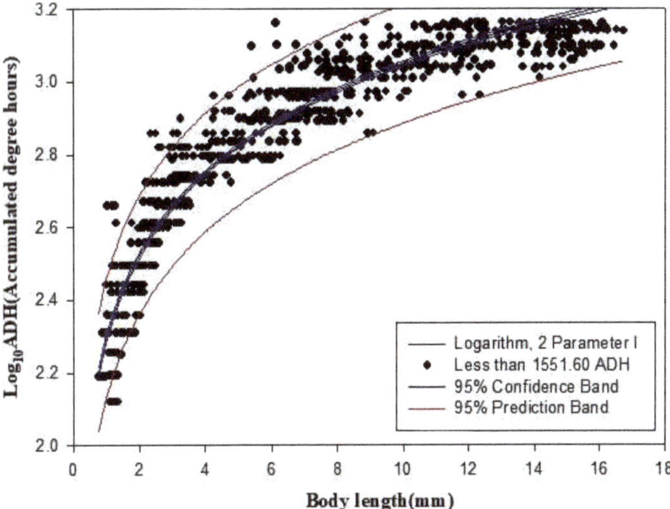

Figure 6. A nonlinear regression analysis was performed from the scatter plot ($n = 973$) of feeding larval stages ($\leq$1551.60 ADH) of *Lucilia sericata* to determine the correlation between ADH values and body lengths. The adjusted R^2 value and mean square error were 0.911 and 0.007, respectively.

4. Discussion

The growth rate of insects is strongly influenced by temperature and is presented as an S-shaped velocity curve at constant temperatures [4]. In the present study, the forensically important *L. sericata* was reared under conditions that aligned with five criteria for controlling variation factors—namely, a food source of fresh pork liver thawed within 24 h [31], a photoperiod of 16 h (light) [25], the placement of rearing bottles in the center of the chamber with a thermometer [32], the number of eggs being limited to 25 to prevent heat generation by friction [33], and the random use of chambers to distinguish the effect of programed temperatures and mechanical errors [32].

In our pilot study, the sampling of entire-age cohorts at 20 °C to produce insect growth models [34] led to the number of bottles exceeding the capacity of a rearing chamber. This problem eventually caused poor ventilation, thereby reducing the effect of the programed temperature. Meanwhile, in the present study, four individuals were sampled from one of the ten shuffled bottles every 12 h, and rearing experiments were repeated five times at seven temperatures to meet the minimum sample size for statistical significance ($n = 318$) [35]. This was done in accordance with the sampling method outlined by Anderson [18]. In addition, because it is difficult to count moving first instar immediately after hatching, the hatch rate for 25 eggs could be estimated through the sum of the number of sampled individuals from egg to adult and the number of left puparia.

It is important to note that the minimum amount of time taken to reach each stage of development was not based on ecologically meaningful 50% transition times but rather on the observed time of the first individual [17–19]. This is because, in forensic science, the existence of each development stage in a scene becomes scientific evidence, and the best standard practice recommends collecting at least 10% of the total population to ensure the collection of the oldest (or the largest) insects [36]. In addition, it was confirmed that the previous growth data were mostly included within 95% prediction intervals based on the growth data of this study (Figure 2), despite differences in geographic region and type and properties of food. These results suggest that the application of a consistent analysis method of developmental stages, based on the observed time of the first individual (minimum), is more important than geographic region or food in insect growth model studies.

Currently, forensic entomologists need to know the minimum growth time of the oldest insect collected at a scene and also need to require information on the optimal temperature for laboratory rearing after sampling [27]. In a study of larval mass effect on *Lucilia sericata*, ambient temperatures between 22 °C and 25 °C were reported as the optimal temperature range for the highest heat emission per larva [37]. These results were consistent with the fact that 22.31 °C ($\pm$1.21 °C, 95% CI) was measured as the optimum temperature for this study, and it was thought to be related to the fact that the growth of insects is dependent on the temperature. Kotzé et al. (2015) found that the body length of *Lucilia cuprina* was greatest near the optimal temperature [38], and this result was also the same as our result, as mean body lengths of third instar were greatest at 24 °C (14.06 $\pm$ 1.85 mm).

For the estimation of base temperature from egg to adult, five temperatures between 20 and 28 °C were used on a linear growth graph, and minimum ADH values were determined using the x-intercept approach at 9.07 °C [39] (Table 1). Notably, this value is similar to the 9.0 °C reported by Marchenko [20]. This similarity suggests that there is little difference by geographical region between Russia and South Korea for *Lucilia sericata* and that methodological factors (food, larval mass effects, etc.) were consistent with this study, including the setting of the temperature range centering on the flection point of the sigmoid growth curve (the optimum temperature) [27].

The minimum ADH model of this study (Figure 4) was produced using the minimum time required to reach each developmental stage, from the initial egg laying phase, for easy identification of minimum ADH values in a scene. Reibe et al. (2010) [40] also published a ADH model similar to this study, but this study used a base temperature of 9.07 °C and growth data made of 25 eggs, whereas they used 8.0 °C and growth data made of 100 eggs by Grassberger & Reiter (2001) [19]. Unlike this study, according to Marchenko (1985) [41], their growth data were estimated to reflect the larval mass effects. Therefore, in order to reduce the estimation error of PMI-min, it was considered essential to select an appropriate ADH model according to the field situation even if the same species of insects were found.

Additionally, the regression curves in Figure 3 demonstrate the delay in puparia-tion due to the extension of the post-feeding larval stage, which resulted from the high programed temperature rather than the crowding of larvae [17]. We excluded any heat generated from the larval population because we had placed only few individuals in each bottle to avoid the heat generated by their bodies [33]. In addition, food and dry sawdust were sufficiently provided [42].

A practical ADH model should include useful parameters such as body length and growth stage, as well as prediction intervals. However, the isomegalen- and isomorphen-diagrams by Grassberger and Reiter [19] have no error values, and the ADH model by Reibe et al. [40] has no data for body length. For these reasons, a new scatter plot was designed to show body length, growth stage, and ADH values (Figure 5). The following is a summary of our findings. First, ADH values and body length during feeding larval stages have a linear relationship. Second, other parameters such as gene expression differences are needed for age prediction during the post-feeding third larval stage and the pupa stage [43]. Third, the minimum ADH values taken to reach each developmental stage can be determined from 20 to 35 °C, and lastly, the first and second instar larvae transitioning to the next developmental stage have potential as forensic indicators due to their relatively narrow ADH ranges.

In the feeding larval stages of the scatter plot, linear and nonlinear regressions were performed to understand the correlation between ADH values and body length. The best fit regression model was the logarithmic model (2 Parameter I) using $Log_{10}ADH$ as the dependent variable (Table 2), considering the R^2_{adj} value and MSE. It was expected that using ADD, rather than ADH, as the dependent variable would result in a low MSE [44]. However, it was excluded from this study because it was thought that the precision was low, even though the measurement of the growth period in units of days had high accuracy. In addition, 95% of the prediction intervals in fitted models, or errors of ADH estimates from body length values, were due to variability within a species [27] and the sampling interval of 12 h. Therefore, shorter sampling intervals were suggested within a growth period of 120 h for a more precise estimation of PMI-min [1,34].

5. Conclusions

The growth models for Korean *L. sericata* showed little difference in this study when compared with the results of previous studies; minimum ADH values at each stage of development could be determined. Based on the scatter plot of ADH values and body length values at immature stages, the logarithmic model was the best fit. In addition, minimum ADH values and 95% prediction intervals at each body length value could be statistically estimated (adjusted R^2 = 0.92). In future studies, it is our intention to rear subdominant necrophagous fly species and develop additional markers for age prediction at post-feeding and pupal stages.

Supplementary Materials: The following are available online at https://www.mdpi.com/article/10.3390/insects12040323/s1, Table S1. Body lengths (mean ± SD, mm) for each stage of Lucilia sericata. Table S2. Minimum developmental times (mean ± SD, hours) for each stage of *Lucilia sericata*. Table S3. Minimum ADH values (mean ± SE, ADH) for each stage of *Lucilia sericata*.

Author Contributions: S.E.S. mainly performed the experiments and wrote the manuscript. J.H.P. participate in the experiments. S.J.J. performed statistical analysis. S.H.P. supervised the whole research and reviewed the final manuscript before submission. All authors have read and agreed to the published version of the manuscript.

Funding: This research was funded by the Basic Science Research Program through the National Research Foundation of Korea (NRF), the Ministry of Science, ICT & Future Planning (NRF-2013R1A1A1012223) and by the Korean National Police Agency (Grant No. PA-G000001).

Institutional Review Board Statement: Not applicable.

Data Availability Statement: Data is contained within the article or Supplementary Material.

Conflicts of Interest: The authors declare no conflict of interest.

References

1. Amendt, J.; Campobasso, C.P.; Goff, M.L.; Grassberger, M. *Current Concepts in Forensic Entomology*; Springer: Dordrecht, The Netherlands, 2010.
2. Byrd, H.; Castner, J.L. *Forensic Entomology: The Utility of Arthropods in Legal Investigations*; CRC Press: Boca Raton, FL, USA, 2010.
3. Shin, S.E.; Jang, M.S.; Park, J.H.; Park, S.H. A forensic entomology case estimating the minimum postmortem interval using the distribution of fly pupae in fallow ground and maggots with freezing injury. *Korean J. Leg. Med.* **2015**, *39*, 17–21. [CrossRef]
4. Davidson, J. On the relationship between temperature and rate of development of insects at constant temperatures. *J. Anim. Ecol.* **1944**, *13*, 26–38. [CrossRef]
5. Gennard, D.E. *Forensic Entomology: An Introduction*, 2nd ed.; John Wiley and Sons: Chichester, UK, 2012.
6. Amendt, J.; Campobasso, C.P.; Gaudry, E.; Reiter, C.; LeBlanc, H.N.; Hall, M.J. Best practice in forensic entomology—Standards and guidelines. *Int. J. Legal Med.* **2007**, *121*, 90–104. [CrossRef] [PubMed]

7.   Tarone, A.M.; Foran, D.R. Generalized additive models and *Lucilia sericata* growth: Assessing confidence intervals and error rates in forensic entomology. *J. Forensic Sci.* **2008**, *53*, 942–948. [CrossRef] [PubMed]

8.   Smith, K.G.V. *A Manual of Forensic Entomology*; British Museum (Natural History): London, UK, 1986.

9.   Benecke, M.; Josephi, E.; Zweihoff, R. Neglect of the elderly: Forensic entomology cases and considerations. *Forensic Sci. Int.* **2004**, *146S*, S195–S199. [CrossRef]

10.   Arnaldos, M.I.; García, M.D.; Romera, E.; Presa, J.J.; Luna, A. Estimation of postmortem interval in real cases based on experimentally obtained entomological evidence. *Forensic Sci. Int.* **2005**, *149*, 57–65. [CrossRef]

11.   Saigusa, K.; Takamiya, M.; Aoki, Y. Species identificaion of the forensically important flies in Iwate prefecture, Japan based on mitochondrial cytochrome oxidase gene subunit I (CO I) sequences. *Legal Med.* **2005**, *7*, 175–178. [CrossRef]

12.   Bugelli, V.; Forni, D.; Bassi, L.A.; Paolo, M.D.; Marra, D.; Lenzi, S.; Toni, C.; Giusiani, M.; Domenici, R.; Gherardi, M.; et al. Forensic entomology and the estimation of the minimum time since death in indoor cases. *J. Forensic Sci.* **2015**, *60*, 525–531. [CrossRef]

13.   Shin, S.E.; Lee, H.J.; Park, J.H.; Ko, K.S.; Kim, Y.H.; Kim, K.R.; Park, S.H. The first survey of forensically important entomofauna collected from medicolegal autopsies in South Korea. *Biomed. Res. Int.* **2015**, *2015*, 606728. [CrossRef]

14.   Choe, S.; Lee, D.; Park, H.; Jeon, H.K.; Kim, H.; Kang, J.H.; Jee, C.H.; Eom, K.S. Canine wound myiasis caused by *Lucilia sericata* (Diptera: Calliphoridae) in Korea. *Korean J. Parasitol.* **2016**, *54*, 667–671. [CrossRef]

15.   Kamal, A.S. Comparative study of thirteen species of sarcosaprophagous Calliphoridae and Sarcophagidae (Diptera) I. Bionomics. *Ann. Entomol. Soc. Am.* **1958**, *51*, 261–271. [CrossRef]

16.   Ash, N.; Greenberg, B. Developmental temperature responses of the sibling species *Phaenicia sericata* and *Phaenicia pallescens*. *Ann. Entomol. Soc. Am.* **1975**, *68*, 197–200. [CrossRef]

17.   Greenberg, B. Flies as forensic indicators. *J. Med. Entomol.* **1991**, *28*, 565–577. [CrossRef] [PubMed]

18.   Anderson, G.S. Minimum and maximum development rates of some forensically important Calliphoridae (Diptera). *J. Forensic Sci.* **2000**, *45*, 824–832. [CrossRef] [PubMed]

19.   Grassberger, M.; Reiter, C. Effect of temperature on *Lucilia sericata* (Diptera: Calliphoridae) development with special reference to the isomegalen- and isomorphen-diagram. *Forensic Sci. Int.* **2001**, *120*, 32–36. [CrossRef]

20.   Marchenko, M.I. Medicolegal relevance of cadaver entomofauna for the determination of the time of death. *Forensic Sci. Int.* **2001**, *120*, 89–109. [CrossRef]

21.   Roe, A.; Higley, L.G. Development modeling of *Lucilia sericata* (Diptera: Calliphoridae). *PeerJ* **2015**, *3*, e803. [CrossRef]

22.   VanLaerhoven, S.L. Blind validation of postmortem interval estimates using developmental rates of blow flies. *Forensic Sci. Int.* **2008**, *180*, 76–80. [CrossRef]

23.   Núñez-Vázquez, C.; Tomberlin, J.K.; Cantú-Sifuentes, M.; García-Martínez, O. Laboratory development and field validation of *Phormia regina* (Diptera: Calliphoridae). *J. Med. Entomol.* **2013**, *50*, 252–260. [CrossRef]

24.   Kanō, R.; Shinonaga, S. *Calliphoridae (Insecta: Diptera), Fauna Japonica*; Biogeographical Society of Japan: Tokyo, Japan, 1968; p. 181.

25.   Tachibana, S.; Numata, H. Effects of temperature and photoperiod on the termination of larval diapause in *Lucilia sericata* (Diptera: Calliphoridae). *Zoolog. Sci.* **2004**, *21*, 197–202. [CrossRef]

26.   Rivers, D.B.; Thompson, C.; Brogan, R. Physiological trade-offs of forming maggot masses by necrophagous flies on vertebrate carrion. *Bull. Entomol. Res.* **2011**, *101*, 599–611. [CrossRef] [PubMed]

27.   Tomberlin, J.K.; Benbow, M.E. *Forensic Entomology: International Dimensions and Frontiers*; CRC Press: Boca Raton, FL, USA, 2015.

28.   Adams, Z.J.; Hall, M.J. Methods used for the killing and preservation of blowfly larvae, and their effect on post-mortem larval length. *Forensic Sci. Int.* **2003**, *138*, 50–61. [CrossRef]

29.   SAS Institute. *PROC User's Manual, Version 9.1*; SAS Institute: Cary, NC, USA, 2002.

30.   Motulsky, H.; Christopoulos, A. *Fitting Models to Biological Data Using Linear and Nonlinear Regression: A Practical Guide to Curve Fitting*; Oxford University Press: Oxford, UK, 2004.

31.   Richards, C.S.; Rowlinson, C.C.; Cuttiford, L.; Grimsley, R.; Hall, M.J.R. Decomposed liver has a significantly adverse affect on the development rate of the blowfly *Calliphora vicina*. *Int. J. Legal Med.* **2013**, *127*, 259–262. [CrossRef] [PubMed]

32.   Moreau, G.; Michaud, J.P.; Schoenly, K.G. Experimental Design, Inferential Statistics, and Computer Modeling. In *Forensic Entomology: International Dimensions and Frontiers*; Tomberlin, J.K., Benbow, M.E., Eds.; CRC Press: Boca Raton, FL, USA, 2015; pp. 205–229.

33.   Johnson, A.P.; Wallman, J.F. Effect of massing on larval growth rate. *Forensic Sci. Int.* **2014**, *241*, 141–149. [CrossRef]

34.   Wells, J.D.; Lecheta, M.C.; Moura, M.O.; LaMotte, L.R. An evaluation of sampling methods used to produce insect growth models for postmortem interval estimation. *Int. J. Legal Med.* **2015**, *129*, 405–410. [CrossRef] [PubMed]

35.   LaMotte, L.R.; Roe, A.L.; Wells, J.D.; Higley, L.G. A statistical method to construct confidence sets on carrion insect age from development stage. *J. Agric. Biol. Environ. Stat.* **2017**. [CrossRef]

36.   Goodbrod, J.R.; Goff, M.L. Effects of larval population density on rates of development and interactions between two species of *Chrysomya* (Diptera: Calliphoridae) in laboratory culture. *J. Med. Entomol.* **1990**, *27*, 338–343. [CrossRef]

37.   Charabidze, D.; Bourel, B.; Gosset, D. Larval-mass effect: Characterisation of heat emission by necrophageous blowflies (Diptera: Calliphoridae) larval aggregates. *Forensic Sci. Int.* **2011**, *211*, 61–66. [CrossRef]

38.   Kotzé, Z.; Villet, M.H.; Weldon, C.W. Effect of temperature on development of the blowfly, *Lucilia cuprina* (Wiedemann) (Diptera: Calliphoridae). *Int. J. Legal Med.* **2015**, *129*, 1155–1162. [CrossRef]

39. Arnold, C.Y. The determination and significance of the base temperature in a linear heat unit system. *Proc. Am. Soc. Hortic. Sci.* **1959**, *74*, 430–445.
40. Reibe, S.; Doetinchem, P.V.; Madea, B. A new simulation-based model for calculating post-mortem intervals using developmental data for *Lucilia sericata* (Dipt.: *Calliphoridae*). *Parasitol. Res.* **2010**, *107*, 9–16. [CrossRef] [PubMed]
41. Marchenko, M.I. Development of *Chrysomyia albiceps* WD. (Diptera, *Calliphoridae*). *Entomol. Rev.* **1985**, *64*, 107–112.
42. Greenberg, B.; Kunich, J.C. *Entomology and the Law: Flies as Forensic Indicators*; Cambridge University Press: Cambridge, MA, USA, 2002.
43. Tarone, A.M.; Foran, D.R. Gene expression during blow fly development: Improving the precision of age estimates in forensic entomology. *J. Forensic Sci.* **2011**, *56*, S112–S122. [CrossRef]
44. Park, J.E.; Shin, S.E.; Park, S.H.; Jeong, S.J.; Park, S.H.; Moon, T.Y.; Lee, J.W. A study for estimating growth time of *Calliphoridae* flies using statistical models. *J. Sci. Crim. Investig.* **2019**, *13*, 89–94. [CrossRef]

Article

Impact of Comingled Heterospecific Assemblages on Developmentally Based Estimates of the Post-Mortem Interval—A Study with *Lucilia sericata* (Meigen), *Phormia regina* (Meigen) and *Calliphora vicina* Robineau-Desvoidy (Diptera: Calliphoridae)

Krystal R. Hans [1,2,*] and Sherah L. Vanlaerhoven [1]

[1] Department of Biology, University of Windsor, 401 Sunset Ave, Windsor, ON N9B 3P4, Canada; vanlaerh@uwindsor.ca
[2] Department of Entomology, Purdue University, 901 W. State St., West Lafayette, IN 47907, USA
* Correspondence: hans3@purdue.edu; Tel.: +1-765-494-1079

Simple Summary: In forensic entomology, blow flies are often the first insects to arrive to decomposing remains. The development rates of blow flies are used to estimate the minimum amount of time between death and discovery of the remains, or post-mortem interval (PMI). When there are multiple species of flies interacting on the same remains, there could be changes in the development of the flies. We tested the development of three different species of blow flies in different combinations at different temperatures and measured the development and the rate of growth. One species (*Lucilia sericata*) grew larger when it developed with the species *Phormia regina* at certain temperatures. The larvae of *Calliphora vicina* gained weight slower when interacting with *P. regina* and *P. regina* grew faster when interacting with these two other species. Due to these differences in the development rates of the flies, depending on the species they are interacting with, more research is needed to further examine other species combinations and temperatures.

Citation: Hans, K.R.; Vanlaerhoven, S.L. Impact of Comingled Heterospecific Assemblages on Developmentally Based Estimates of the Post-Mortem Interval—A Study with *Lucilia sericata* (Meigen), *Phormia regina* (Meigen) and *Calliphora vicina* Robineau-Desvoidy (Diptera: Calliphoridae). *Insects* **2021**, *12*, 280. https://doi.org/10.3390/insects12040280

Academic Editors: Damien Charabidze and Daniel Martín-Vega

Received: 27 February 2021
Accepted: 23 March 2021
Published: 25 March 2021

Publisher's Note: MDPI stays neutral with regard to jurisdictional claims in published maps and institutional affiliations.

Abstract: Estimates of the minimum post-mortem interval (mPMI) using the development rate of blow flies (Diptera: Calliphoridae) are common in modern forensic entomology casework. These estimates are based on single species developing in the absence of heterospecific interactions. Yet, in real-world situations, it is not uncommon to have 2 or more blow fly species developing on a body. Species interactions have the potential to change the acceptance of resources as suitable for oviposition, the timing of oviposition, growth rate, size and development time of immature stages, as well as impacting the survival of immature stages to reach adult. This study measured larval development and growth rate of the blow flies *Lucilia sericata* (Meigen, 1826), *Phormia regina* (Meigen, 1826) and *Calliphora vicina* Robineau-Desvoidy (Diptera: Calliphoridae) over five constant temperatures (15, 20, 25, 30, 35 °C), in the presence of conspecifics or two-species heterospecific assemblages. Temperature and species treatment interacted such that *L. sericata* larvae gained mass more rapidly when in the presence of *P. regina* at 20 and 30 °C, however only developed faster at first instar. At later stages, the presence of *P. regina* slowed development of *L. sericata* immatures. Development time of *C. vicina* immatures was not affected by the presence of *P. regina*, however larvae gained mass more slowly. Development time of *P. regina* immatures was faster in the presence of either *L. sericata* or *C. vicina* until third instar, at which point, the presence of *L. sericata* was neutral whereas *C. vicina* negatively impacted development time. *Phormia regina* larvae gained mass more rapidly in the presence of *L. sericata* at 20 °C but were negatively impacted at 25 °C by the presence of either *L. sericata* or *C. vicina*. The results of this study indicate that metrics such as development time or larval mass used for estimating mPMI with blow flies are impacted by the presence of comingled heterospecific blow fly assemblages. As the effects of heterospecific assemblages are not uniformly positive or negative between stages, temperatures or species combinations, more research into these effects is vital. Until then, caution should be used when estimating mPMI in cases with multiple blow fly species interacting on a body.

Keywords: medico-legal entomology; time of colonization; accumulated degree day estimates; length-weight estimates; species interactions

1. Introduction

To estimate the minimum Post-mortem Interval (mPMI) using insect evidence, there are two general approaches; one method uses the predictable rates of development of blow flies (Diptera: Calliphoridae) [1–3] and the other uses the predictable changes in community composition in the succession of insects through decomposition [4,5]. Blow fly species have different growth and developmental rates, which have been measured for numerous species including: *Lucilia sericata* Meigen [6–11], *Calliphora vicina* Robineau-Desvoidy [6,8,10,12,13], and *Phormia regina* Meigen [6,8,14,15]. Using these species-specific development rates, combined with temperature conditions that the larvae experience during development, the mPMI estimate can be calculated. Numerous studies provide examples of variable development within species of blow flies. This may be a result of different methods such as food source, density, constant versus fluctuating temperatures, or different geographic populations [7,11,16–18]. Greater awareness of this issue is being addressed in the forensic entomology community through validation studies that compare estimates of mPMI using different developmental data [19].

Developmentally based estimates of mPMI must first determine how far along the insect evidence has progressed through immature larval development. This is done by examining the larval spiracular slits for instar determination [2,20], and/or by measuring larval size using either length or weight [3,21]. Any factors that impact the timing of larval development between instars or impact the rate at which larvae grow or gain mass will impact mPMI estimates based on these metrics. It is well known that growth rate, larval weight and adult size may all be influenced by temperature [22] and density of larvae on the food source [23–25]. Less well studied is the effect of species interactions, including competition, which can alter body size and fecundity of blow flies [25–30]. Yet, understanding the role of species interactions and temperature on the behavior and development of blow flies should allow for more accurate estimates of mPMI [31].

Although many development studies have examined the influence of temperature on blow flies, many of these studies examine only conspecifics developing, with species interactions noticeably absent [14,15,32–34] and it is likely that species interactions together with environmental effects result in variable blow fly development. Group oviposition sets up a potential scenario for different species interactions among larvae, such as intraspecific and interspecific competition [24,35,36]. The blow flies *L. sericata*, *C. vicina* and *P. regina* are three widely distributed species that are frequently encountered on decomposing remains [37]. These blow fly species often arrive to carrion and oviposit large aggregations of eggs. *Lucilia sericata* and *P. regina* have the same developmental range of 10–35 °C [14,38–40], and often colonize the same carrion resource [41–44]. *Calliphora vicina* has a colder developmental range of between 3.5–30 °C [13,45]. In southern Ontario (Canada), *C. vicina* and *P. regina* co-occur during the spring and late fall, whereas *L. sericata* and *P. regina* co-occur during the late spring through to early fall (VanLaerhoven, personal observations).

It is generally believed that *L. sericata* is a poor competitor, exhibiting negative effects such as reduced body size and reduced survival due to intraspecific competition [35,46,47]. However, when competing with *P. regina*, *L. sericata* development and survival were not reduced, but *P. regina* larvae were completely eliminated from resources when competing with *L. sericata* [48].

The growth rate of *C. vicina* is often greater than that of other species, due to the larger size of this species [35]. This species experiences increased mortality and reduced adult size due to intraspecific competition during development, indicating that the growth of *C. vicina* should be reduced during intraspecific interactions [24,35]. Due to their large size, *C. vicina* may not be as heavily impacted by interspecific competition when competing with smaller larvae and may have the ability to outcompete and exclude a smaller species. Conversely, the larger size of *C. vicina* may result in increased competition with other species due to their greater resource requirements.

The objective of this study was to measure the effect of conspecific or heterospecifics rearing conditions at five different constant temperatures between 15–35 °C on development time and growth rate of the blow flies *L. sericata*, *C. vicina* and *P. regina*.

2. Materials and Methods

2.1. Colony Maintenance

All adult flies were maintained in colonies at the University of Windsor and were housed in 46 × 46 × 46 cm aluminum cages (Bioquip 1450C aluminum collapsible cage) at 25 °C, 60% RH and a 12:12 L:D diel cycle. Colonies originated from wild-caught females, collected in Windsor, Ontario, Canada with new wild type flies trapped and added to the colonies every year. Adult flies were provided with sugar and water ad libitum. Pork liver was used as an oviposition substrate for gravid females within the colony cages. Larvae were given fresh liver throughout development and were monitored until emergence of adults. After emergence, flies were transferred to clean colony cages.

2.2. Experimental Design

The species treatments for this study were as follows: (1) *P. regina* only, (2) *L. sericata* only, (3) *C. vicina* only, and (4) *P. regina* and *L. sericata* mixed (5) *P. regina* and *C. vicina* mixed. In treatments 1–3, each species developed with conspecifics and therefore experienced intraspecific interactions only. Treatments 4–5 represented mixed species treatment, with the species experiencing both intra and interspecific interactions. All species and temperature treatments were replicated five times. Prior to setting up each treatment replicate, colony cages of adult flies were supplied with pork liver as an oviposition media to obtain egg masses of up to 2000 eggs. Eggs were held at 25 °C until eclosion, at which point, they were transferred to treatment conditions within an hour of eclosion. For each temperature and species treatment, 40 cups were prepared and were composed of 20 first stage larvae were transferred to 59 mL polystyrene cups using a dampened paintbrush. Density of larvae was maintained at 20 individuals regardless of treatment, thus in mixed species treatments, 10 individuals of each species were placed into the cup. Each cup contained 20 g of pork liver to ensure that excess liver would be present, as each larva requires between 0.5–1 g liver [49,50] and 1.5 cm of sawdust, to act as a pupation medium. All cups were covered with landscape tarp and secured with a plastic lid. Cups were placed into growth chambers (Conviron Adaptis A1000) that were programmed to a constant temperature, with 50% (±0.41–2.59) RH and a 12:12 L:D cycle. Temperatures within the growth chambers were programmed to be one of five temperatures (15 °C, 20 °C, 25 °C, 30 °C, 35 °C (±0.66)). Dataloggers (HOBO U-12 data loggers, Onset, Pocasset, MA, USA) were placed into the growth chambers to record hourly temperature and relative humidity.

2.3. Sampling

For development up to pupation, cups were checked every 8–12 h for developmental stage based on the number of spiracular slits or visual observation of pupation. Following pupation, cups were checked every 20–24 h for adult emergence. At each check, one cup from each species and temperature treatment was removed entirely from the growth chambers. All larvae present in that cup were immediately placed into boiling water for 30 s to prevent shrinkage and were then preserved in 70% ethanol. This continued until pupation was observed for all larvae in the polystyrene cups. In heterospecific treatments, larvae were identified to species using the peritreme and accessory oral sclerite [51] prior to determination of stage or larval mass. The wet mass of individual larvae was measured to the nearest 0.1 mg using an analytical scale (Denver Instruments M-120).

2.4. Statistical Analyses

Analyses were conducted for each mixed species combination (*L. sericata* with *P. regina*, *C. vicina* with *P. regina*, *P. regina* with *L. sericata* and *P. regina* with *C. vicina*) to determine if species composition influenced growth rate, minimum development time and pupal mass. To satisfy the assumptions of normality and homogeneity of variance for ANOVA, variables were square-root (larval weight) transformed.

Given the strong known effect of temperature on development time, instead of a two-way ANOVA to analyze interactions between temperature and species composition treatment, Bonferonni corrected t-tests were used to compare individual pairwise comparisons within stage and temperature combinations that differed between of a species on its own compared to development time in the presence of the second species. Thus, $\alpha = 0.003$ was used to distinguish significant differences.

To analyze if growth rate (mg larval mass/sampling time) differed between larvae in conspecific or heterospecific conditions, the relationship of larval mass over time was transformed into a linear relationship for each species treatment at each temperature using the equation:

$$HM = wH + k \tag{1}$$

HM = product of duration of growth (in hours) and larval mass (in mg)
k = y intercept or cumulative mass acquisition during development
w = rate of change in mass or slope
H = duration of growth (in hours)

Within each temperature and pairwise species treatment of conspecific to heterospecific, an indicator variable linear regression with interaction analysis was conducted to determine if slope of the conspecific treatment regression differs from that of the heterospecific treatment regression (indicated by an interaction term with $p \leq 0.05$). In addition, evidence of an overall mean difference between conspecific and heterospecific growth rate is indicated by a significant species treatment term ($p \leq 0.05$).

3. Results

3.1. Development Time

As expected, temperature affected the minimum development time to each stage of all three species in this study, with faster development times as temperature increased (Tables 1–3). *Phormia regina* took less time to develop at all temperatures tested when compared to *L. sericata* and *C. vicina*, whether developing with conspecifics or heterospecifics.

The presence of *p. regina* reduced the developmental time of *L. sericata* compared to developing with conspecifics for the first larval stages at some temperatures, but increased the developmental time of *L. sericata* for the second, third and pupal stages at some temperatures (Table 1). Development time to second instar was shorter when developing with *P. regina* than when developing with conspecifics at 20 and 30 °C with development time either not different or not significantly shorter but trending shorter at other temperatures. In contrast, development time to subsequent stages was longer when *L. sericata* developed with *P. regina*, particularly for time to third instar at 25 °C, time to pupation at 15 °C and time to adult emergence at 25 °C. Development time between *L. sericata* raised alone or with *P. regina* either did not differ or was not significantly longer but trending longer at other stages and temperatures for those reared with *P. regina*.

The presence of *L. sericata* reduced the developmental time of *P. regina* compared to developing with conspecifics for the first and second larval stages at some temperatures (Table 2). Development time to second instar was shorter when developing with *L. sericata* than when developing with conspecifics, except at 30 °C. Development time to third instar was only shorter when developing with *L. sericata* than when developing with conspecifics at 15 °C. Development time to pupal and adult emergence did not differ between those *P. regina* with conspecifics compared to those with *L. sericata*.

The presence of *C. vicina* reduced the developmental time of *P. regina* compared to developing with conspecifics for the first and second larval stages at some temperatures (Table 2). Development time to second and third instar was shorter when *P. regina* developed with *C. vicina* than when developing with conspecifics at 15 and 25 °C, and either not different or not significantly shorter but trending shorter at other temperatures. Development time to pupation was only longer when *P. regina* developed with *C. vicina* than when developing with conspecifics at 20 °C. Development time to adult emergence was longer for *P. regina* developing with *C. vicina* compared to those with conspecifics at 15–25 °C and not significantly longer but trending longer at 30 and 35 °C.

The presence of *P. regina* had no effect on the developmental time of *C. vicina* compared to developing with conspecifics at any stage or temperature (Table 3).

Table 1. Mean minimum development times (h) to reach each life stage starting from 1 h old first instar larvae for *L. sericata* when developing with conspecifics (*L. sericata* alone) and heterospecifics (*L. sericata* with *P. regina*). Variation around the mean is presented as the greater of either half the sampling interval (6 h for second instar to pupal stages, 12 h for adult stage) or the calculated standard error. Asterisk * indicate differences ($p < 0.003$) between *L. sericata* alone or with *P. regina* combinations within a single stage and temperature pairwise comparison.

Species	*L. sericata* **Alone**				*L. sericata* **with** *P. regina*			
Temp (°C)	**L2**	**L3**	**Pupae**	**Adult**	**L2**	**L3**	**Pupae**	**Adult**
15	92 ± 6	125 ± 6	437 ± 6	1119 ± 12	87 ± 6	147 ± 16	498 ± 30 *	1122 ± 55
20	49 ± 6	65 ± 6	221 ± 6	533 ± 12	24 ± 6 *	72 ± 6	228 ± 17	524 ± 30
25	16 ± 6	30 ± 6	158 ± 9	325 ± 12	16 ± 6	61 ± 6 *	139 ± 6	385 ± 12 *
30	25 ± 6	49 ± 6	111 ± 6	236 ± 12	16 ± 6 *	35 ± 6	128 ± 8	282 ± 17
35	17 ± 6	32 ± 6	145 ± 10	248 ± 18	12 ± 6	30 ± 6	138 ± 12	244 ± 12

Table 2. Mean minimum development times (h) to reach each life stage starting from 1 h old first instar larvae for *P. regina* when developing with conspecifics (*P. regina* alone) and heterospecifics (*P. regina* with *L. sericata* or with *C. vicina*). Variation around the mean is presented as the greater of either half the sampling interval (6 h for second instar to pupal stages, 12 h for adult stage) or the calculated standard error. Asterisk * indicate differences ($p \leq 0.003$) between *P. regina* alone and with *L. sericata* combinations, or with *C. vicina* combinations within a single stage and temperature pairwise comparison.

Species	*P. regina* Alone				*P. regina* with *L. sericata*				*P. regina* with *C. vicina*			
Temp (°C)	L2	L3	Pupae	Adult	L2	L3	Pupae	Adult	L2	L3	Pupae	Adult
15	156 ± 7	209 ± 6	439 ± 42	869 ± 51	87 ± 6 *	147 ± 16 *	488 ± 40	972 ± 47	26 ± 6 *	101 ± 6 *	370 ± 6	1202 ± 12 *
20	51 ± 6	73 ± 6	217 ± 6	409 ± 12	24 ± 6 *	65 ± 6	212 ± 10	420 ± 20	43 ± 6	72 ± 6	298 ± 10 *	537 ± 27 *
25	50 ± 6	65 ± 6	139 ± 6	253 ± 12	16 ± 6 *	61 ± 6	133 ± 6	259 ± 12	24 ± 6 *	48 ± 6*	158 ± 6	277 ± 12 *
30	19 ± 6	35 ± 6	95 ± 6	201 ± 12	16 ± 6	35 ± 6	121 ± 10	227 ± 12	12 ± 6	24 ± 6	96 ± 6	216 ± 12
35	16 ± 6	31 ± 6	93 ± 6	174 ± 12	10 ± 6 *	27 ± 6	99 ± 6	195 ± 12	12 ± 6	24 ± 6	86 ± 10	182 ± 12

Table 3. Mean minimum development times (h) to reach each life stage starting from 1 h old first instar larvae for *C. vicina* when developing with conspecifics (*C. vicina* alone) and heterospecifics (*C. vicina* with *P. regina*). Variation around the mean is presented as the greater of either half the sampling interval (6 h for second instar to pupal stages, 12 h for adult stage) or the calculated standard error. N/A indicates all individuals died prior to reaching this stage.

Species	*C. vicina* alone				*C. vicina* with *P. regina*			
Temp (°C)	L2	L3	Pupae	Adult	L2	L3	Pupae	Adult
15	33 ± 6	78 ± 6	342 ± 6	798 ± 12	26 ± 6	72 ± 6	322 ± 6	787 ± 12
20	15 ± 6	53 ± 6	197 ± 6	505 ± 12	23 ± 6	53 ± 6	178 ± 6	432 ± 25
25	15 ± 6	27 ± 6	140 ± 8	413 ± 13	12 ± 6	24 ± 6	130 ± 10	451 ± 12
30	12 ± 6	24 ± 6	154 ± 6	366 ± 12	12 ± 6	24 ± 6	146 ± 6	N/A
35	29 ± 6	50 ± 6	216 ± 6	N/A	26 ± 6	55 ± 6	124 ± 33	N/A

3.2. Growth Rate

As expected, temperature affected growth rate (larval mass/time) of all three species in this study, with faster mass acquisition as temperature increased through larval stages until the start of the post-feeding period prior to pupation (Figures 1–3).

The slope of the regression line for larval growth rate of *L. sericata* differs between developing on its own compared to with *P. regina* at 30 °C such that *L. sericata* is slower in gaining mass over time when developing on its own (Figure 1; $r^2 = 0.96$, $F_{1,14} = 4.6$, $p = 0.05$). In addition, *L. sericata* acquires less overall mass when on its own compared to with *P. regina* at 20 and 30 °C ($r^2 = 0.86$, $F_{1,48} = 6.7$, $p = 0.01$; $r^2 = 0.96$, $F_{1,14} = 10.8$, $p = 0.005$, respectively).

At 20 °C, *P. regina* acquires more overall mass when with *L. sericata* than when on its own or with *C. vicina* ($r^2 = 0.89$, $F_{2,56} = 24.2$, $p < 0.0001$). At 25 °C, larval growth rate of *P. regina* differs between developing on its own compared to with *L. sericata* or *C. vicina* at 25 °C such that *P. regina* is initially slower in gaining mass but as development progresses, becomes faster when developing on its own (Figure 2; $r^2 = 0.92$, $F_{2,30} = 4.3$, $p = 0.02$). In addition, at 30 °C, both the slope of the regression line and the overall growth rate differ such that *P. regina* larvae developing with *C. vicina* gain more overall mass at a faster rate than *P. regina* alone or with *L. sericata* ($r^2 = 0.89$, slope: $F_{2,20} = 3.7$, $p = 0.04$, overall: $F_{2,20} = 5.4$, $p = 0.01$).

The slope of the regression line for larval growth rate of *C. vicina* differs between developing on its own compared to with *P. regina* at 15 and 25 °C such that *C. vicina* is initially slower at gaining mass when developing on its own but becomes faster when developing on its own over time (Figure 3; $r^2 = 0.89$, $F_{1,53} = 6.46$, $p = 0.01$; $r^2 = 0.93$, $F_{1,19} = 13.5$, $p = 0.002$, respectively). In addition, *C. vicina* acquires more overall mass when on its own compared to with *P. regina* at 25 and 30 °C ($r^2 = 0.93$, $F_{1,19} = 14.1$, $p = 0.001$; $r^2 = 0.94$, $F_{1,23} = 5.44$, $p = 0.03$, respectively).

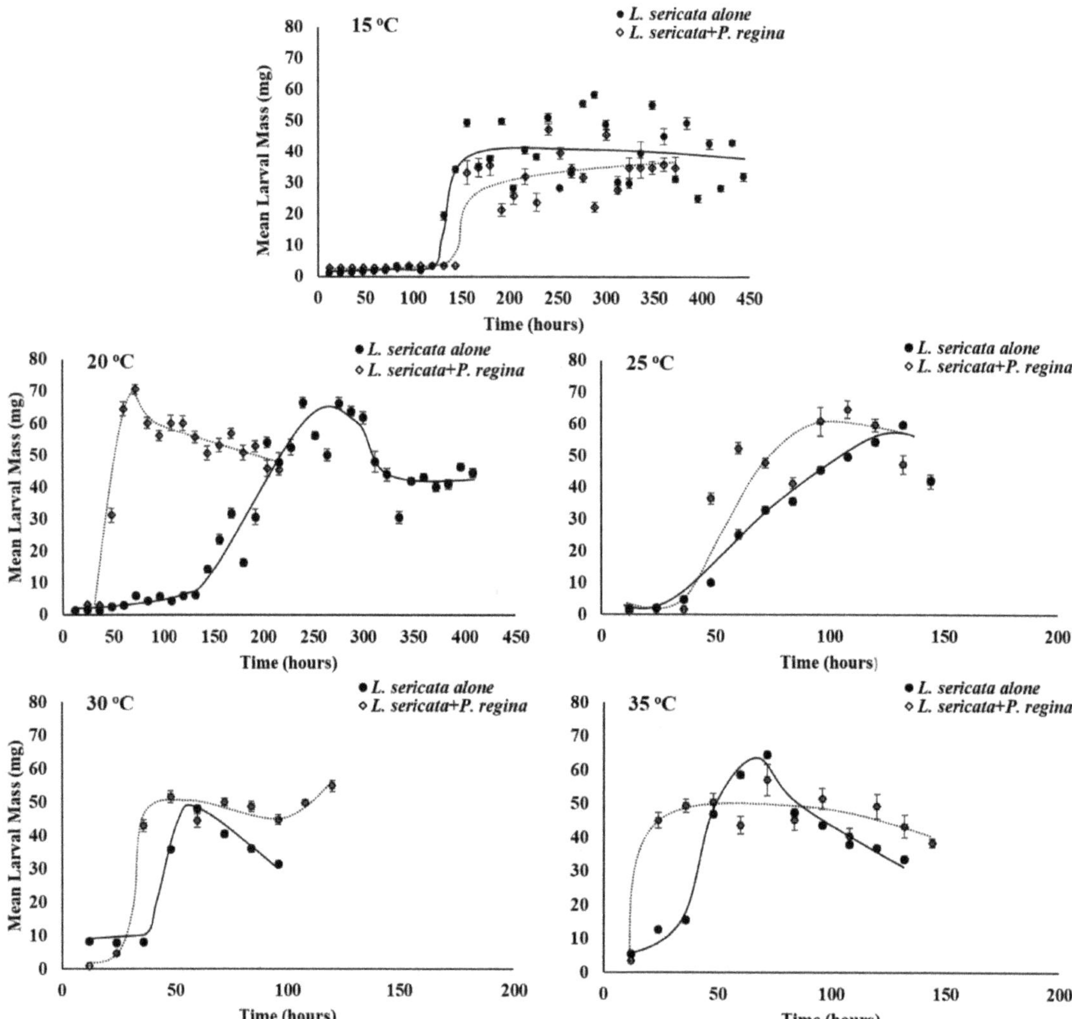

Figure 1. Mean (±SE) larval growth rate (mg/time) of *L. sericata* larvae reared with conspecifics, or reared with *P. regina* larvae at various temperatures. Indicator variable linear regression with interaction was conducted on the product of larval mass (mg) and time (in h) over time (not shown) for each species treatment and temperature combination. Slope (growth rate) differs at 30 °C ($r^2 = 0.96$, $F_{1,14} = 4.6$, $p = 0.05$). Overall growth rate differs at 20 and 30 °C ($r^2 = 0.86$, $F_{1,48} = 6.7$, $p = 0.01$; $r^2 = 0.96$, $F_{1,14} = 10.8$, $p = 0.005$, respectively).

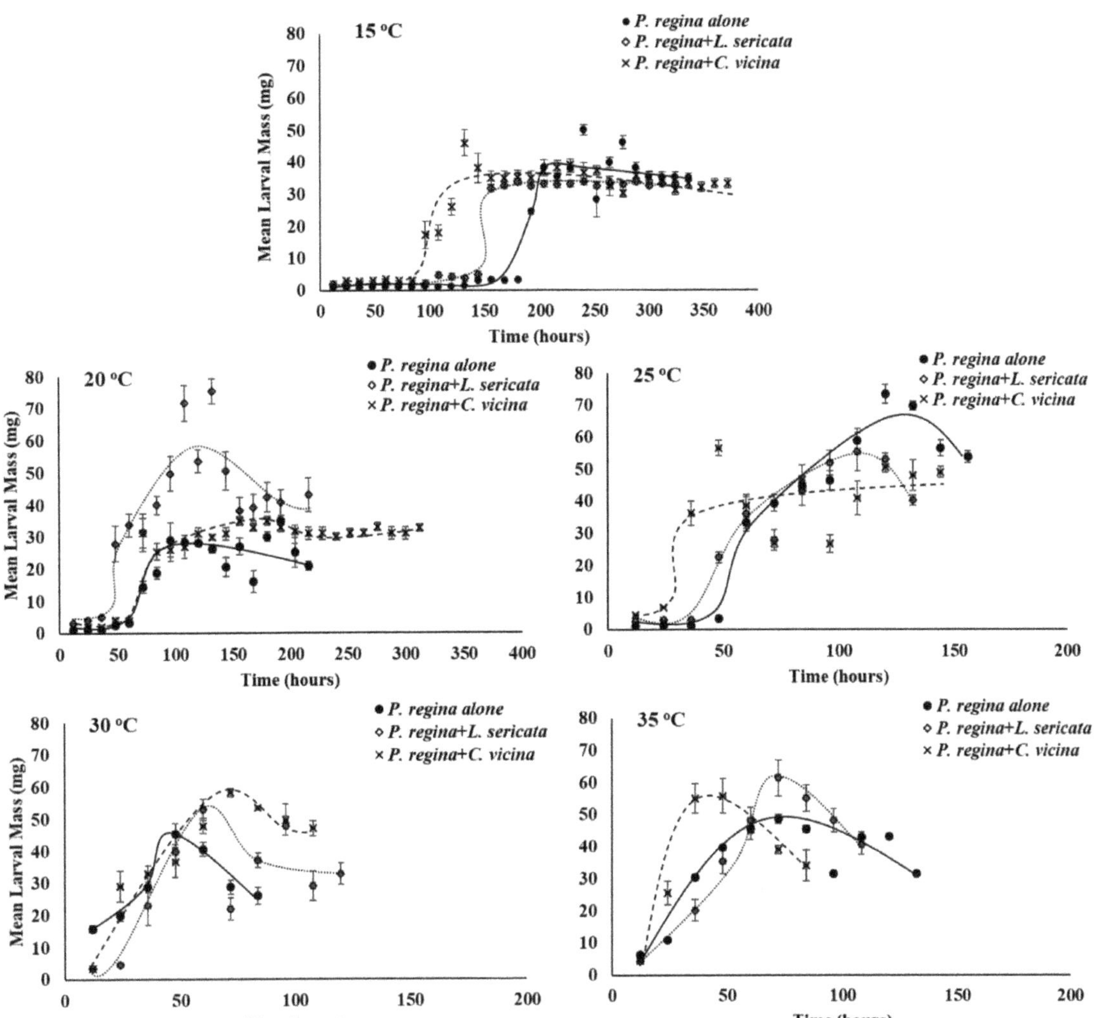

Figure 2. Mean (±SE) larval growth rate (mg/time) of *P. regina* larvae reared with conspecifics, or reared with *L. sericata or C. vicina* larvae at various temperatures. Indicator variable linear regression with interaction was conducted on the product of larval mass (mg) and time (in h) over time (not shown) for each species treatment and temperature combination. Slope (growth rate) differs at 25 °C for *P. regina* alone ($F_{2,30} = 4.3$, $p = 0.02$) and at 30 °C for *P. regina* with *C. vicina* ($r^2 = 0.89$, $F_{2,20} = 3.7$, $p = 0.04$). Overall growth rate differs at 20 °C for *P. regina* with *L. sericata* ($r^2 = 0.89$, $F_{2,56} = 24.2$, $p < 0.0001$) and at 30 °C for *P. regina* with *C. vicina* ($r^2 = 0.89$, $F_{2,20} = 5.4$, $p = 0.01$).

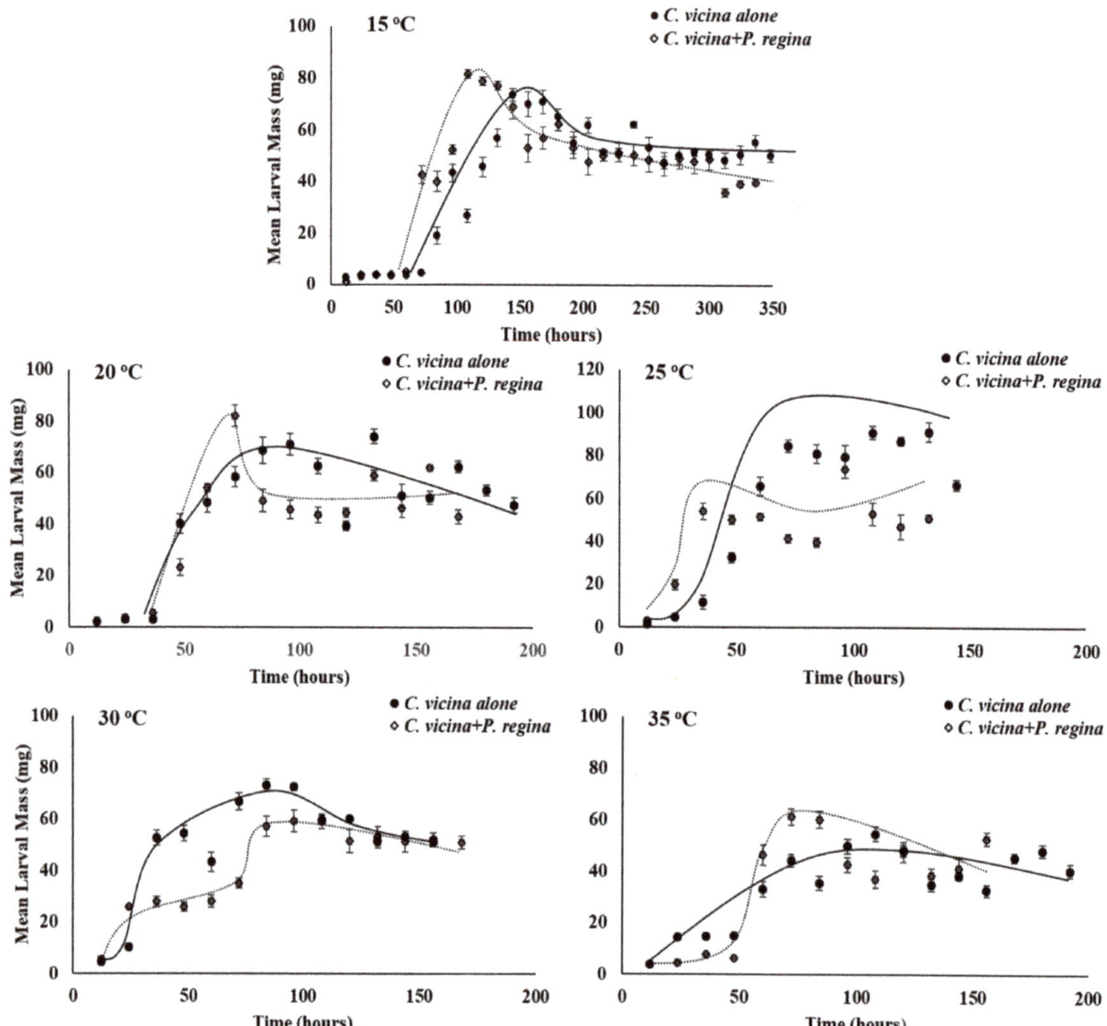

Figure 3. Mean (±SE) larval growth rate (mg/time) of *C. vicina* larvae reared with conspecifics, or reared with *P. regina* larvae at various temperatures. Indicator variable linear regression with interaction was conducted on the product of larval mass (mg) and time (in h) over time (not shown) for each species treatment and temperature combination. Slope (growth rate) differs at 15 and 25 °C ($r^2 = 0.89$, $F_{1,53} = 6.46$, $p = 0.01$; $r^2 = 0.93$, $F_{1,19} = 13.5$, $p = 0.002$, respectively). Overall growth rate differs at 25 and 30 °C ($r^2 = 0.93$, $F_{1,19} = 14.1$, $p = 0.001$; $r^2 = 0.94$, $F_{1,23} = 5.44$, $p = 0.03$, respectively).

4. Discussion

This study has demonstrated that metrics such as development time or larval mass used for estimating mPMI with blow flies are impacted by the presence of comingled heterospecific blow fly assemblages and the effects are not uniformly positive or negative between stages, temperatures or species combinations (Table 4). Indeed, this complexity in response to heterospecifics is expected as co-existence of these three blow fly species within overlapping temporal and spatial scales requires evolution of fluctuation dependent and independent stabilizing mechanisms such as facilitation, environmental fluctuation and resource partitioning [52,53]. Similarly, it has been argued that larval aggregation may be costly or beneficial [54] but it is expected that the outcome depends on the specific species, stages, densities, amount of resources and temperatures [52].

Table 4. Summary of heterospecific species treatment effects on development time and growth rate (larval mass/time) when compared to species reared alone. + indicates a positive effect; − indicates a negative effect; = indicates a neutral effect.

Species Treatment	Development Time	Growth Rate
L. sericata with *P. regina*	+ until 2nd instar, particularly at 20 & 30 °C − 2nd instar onward	+ at 20 & 30 °C
C. vicina with *P. regina*	=	−
P. regina with *L. sericata*	+ until 3rd instar = 3rd instar onward	+ at 20 °C − at 25 °C
P. regina with *C. vicina*	+ until 3rd instar − 3rd instar onward	− at 25 °C

Previous studies have found that *L. sericata* is a poor competitor, by displaying negative effects on survival and adult size due to interspecific competition [35,46,47]. When reared together with *C. vicina*, *L. sericata* developed slower than when developing with conspecifics at 25 °C [55]. It has been suggested that costs of interspecific aggregation may be from generating temperatures closer to thermal maximum of a species, decreased quantity of food resource and availability of nutrients, and increased risk of pathogens and disease [54].

Other studies have indicated that *L. sericata* may be facilitated by the presence of other species, such as *P. regina* [56]. In our study, temperature and species treatment interacted such that *L. sericata* larvae gained mass more rapidly when in the presence of *P. regina* at 20 and 30 °C, providing support for the idea that *L. sericata* may benefit by the presence of *P. regina*. Indeed, *L. sericata* also developed faster with *P. regina*, but only during the first instar. However, at later stages the presence of *P. regina* slowed development of *L. sericata* immatures. In studies with *C. vicina*, *L. sericata* pupae were larger when reared with heterospecifics at 15 °C [57] and at 25 °C [55]. In a study by Fouche et al. [58], *L. sericata* adult size was larger when larvae were reared with *C. vicina*, than when reared separately at 25 °C on 7 day rotted liver but not on fresh liver. Similarly, *L. sericata* had reduced mortality when developing with *C. vicina*, than when developing with conspecifics at 25 °C [55] but not at 15 or 28 °C [57].

Lucilia sericata demonstrates considerable plasticity in growth and development [17,55,59,60] and adaptations to combat adverse conditions, either environmental or due to species interactions, consist of rapid growth during larval stages to reach their critical weight for pupation in order to promote greater survival [17,59]. Developmental plasticity is also observed in *C. vicina*, when developing in heterospecific combinations with *L. sericata*, developed faster and migrated earlier at higher temperatures [57,60].

At temperatures above 30 °C, Reiter [12] reported that *C. vicina* larvae exhibit inhibited growth, with high mortality and few surviving to pupation. This is not surprising given that *C. vicina* is considered a cold weather species [2,13]. At 35 °C, *C. vicina* did not successfully emerge from pupation regardless of species treatment. However, when developing with *P. regina*, *C. vicina* larvae failed to successfully emerge from pupation at 30 °C, an indication that *P. regina* negatively impacted *C. vicina* at this temperature. Given that development time of *C. vicina* immatures was not affected by the presence of *P. regina*, it is possible that temperature and the presence of *P. regina* interacted to reduce the ability of *C. vicina* to achieve its critical weight for successful pupation and emergence to adult.

Critical weights required for pupation have been studied in many insect systems [61–63]. Saunders and Bee [24] found that the minimum pupal weight for *C. vicina* was 30 mg when developing in low densities of less than 50 larvae, but this critical weight was reduced to 15–20 mg when *C. vicina* was developing at higher densities of 150 conspecific larvae or more. These results indicate that critical weight may fluctuate due to the influence of species interactions and competition. In this study, *C. vicina* larvae gained mass more slowly in the presence of *P. regina*, which may explain the lack of survival of *C. vicina* to adult emergence at 30 and 35 °C.

The presence of proteolytic enzymes released by *L. sericata* may facilitate more efficient feeding by other blow fly species [25,30,47,64]. Scanvion et al. [65] found that exodigestion with enzymes aided in the consumption of the carrion resource and enhanced development of *L. sericata*. When developing in higher densities, *L. sericata* developed faster and had decreased mortality, due to enzymatic activity [65]. Development time of *P. regina* immatures was faster in the presence of either *L. sericata* or *C. vicina* until third instar, at which point, the presence of *L. sericata* was neutral whereas *C. vicina* negatively impacted development time. The release of these enzymes may be a mechanism that facilitates feeding by *P. regina*, resulting in larger larvae at some temperatures and faster initial development.

Temperature can also have an effect, as there is a direct relationship between temperature and feeding rates; as temperatures increase, feeding rates increase and larvae may not be able to metabolize as quickly, leading to smaller individuals [17,66]. This may explain why *P. regina* larval growth rate was negatively impacted at 25 °C by the presence of either *L. sericata* or *C. vicina* but *P. regina* larvae gained mass more rapidly in the presence of *L. sericata* at 20 °C.

5. Conclusions

As presented in this study, there is complexity to understanding the impact of species interaction on the metrics used to estimate immature development, whether it be growth rate measured by larval mass or length over time, or time to each developmental stage. Clearly it depends on the particular species involved and the temperatures.

Fluctuating temperatures exemplify temperatures blow flies experience in natural conditions, but the results of fluctuating temperatures on larval development are mixed. Some species develop faster under fluctuating temperatures, such as *Calliphora vomitoria* L. [67], whereas development of *L. sericata*, *P. regina* and *C. vicina* are delayed under periods of fluctuating temperatures [8,68]. In addition, larvae may balance thermoregulation and social behavior, trading off aggregation in suboptimal temperatures with thermal optimization [69]. Since temperature interacts with development to mediate species interactions, fluctuating temperatures have the potential to change outcomes of species interactions.

It is likely that density of individuals will impact these outcomes. The density of 20 larvae utilized in this study limits the amount of competition/facilitation that the larvae can experience while developing. On carrion, much higher densities of larvae occur which is influenced by the recruitment of gravid females and larval aggregation behavior. Benefits of aggregation include increased heat production, enzyme activity and cooperative feeding, reduced risk of predation and parasitism and protection from fluctuations in environmental factors which overall may result in faster development and increased survival [3,54,70]. However, as density of individuals increases, the cost associated with aggregation are generating temperatures closer to thermal maximum of a species, decreased quantity of food resource and availability of nutrients, and increased risk of pathogens and disease which may result in smaller sizes, decreased weight and decreased survival of the species involved [54]. Presumably, this would also exacerbate species interactions.

However, it is not only the overall density that is important. It is likely that initial densities of each species' population affect the outcome of the interaction, where the more abundant species has the greatest probability of dominating a resource [52,71]. Known as founder control, at each carrion resource, a different species could arrive to and colonize the resource at a greater abundance ultimately resulting in different potential outcomes of species interactions on each of these carrion patches [52].

Ultimately, research that incorporates species interactions and ecological theory should allow forensic entomologists to better model insect development for use in estimating mPMI. As we account for more of the factors affecting insect development of each forensically relevant species we will improve our estimates, thereby increasing confidence in the interpretation of insect evidence in the judicial system [31].

Author Contributions: Conceptualization, K.R.H. and S.L.V.; Methodology, K.R.H. and S.L.V.; Formal Analysis, K.R.H.; Investigation: K.R.H.; Resources, S.L.V.; Data Curation, K.R.H.; Writing—Original Draft Preparation, K.R.H.; Writing—Review and Editing, K.R.H. and S.L.V.; Visualization, K.R.H. and S.L.V.; Supervision, S.L.V.; Funding Acquisition, S.L.V. All authors have read and agreed to the published version of the manuscript.

Funding: This work was supported by the Government of Ontario Early Researcher Award; Canadian Foundation for Innovation; Ontario Innovation Trust; and the University of Windsor.

Institutional Review Board Statement: Not applicable.

Data Availability Statement: The data presented in this study are openly available in FigShare at 10.6084/m9.figshare.14294699.

Acknowledgments: The authors would like to acknowledge Alexandra Basden, Eve Deck, Celia Li, Vincenzo Pacheco, and Sydney Wardell for their assistance with this project.

Conflicts of Interest: The authors have no conflict of interest to declare.

References

1. Nuorteva, P. Sarcosaprophagous insects as forensic indicators. In *Forensic Medicine: A Study in Trauma and Environmental Hazards*; Tedeschi, C.G., Tedeschi, L.G., Eckert, W.G., Eds.; Saunders: Philadelphia, PA, USA, 1977; Volume 2, pp. 1080–1084, ISBN 0721687725.
2. Smith, K.G.V. *A Manual of Forensic Entomology*; British Museum: Oxford, UK, 1986.
3. Catts, E.P.; Goff, M.L. Forensic Entomology in Criminal Investigations. *Ann. Rev. Entomol.* **1992**, *9*, 253–273. [CrossRef]
4. Payne, J.A. A Summer Carrion Study of the Baby Pig *Sus Scrofa* Linnaeus. *Ecol. Wash. C* **1965**, *45*, 592–602. [CrossRef]
5. Schoenly, K.; Reid, W. Dynamics of heterotrophic succession in carrion arthropod assemblages: Discrete series or a continuum of change? *Oecologia* **1987**, *73*, 192–202. [CrossRef] [PubMed]
6. Kamal, A.S. Comparative Study of Thirteen Species of Sarcosaprophagous Calliphoridae and Sarcophagidae (Diptera). 1. Bionomics. *Ann. Entomol. Soc. Am.* **1958**, *51*, 261–271. [CrossRef]
7. Ash, N.; Greenberg, B. Developmental temperature responses of the sibling species *Phaenicia sericata* and *Phaenicia pallescens*. *Ann. Entomol. Soc.* **1975**, *68*, 197–200. [CrossRef]
8. Greenberg, B. Flies as Forensic Indicators. *J. Med. Entomol.* **1991**, *28*, 565–577. [CrossRef] [PubMed]
9. Wall, R.; French, N.; Morgan, K.L. Effects of temperature on the development and abundance of the sheep blowfly *Lucilia sericata* (Diptera: Calliphoridae). *Bull. Entomol. Res.* **1992**, *82*, 125–131. [CrossRef]
10. Davies, L.; Ratcliffe, G.G. Development Rates of Some Pre-Adult Stages in Blowflies with References to Low Temperatures. *Med. Vet. Entomol.* **1994**, *8*, 245–254. [CrossRef]
11. Grassberger, M.; Reiter, C. Effect of temperature on *Lucilia sericata* (Diptera: Calliphoridae) development with special reference to the isomegalen- and isomorphen-diagram. *Forensic Sci. Int.* **2001**, *120*, 32–36. [CrossRef]
12. Reiter, E. Zum Wachstumsverhalten der Maden der blauen Sehmeissfliege *Calliphora vicina*. *Z. Rechtsmedizin* **1984**, *91*, 295–308. [CrossRef]
13. Donovan, S.E.; Hall, M.J.; Turner, B.D.; Moncrieff, C.B. Larval Growth Rates of the Blowfly, *Calliphora vicina*, over a Range of Temperatures. *Med. Vet. Entomol.* **2006**, *20*, 106–114. [CrossRef] [PubMed]
14. Byrd, J.H.; Allen, J.C. The Development of the Black Blow Fly, *Phormia regina* (Meigen). *Forensic Sci. Int.* **2001**, *120*, 79–88. [CrossRef]
15. Nabity, P.D.; Higley, L.G.; Heng-Moss, T.M. Effects of temperature on development of *Phormia regina* (Diptera: Calliphoridae) and use of developmental data in determining time intervals in forensic entomology. *J. Med. Entomol.* **2006**, *43*, 1276–1286. [CrossRef] [PubMed]
16. Anderson, G.S. Minimum and Maximum Development Rates of Some Forensically Important Calliphoridae (Diptera). *J. Forensic Sci.* **2000**, *45*, 824–832. [CrossRef]
17. Tarone, A.M.; Foran, D.R. Components of developmental plasticity in a Michigan population of *Lucilia sericata* (Diptera: Calliphoridae). *J. Med. Entomol.* **2006**, *43*, 1023–1033. [CrossRef] [PubMed]

18. Gallagher, M.B.; Sandhu, S.; Kimsey, R. Variation in Developmental Time for Geographically Distinct Populations of the Common Green Bottle Fly, *Lucilia sericata* (Meigen). *J. Forensic Sci.* **2010**, *55*, 438–442. [CrossRef]
19. VanLaerhoven, S.L. Blind validation of postmortem interval estimates using developmental rates of blow flies. *Forensic Sci. Int.* **2008**, *180*, 76–80. [CrossRef]
20. Gennard, D.E. *Forensic Entomology: An Introduction*; John Wiley & Sons Ltd.: Chichester, UK, 2007.
21. Wells, J.D.; LaMotte, L.R. Estimating Maggot Age from Weight Using Inverse Prediction. *J. Forensic Sci.* **1995**, *40*, 585–590. [CrossRef]
22. Kingsolver, J.G.; Huey, R.B. Size, temperature, and fitness: Three rules. *Evol. Ecol. Res.* **2008**, *10*, 251–268.
23. Goodbrod, J.R.; Goff, M.L. Effects of larval population density on rates of development and interactions between two species of *Chrysomya* (Diptera: Calliphoridae) in laboratory culture. *J. Med. Entomol.* **1990**, *23*, 338–343. [CrossRef]
24. Saunders, D.S.; Bee, A. Effects of larval crowding on size and fecundity of the blow fly, *Calliphora vicina* (Diptera: Calliphoridae). *Eur. J. Entomol.* **1995**, *92*, 615–622.
25. Ireland, S.; Turner, B. The Effects of Larval Crowding and Food Type on the Size and Development of the Blowfly, *Calliphora vomitoria*. *Forensic Sci. Int.* **2006**, *159*, 175–181. [CrossRef]
26. Ives, A.R. Aggregation and the coexistence of competitors. *Ann. Zool. Fenn.* **1988**, *25*, 75–88.
27. Catts, E.P. Problems in Estimating the Postmortem Interval in Death Investigations. *J. Agric. Entomol.* **1992**, *9*, 245–255.
28. Von Zuben, C.J.; Bassanezi, R.C.; Reis, S.F.; Godoy, W.A.C.; von Zuben, F.J. Theoretical approaches to forensic entomology: I. Mathematical model of postfeeding larval dispersal. *J. Appl. Entomol.* **1996**, *120*, 379–382. [CrossRef]
29. Faria, L.D.B.; Orsi, L.; Trinca, L.A.; Godoy, W.A.C. Larval predation by *Chrysomya albiceps* on *Cochliomyia macellaria*, *Chrysomya megacephala* and *Chrysomya putoria*. *Entomol. Exp. Appl.* **1999**, *90*, 149–155. [CrossRef]
30. Dos Reis, S.F.; Von Zuben, C.J.; Godoy, W.A.C. Larval aggregation and competition for food in experimental populations of *Chrysomya putoria* (Wied.) and *Cochliomyia macellaria* (F.) (Dipt., Calliphoridae). *J. Appl. Entomol.* **1999**, *123*, 485–489. [CrossRef]
31. VanLaerhoven, S.L. Ecological theory and its application in forensic entomology. In *Forensic Entomology: The Utility of Arthropods in Legal Investigations*, 2nd ed.; Byrd, J.H., Castner, J.L., Eds.; CRC Press: Boca Raton, FL, USA, 2010; pp. 493–517.
32. Byrd, J.H.; Butler, J.F. Effects of temperature on *Cochliomyia macellaria* (Diptera: Calliphoridae) development. *J. Med. Entomol.* **1996**, *33*, 901–905. [CrossRef]
33. Byrd, J.H.; Butler, J.F. Effects of temperature on *Chrysomya rufifacies* (Diptera: Calliphoridae) development. *J. Med. Entomol.* **1997**, *34*, 353–358. [CrossRef]
34. Grassberger, M.; Reiter, C. Effect of Temperature on Development of the Forensically Important Holarctic Blow Fly *Protophormia terraenovae* (Robineau-Desvoidy) (Diptera: Calliphoridae). *Forensic Sci. Int.* **2002**, *128*, 177–182. [CrossRef]
35. Smith, K.E.; Wall, R. Asymmetric competition between larvae of the blowflies *Calliphora vicina* and *Lucilia sericata* in carrion. *Ecol. Entomol.* **1997**, *22*, 468–474. [CrossRef]
36. Von Zuben, C.J.; Von Zuben, F.L.; Godoy, W.A.C. Larval competition for patchy resources in *Chrysomya megacephala* (Dipt., Calliphoridae): Implications of the spatial distribution of immatures. *J. Appl. Entomol.* **2001**, *125*, 537–541. [CrossRef]
37. Byrd, J.H.; Castner, J.L. Insects of forensic importance. In *Forensic Entomology: The Utility of Arthropods in Legal Investigations*, 2nd ed.; Byrd, J.H., Castner, J.L., Eds.; CRC Press: Boca Raton, FL, USA, 2010; pp. 39–126.
38. Deonier, C.C. Carcass temperatures and their relation to winter blowfly populations and activity in the southwest. *J. Econ. Entomol.* **1940**, *33*, 166–170. [CrossRef]
39. Gosselin, M.; Charabidze, D.; Frippiat, C.; Bourel, B.; Gosset, D. Development Time Variability: Adaptation of Régnière's Method to the Intrinsic Variability of Belgian *Lucilia sericata* (Diptera, Calliphoridae) Population. *J. Forensic Res.* **2010**, *1*. [CrossRef]
40. Roe, A.; Higley, L.G. Development Modeling of *Lucilia sericata* (Diptera: Calliphoridae). *PeerJ* **2015**, *3*, e803. [CrossRef] [PubMed]
41. Anderson, G.S.; VanLaerhoven, S.L. Initial Studies on Insect Succession on Carrion in Southwestern British Columbia. *Forensic Sci.* **1996**, *41*, 617–625. [CrossRef]
42. VanLaerhoven, S.L.; Anderson, G.S. Insect succession on buried carrion in two biogeoclimatic zones of British Columbia. *J. Forensic Sci.* **1999**, *44*, 32–43. [CrossRef] [PubMed]
43. Sharanowski, B.J.; Walker, E.G.; Anderson, G.S. Insect Succession and Decomposition Patterns on Shaded and Sunlit Carrion in Saskatchewan in Three Different Seasons. *Forensic Sci. Int.* **2008**, *179*, 219–240. [CrossRef] [PubMed]
44. Vanin, S.; Zanotti, E.; Gibelli, D.; Taborelli, A.; Andreola, S.; Cattaneo, C. Decomposition and entomological colonization of charred bodies—A pilot study. *Croat. Med. J.* **2013**, *54*, 387–393. [CrossRef]
45. Hwang, C.C.; Turner, B.D. Small-Scaled Geographical Variation in Life-History Traits of the Blowfly *Calliphora vicina* between Rural and Urban Populations. *Entomol. Exp. Appl.* **2009**, *132*, 218–224. [CrossRef]
46. Prinkkilä, M.; Hanski, I. Complex competitive interactions in four species of *Lucilia* blowflies. *Ecol. Entomol.* **1995**, *20*, 261–272. [CrossRef]
47. Kheirallah, A.M.; Tantawi, T.I.; Aly, A.H.; El-Moaty, Z.A. Competitive interaction between larvae of *Lucilia sericata* (Meigen) and *Chrysomya albiceps* (Weidemann) (Diptera: Calliphoridae). *Pak. J. Biol. Sci.* **2007**, *10*, 1001. [CrossRef]
48. Hutton, G.F.; Wasti, S.S. Competitive interactions between larvae of the green bottle fly, *Phaenicia sericata* (Meig.) and the black blow fly, *Phormia regina* (Meig.). *Comp. Physiol. Ecol.* **1980**, *5*, 1–4.
49. Ives, A.R. Aggregation and coexistence in a carrion fly community. *Ecol. Monogr.* **1991**, *61*, 75–94. [CrossRef]

50. Reid, C.M. The Role of Community Composition and Basal Resources in a Carrion Community (Diptera: Calliphoridae). Master's Thesis, University of Windsor, Windsor, ON, Canada, 2012. Available online: scholar.uwindsor.ca (accessed on 21 February 2021).
51. Szpila, K. Key for the identification of third instars of European Blowflies (Diptera: Calliphoridae) of forensic importance. In *Current Concepts in Forensic Entomology*; Amendt, J., Campobasso, C.P., Goff, M.L., Grassberger, M., Eds.; Springer: Berlin/Heidelberg, Germany, 2010; pp. 43–56.
52. VanLaerhoven, S.L. Modeling species interactions within carrion food webs. In *Carrion Ecology, Evolution, and Their Applications*; Benbow, M.E., Tomberlin, J.K., Tarone, A.M., Eds.; CRC Press: Boca Raton, FL, USA, 2015; pp. 231–245.
53. VanLaerhoven, S.L. Ecological theory of community assembly and its application in forensic entomology. In *Forensic Entomology: The Utility of Arthropods in Legal Investigations*, 3rd ed.; Byrd, J.H., Tomberlin, J.K., Eds.; CRC Press: Boca Raton, FL, USA, 2020; pp. 387–404.
54. Rivers, D.; Thompson, C.; Brogan, R. Physiological trade-offs of forming maggot masses by necrophagous flies on vertebrate carrion. *Bull. Ent. Res.* **2011**, *101*, 599–611. [CrossRef]
55. Komo, L.; Scanvion, Q.; Hedouin, V.; Charabidze, D. Facing death together: Heterospecifics aggregations of blowfly larvae evince mutual benefits. *Behav. Ecol.* **2019**, *30*, 1113–1122. [CrossRef]
56. Pacheco, V.A. Served Medium Rare: The Effect of Burnt Remains on Oviposition, Survival and Fitness of the Local Blow Fly (Diptera: Calliphoridae) Community. Master's Thesis, University of Windsor, Windsor, ON, Canada, 2015. Available online: scholar.uwindsor.ca (accessed on 20 February 2021).
57. Komo, L.; Hedouin, V.; Charabidze, D. Benefits of heterospecific aggregation on necromass: Influence of temperature, group density, and composition on fitness-related traits. *Insect Sci.* **2020**, *28*, 144–152. [CrossRef]
58. Fouche, Q.; Hedouin, V.; Charabidze, D. Effect of density and species preferences on collective choices: An experimental study on maggot aggregation behaviours. *J. Exp. Biol.* **2021**, *224*, jeb233791. [CrossRef]
59. Ullyett, G.C. Competition for food and allied phenomena in sheep-blowfly populations. *Philos. Trans. R. Soc. Lond. B Biol. Sci.* **1950**, *234*, 77–174.
60. Komo, L.; Hedouin, V.; Charabidze, D. Quickie well done: No evidence of physiological costs in the development race of Lucilia sericata necrophagous larvae. *Physiol. Entomol.* **2019**, *45*, 30–37. [CrossRef]
61. Kingsolver, J.G. Feeding, growth, and the thermal environment of cabbage white caterpillars, *Pieris rapae* L. *Physiol. Biochem. Zool.* **2000**, *73*, 621–628. [CrossRef]
62. Davidowitz, G.; D'Amico, L.J.; Nijhout, H.F. Critical weight in the development of insect body size. *Evol. Dev.* **2003**, *5*, 188–197. [CrossRef]
63. Mirth, C.K.; Riddiford, L.M. Size assessment and growth control: How adult size is determined in insects. *Bioessays* **2007**, *29*, 344–355. [CrossRef] [PubMed]
64. Baxter, J.A.; Morrison, P.E. Dynamics of growth modified by larval population density in the flesh fly Sarcophaga bullata. *Can. J. Zool.* **1983**, *61*, 512–517. [CrossRef]
65. Scanvion, Q.; Hedouin, V.; Charabidze, D. Collective exodigestion favours blow fly colonization and development on fresh carcasses. *Anim. Behav.* **2018**, *141*, 221–232. [CrossRef]
66. Atkinson, D. On the solutions to a major life-history puzzle. *Oikos* **1996**, *77*, 359–365. [CrossRef]
67. Hagstrum, D.W.; Hagstrum, W.R. A simple device for producing fluctuating temperatures, with an evaluation of the ecological significance of fluctuating temperatures. *Ann. Entomol. Soc.* **1970**, *63*, 1385–1389. [CrossRef] [PubMed]
68. Niederegger, S.; Pastuschek, J.; Mall, G. Preliminary Studies of the Influence of Fluctuating Temperatures on the Development of Various Forensically Relevant Flies. *Forensic Sci. Int.* **2010**, *199*, 72–78. [CrossRef] [PubMed]
69. Aubernon, C.; Hedouin, V.; Charabidze, D. The maggot, the ethologist and the forensic entomologist: Sociality and thermoregulation in necrophagous larvae. *J. Adv. Res.* **2019**, *16*, 67–73. [CrossRef]
70. Turner, B.; Howard, T. Metabolic heat generation in Dipteran larval aggregations: A consideration for forensic entomology. *Med. Vet. Entomol.* **1992**, *6*, 179–181. [CrossRef]
71. Mittelbach, G.G. *Community Ecology*; Sinauer Associates: Sunderland, MA, USA, 2012.

Article

Comparison of Accumulated Degree-Days and Entomological Approaches in Post Mortem Interval Estimation

Lorenzo Franceschetti [1,2], Jennifer Pradelli [3], Fabiola Tuccia [3], Giorgia Giordani [4], Cristina Cattaneo [2] and Stefano Vanin [5,6,*]

1   Forensic Medicine Unit, Department of Medical and Surgical Specialties, Radiological Sciences and Public Health, University of Brescia, Piazzale Spedali Civili, 1, 25123 Brescia, Italy; lorenzofranceschetti@gmail.com
2   LABANOF (Laboratorio di Antropologia e Odontologia Forense), Sezione di Medicina Legale Dipartimento di Scienze Biomediche per la Salute, Università degli Studi di Milano, Milano Via Luigi Mangiagalli, 37, 20133 Milano, Italy; cristina.cattaneo@unimi.it
3   School of Applied Sciences, University of Huddersfield, Queensgate, Huddersfield HD1 3DH, UK; jennifer.pradelli@gmail.com (J.P.); tuccia.fabiola@gmail.com (F.T.)
4   Dipartimento di Farmacia e Biotecnologie (FABIT), Alma Mater Studiorum Università di Bologna, 40126 Bologna, Italy; giorgia.giordani.gg@gmail.com
5   Dipartimento di Scienze della Terra dell'Ambiente e della Vita (DISTAV), Università di Genova, Corso Europa 26, 16132 Genova, Italy
6   National Research Council, Institute for the Study of Anthropic Impact and Sustainability in the Marine Environment (CNR-IAS), Via de Marini 6, 16149 Genova, Italy
*   Correspondence: stefano.vanin@unige.it

Citation: Franceschetti, L.; Pradelli, J.; Tuccia, F.; Giordani, G.; Cattaneo, C.; Vanin, S. Comparison of Accumulated Degree-Days and Entomological Approaches in Post Mortem Interval Estimation. *Insects* **2021**, *12*, 264. https://doi.org/10.3390/insects12030264

Academic Editors: Damien Charabidze and Daniel Martín-Vega

Received: 3 March 2021
Accepted: 16 March 2021
Published: 21 March 2021

Publisher's Note: MDPI stays neutral with regard to jurisdictional claims in published maps and institutional affiliations.

Simple Summary: Among the investigative questions to define the temporal frame of a criminal event, the time since death plays a fundamental role. After death, the body goes through a series of physical and chemical transformations—known as decomposition. The way in which different parts of the body undergo these transformations can be quantified with a scale of scores (TBS, the total body score) and used for the time since death evaluation using the accumulated degree-days (ADDs) parameter, which accounts for time and temperature. This method is reported as TBS/ADD. Another way to estimate the time since death is based on the insect development on the body. Flies represent the first body coloniser and the development of their immature stages is used to define the time of colonisation that is temperature dependent and species specific. In this study, the two methods were compared based on 30 forensic cases occurring in northern Italy. The results highlighted the limits of the TBS/ADD method and the importance of the entomological approach, keeping in mind that with insects the colonization time is evaluated. This time is the minimum time since death.

Abstract: Establishing the post mortem interval (PMI) is a key component of every medicolegal death investigation. Several methods based on different approaches have been suggested to perform this estimation. Among them, two methods based their evaluation on the effect of the temperature and time on the considered parameters: total body score (TBS)/accumulated degree-days (ADDs) and insect development. In this work, the two methods were compared using the results of minPMI and PMI estimates of 30 forensic cases occurring in northern Italy. Species in the family Calliphoridae (*Lucilia sericata*, *Calliphora vomitoria* and *Chrysomya albiceps*) were considered in the analyses. The results highlighted the limits of the TBS/ADD method and the importance of the entomological approach, keeping in mind that the minPMI is evaluated. Due to the fact that the majority of the cases occurred in indoor conditions, further research must also be conducted on the different taxa to verify the possibility of increasing the accuracy of the minPIM estimation based on the entomological approach.

Keywords: ADD; TBS; Diptera; PMI; colonisation; temperature

1. Introduction

Establishing the post mortem interval (PMI) is a key component of every medicolegal death investigation. The longer the time since death, the more imprecise the chronological or sequential indicators detectable on the remains are [1]. A reliable PMI estimation is extremely important for the reconstruction of events surrounding the death. Testimonial statements provided by neighbours, family members and co-workers are not always reliable, thus, crime investigators normally base the PMI assessment on ante mortem physical evidence when available (e.g., video recording, cell phone records, bank activities, calendars, social networking, medicine reminders, medical records, etc.). However, scientifically, the most accurate and precise PMI estimate relies on post mortem changes and decomposition processes, which are the prerogative of forensic pathologists or, depending on the country, of other forensic medical specialists (e.g., coroner) [2].

Human decomposition is a complex biological and chemical process that begins immediately after death and involves the interaction of cadaver enzymes, bacteria, fungi, and protozoa [3,4]. Body decomposition is characterised by stages of gradual physical decay, from the fresh stage to skeletonization, through a bloated stage, followed by an active and then an advanced stage [5–7]. Vass (2011) described the four main abiotic factors that affect the decomposition rate: temperature, moisture, pH and the partial pressure of oxygen [8]. In addition, other parameters influence the decomposition such as the cause of death, wounds or trauma, bodyweight, degree of exposure to sunlight, body coverage, insect activity and subsoil parameters [9–15].

Temperature is the most important factor affecting decomposition and for this reason, meteorological information is fundamental to estimate the PMI [9]. Obtaining measurements with the best possible confidence of the actual temperatures the body experienced at the potential crime scene is crucial for an accurate estimation [16]. After death, the environmental temperature affects the body's chemical and biochemical processes, impacting the decomposition and at the same time, affecting the entomological colonisation, both in terms of body search and insect development. Moreover, the temperature and moisture also have great influence on other organisms that can develop on or around the body, such as plants, fungi, bacteria and other microorganisms [17].

In forensic practice, methods for PMI estimation are based on the macroscopic examination of the soft tissues' decomposition degree. However, when the skeletal stage is reached, few markers for this purpose exist [18,19]. In the literature, the morphological changes that take place during decomposition have been described in detail [5,20]. Most of these descriptions are qualitative, based on personal opinions and experience, and not applicable to all geographic and environmental conditions [10,21,22]. Furthermore, many decomposition studies have been conducted in different seasons and climatic conditions using varying methods [15,22,23]. To bypass the limitation mentioned above, a method based on the observation of decomposition stages of different body districts have been developed in forensic anthropology. The body decomposition degree can be quantified using the total body score (TBS) [10]. It is a scale that distinguishes the different stages of decomposition, allowing to assign points to specific categories and eventually to score overall decomposition [24,25]. Numerous studies utilised the TBS method either with or without some modification [8,24–32].

Due to the relationship between temperature and decomposition, a semi-quantitative model to estimate PMI was based on accumulated degree-days (ADD). ADD represent the combination between chronological time and temperature. They are defined as heat-energy units representing the accumulation of thermal energy needed for chemical and biological reactions to take place in soft tissue during decomposition [10]. This method takes into consideration the overall body decomposition (evaluated by a score like the TBS) and the number of environmental degrees recorded since the death. Megyesi et al. (2005) proposed a retrospective study in which TBS and ADD could be used to quantitatively estimate PMI [10]. A few authors have tested Megyesi's research in their countries and have found positive correlation [33–35]. However, worldwide there is an increasing number of authors that did not manage to corroborate Megyesi's equation at their latitude (such as South Africa [22], Netherlands [32], Australia [36], Canada [37]). An alternative method for PMI estimation is based on the study of the insects developing on a body, taking in account that insect development is temperature dependent and species specific [38–42]. Forensic entomology uses, depending on the case, the development of necrophagous insects and the composition of insect communities present on the body after the death for the minimum PMI (minPMI) estimation defined as the colonisation time [41–51].

Despite several attempts and new technologies, medicolegal death investigation still lacks a universal and reliable method to estimate PMI that allows an investigation to proceed appropriately and without delay, while providing time for more complex analyses [2]. Currently, there is no scientifically recognised PMI estimation method for any specific or general geographic region. As more and more knowledge is gained from thanatological experimental studies, a working model that encompasses the factors affecting decomposition is becoming more and more plausible. Furthermore, since the environment plays a large role in the rate of decomposition, the applicability of current PMI models needs to be tested and validated at a regional level.

For these reasons, this study was aimed to test the efficiency and reliability of the semi-quantitative method of TBS and ADD applying the Megyesi mathematical formula [10] on real forensic cases and to compare it with the entomological evaluations of the PMI to verify the complementarity of the two methods in death investigations.

2. Materials and Methods

2.1. Accumulated Degree-Days Analysis

Thirty human corpses in an advanced decomposition stage from the Lombardia region (Northern Italy), subjected to on-site external examination and then to autopsy at the Institute of Legal Medicine, University of Milan, between 2016 and 2018, were included in this study. The bodies were stored under refrigerated conditions in adherence to local laws between the scene recovery and the autopsy. Photographs shot during the crime scene inspection and during the autopsy were used to quantify/evaluate the level of decomposition of each body using the TBS as suggested by Galloway et al. (1989) [5] and Nawrocka et al. (2016) [24]. The ADD for the PMI estimation were calculated using Megyesi's formula [10].

2.2. Entomological Analysis

The insects were sampled from the bodies before their removal from the crime scene and later during the autopsy following the EAFE (European Association for Forensic Entomology) guidelines [51] and stored at −20 °C until analysis as per GIEF (Gruppo Italiano di Entomologia Forense) protocol (GIEF, 2016) [52]. After defrosting, the larvae were fixed with hot water (80 °C, 1 min) and finally stored in an 80% ethanol solution.

Entomological samples were observed and photographed using a Leica M60 stereomicroscope equipped with a DFC425C camera and the LAS software (Leica, Germany).

Species were identified using morphological keys [53,54] and confirmed by molecular analysis after dissection as described by Tuccia et al. (2016) [55]. DNA was extracted from a fraction of larval tissue using the QIAamp DNA Investigator Kit (QIAGEN, Redwood City, CA, USA). The manufacture protocol was partially modified in order to increase the quality of the reaction by adding 4 µL of RNAse A (4 mg/mL) after over-night incubation with the Proteinase K (100 µg/mL) (PROMEGA, Madison, WI, USA). DNA was eluted in 100 µL of elution buffer and quantified via Qubit 3.0 fluorimeter (Thermo Scientific, Waltham, MA, USA). Polymerase Chain Reaction was carried out on mitochondrial COI gene 658 bp long and commonly used as a molecular target for insects' barcoding [56]. Universal LCO-1490 Forward primer (5′-GGTCAACAAATCATAAAGATATTGG-3′) and HCO-2198 Reverse primer (5′-TAAACTTCAGGGTGACCAAAAAATCA-3′) were used as described by Folmer and co-workers (1994) [57]. PROMEGA GoTaq® Flexi Polymerase protocol (PROMEGA, Madison, WI, USA) was followed in order to prepare a master mix reaction of 20 µL final volume: 4 µL of Colourless GoTaq Flexi Buffer (5×), 2 µL of MgCl$_2$ (25 mM), 0.5 µL of each primer (10 pmol/µL), 0.5 µL of Nucleotide Mix (10 mM), 0.25 µL GoTaq DNA Polymerase (5 u/µL) and 2 µL of DNA template. The following amplification program was set up on BioRad C1000 Thermal Cycler (Bio-Rad Laboratories, Inc., Hercules, California, USA): 95 °C for 10 min, 35 cycles of 95 °C for 1 min, 49.8 °C for 1 min, 72 °C for 1 min and a final elongation step at 72 °C for 1 min. Each reaction was confirmed by standard gel electrophoresis in 1.5% agarose gel previously stained with Midori Green Advanced DNA Stain (Geneflow, Elmhurst, UK). Thirty-five microliters of PCR products was purified using QIAquick PCR Purification Kit® (QIAGEN, Hilden, Germany) following the manufacturer's instructions. Purified amplicons were eluted in 40 µL of sterile/deionised water and sequenced by an external company (Eurofins Operon MWG, Ebersberg, Germany).

2.3. PMI Estimation and Statistical Analysis

Insect development is temperature dependent and species/population specific in the range between the minimal and the maximal developmental thresholds. Depending on the nature of the sample (already fixed or living specimens, larvae, close or empty puparia) two different methods can be applied to estimate the age of an insect. ADD is currently the most common method used for this estimation when living specimens can breed until the adult stage or when empty puparia are collected from the crime scene. Because of the minimal developmental threshold, the ADDs used in entomology differ from the ADDs applied in the anthropological method. For this reason, ADDs were derived from the average temperature as absolute values for the morphological evaluation, whereas if needed for the entomological evaluation, the minimal developmental threshold was subtracted from the absolute value.

minPMI was also evaluated using the larvae measurements applying the data available in the literature [58] when the samples were already fixed. In this case, the size of the larvae was compared with diagrams reporting the relations between size, temperature and time. Whereas some authors only report the size as a function of the temperature and time [58–60], others also provide the formulas and the interval of confidence of the larval length based on the ADDs [61]. Larvae were counted and then measured using Leica M60 stereomicroscope equipped with a CCD camera (Leica, Wetzlar, Germany) and the automatic measurement tool. When the numbers of larvae were under fifty, measurements were performed on all the specimens. When the count exceeded fifty, only fifty larvae were considered. Measurements were expressed in mm as the average and standard deviation (SD).

The estimations derived from the two methods were compared using an interclass correlation according to Cicchetti (1994) [62] and Koo and Li (2016) [63]. Statistical analyses were performed using the SPSS Statistics software (IBM, Armonk, NY, USA).

3. Results

Twenty-four of the deceased cases were males (80%) whereas 6 were females (20%). At the time of death, the mean age was 63 years (range of 37–88 years). According to the autopsy's reports, the main cause of death was cardiovascular diseases (22 cases, 73.5%), followed by cerebrovascular diseases and acute drugs intoxication (three cases, 10%). Pneumoniae was the cause of death in two cases (6.5%). The last time the person was seen alive ranged between 2 and 10 days. The average temperature for the period before body recovery is summarised per each case in Table 1.

3.1. Accumulated Degree-Days Estimate

The decomposition scores were evaluated for every region, and then added up to calculate the TBS. Data are summarised in Table 1, whereas the derived ADD with the PMI based on circumstantial data and entomological data are reported in Table 2. One case was removed from the PMI evaluation because the body conditions did not allow the application of the morphological method.

Table 1. Summary of the circumstantial, environmental temperature and stage of decomposition for each case (TBS: Total Body Score).

Case	Month and Year of Discovery	Environmental Temperature (Average ± SD, °C)	Stage of Decomposition			
			Head/Neck Score	Trunk Score	Limbs Score	TBS
1	X.2016	16.5 ± 1.0	4	3	2	9
2	III.2017	8.0 ± 2.0	6	3	4	13
3	II.2017	7.5 ± 1.5	4	3	3	10
4	V.2017	20.5 ± 3.0	5	3	4	12
5	IV2017	14.0 ± 1.5	4	3	4	11
6	III.2017	8.5 ± 2.5	4	4	4	12
7	X.2017	14.5 ± 1.5	4	3	3	10
8	I.2018	6.0 ± 2.5	4	4	3	11
9	II.2018	4.0 ± 2.5	5	3	4	12
10	II.2018	5.5 ± 1.0	5	4	4	13
11	XII.2018	3.5 ± 2.5	5	4	4	13
12	VII.2016	28.5 ± 1.5	5	4	3	12
13	VIII.2016	28.5 ± 1.5	5	4	3	12
14	VIII.2017	28.5 ± 2.5	5	3	4	12
15	IV.2018	18.0 ± 2.5	5	4	3	12
16	VIII.2018	27.0 ± 2.0				
17	IV.2016	20.5 ± 1.0	4	3	2	9
18	X.2018	16.5 ± 2.0	8	3	3	14
19	VII2018	26.0 ± 2.5	6	4	3	13
20	X.2016	20.0 ± 4.0	5	3	3	11
21	VIII.2018	29.0 ± 1.0	6	4	3	13
22	VII.2018	29.0 ± 1.0	7	5	4	16
23	VII.2018	26.0 ± 2.0	4	3	3	10
24	VIII.2016	26.0 ± 20	5	4	4	13
25	VIII.2018	29.0± 0.5	5	5	3	13
26	VII.2018	28.0± 1.5	7	5	3	15
27	IV.2017	15.0 ± 3.0	2	3	2	7
28	VIII.2017	25.5 ± 2.0	6	5	4	15
29	VI.2017	28.5 ± 2.5	5	3	4	12
30	VII.2018	26.5 ± 1.5	6	4	3	13

Table 2. Comparison between the circumstantial data estimation and technical assessment using accumulated degree-days (ADDs) converted in a 24 h range.

Taxon	Case Number	PMI from Circumstantial Data		TBS PMI Estimation		Entomological min PMI Estimation	
C. vicina	1	120	144	120	144	120	270
	2	168	192	168	192	120	200
	3	120	144	120	144	96	170
	4	120	144	120	144	96	180
	5	144	168	144	168	120	180
	6	144	168	144	168	96	144
	7	96	120	144	168	96	170
	8	96	120	144	168	96	170
	9	144	168	144	168	120	200
	10	96	120	168	192	72	110
	11	168	192	168	192	110	180
L. sericata	12	72	96	96	120	48	72
	13	72	96	96	120	48	72

Table 2. *Cont.*

Taxon	Case Number	PMI from Circumstantial Data		TBS PMI Estimation		Entomological min PMI Estimation	
	14	72	96	96	120	48	72
	15	168	192	168	192	84	180
	16	72	96	-	-	48	84
	17	96	120	96	120	78	108
	18	192	216	240	264	132	204
	19	96	120	168	192	54	90
	20	168	192	168	192	96	216
	21	48	72	72	96	48	54
	22	48	72	72	96	48	54
	23	216	240	96	120	48	84
	24	216	240	96	120	54	84
	25	48	72	144	168	48	60
	26	48	72	144	168	48	60
	27	144	168	120	144	168	204
C. albiceps	28	144	168	144	168	84	144
	29	72	96	96	120	60	108
	30	96	120	120	144	60	102

(green = positive concordance; yellow = concordance at ranges' limits; orange = no concordance) (PMI: post mortem interval; TBS: TotalBody Score).

The PMI based on the ADD estimates in 12 cases (40%) was in good agreement with the circumstantial data; in the remaining cases, it was overestimated (15 cases, 50%) or underestimated (three cases, 10%).

All cases resulting as positive outcomes were above 5 days for PMI, with the exception of one case, which had a PMI estimate in 4 days (Table 2).

3.2. Entomological Estimate

Calliphoridae sampled from the cases and used for the estimation were: *Lucilia sericata* (Meigen, 1826) in 16 cases (53%), *Calliphora vicina* Robineau-Desvoidy, 1830 in 11 cases (37%) and *Chrysomya albiceps* (Wiedemann, 1819) in three cases (10%).

In 21 cases (70%), the minimum entomological PMI correlates positively with the time of death from the circumstantial data; in the other cases there is an underestimation (9–30%) of the PMI provided by circumstantial information (Table 2).

3.3. Comparison

In 10 cases (33%), both methods allowed to have an estimate of the post-mortal interval in agreement to what was given by the circumstantial data. In 9 out of the 10 cases, the PMI exceeded over 5 days. In the tenth case, the PMI was included between 4 and 5 days (Table 2).

3.4. Statistical Analysis

Interclass correlation for the absolute value based on the central tendency value which resulted to be 0.58 (95% CI: 0.11–0.80) for circumstantial data vs. the ADD estimate and 0.71 for circumstantial data vs. the entomological evaluation (95% CI: 0.39–0.86).

Due to the fact that the ADD method estimates the PMI whereas the entomological one estimates the minimum PMI, the interclass correlation of these two variables was estimated in terms of consistency and not, as in the previous analysis, in terms of absolute value. The results indicate a value around 0.65–0.69 (95% CI: 0.25–0.85) for the lower, upper and central tendency values of the estimates, with an average underestimation of 49 h (95% CI: 36.4–61.5) for the lower limit and of 19.7 h (95% CI: 1.1–38.2) for the upper limits of the estimate.

4. Discussion

In this study, the PMI provided by the circumstantial data showed high variability, from a minimum of two days to a maximum of ten days. The ADD estimates did not allow to positively correlate the derived TBS with the circumstantial data in 60% of the cases. In these cases, the TBS scale presented a higher score of decomposition for the head/neck region, compared to the other body regions. These cases are distributed throughout the whole year and in no case was the body naked, the only parts exposed to the air were the head and the extremities of the limbs. In the remaining cases (40%), a positive concordance between the ADD estimate and the PMI derived from circumstantial data was found. The positive outcomes were shown for PMI above 5 days. This concordance can be explained by the fact that the ADD conversions and subsequent estimations became more accurate during the later stages of decomposition, as suggested by Parsons (2009) [34].

For shorter intervals, the ADD derived from the TBS scale lost effectiveness, while forensic entomology remains the most reliable method. This aspect should be considered as an important tool for the forensic pathologist who approaches a corpse in an advanced state of decomposition. In fact, anomalies of decomposition, affecting the TBS, can easily occur because of ante mortem injuries or post mortem scavenging of human remains due to insects or larger animals such as wolfs, dogs, birds, etc. [64].

Regarding the entomological estimate, our study showed more than two-thirds of the estimations positively arranged the PMI derived from circumstantial information. In other cases, an underestimation of minPMI was found. This observation does not surprise—despite it is, very often, a matter of discussion in the court of law—due to the fact that with insects the minimum time since death is calculated. In fact, for estimating PMI from immature insects it has to be considered also the development interval and the pre-appearance interval (PAI), which is currently scientifically invaluable [65–68]. The "underestimation" is explained as the time of insects' occurrence on a corpse. The casuistry considered for the study was composed of indoor cases, whereas the colonisation of bodies can record delays, depending on how easy it is for gravid females to access the body. Access of the body to insects is the second most important variable affecting the decomposition rate of the human body after temperature [9]. Since blowflies (Diptera: Calliphoridae) are usually the first necrophagous fauna to find a cadaver, the PMI estimation needs to take into account the factors that may delay the arrival of adult flies and subsequent oviposition by the females [68–70]. Further works, based on real cases have, however, to consider the time of colonization in indoor cases by other taxa (e.g.: Phoridae), and their informative potentiality, as already suggested by Bugelli and other authors [71,72]. The interclass correlation (0.71) calculated turned out to be "good" (very close to "excellent"). So, these results emphasise the applicability of forensic entomology in legal and medical cases, in accordance with the international guidelines and standards that help and strengthen forensic practitioners' assessment. Nevertheless, these real forensic cases could be an important factor in improving data collection, and thus optimising scientific results and inferences.

Further retrospective studies and experimentation research should indeed be conducted using cases with a known and large PMI so that not only the precision of the estimation could be measured but also the accuracy. Moreover, other PMI estimation assessments could be taken into account.

5. Conclusions

In conclusion, forensic pathologists must consider every aspect before giving a statement about PMI. They always have to be aware of the possible source of errors and variability, particularly when dealing with cadavers in active-advanced decomposition. If the decomposition stage of the corpse is highly developed, additional pieces of information can be provided by ADD and entomological assessments of PMI.

In our study, the ADD method has shown some limitations, probably due to the experimental morphological approach still not being validated. The TBS and the derived ADD tend to overestimate the degree of the decomposition. The high variability can probably be due to the different stages of decomposition, especially for the head district, even if all the body came from the same geographical area and death occurred in indoor environment [29,73].

The results of our study, as observed by other authors in the literature [31,74], showed that scoring scales and regress equations derived for predicting ADD seem to be of little help in forensic practice, because of the so many factors affecting human decomposition and leading to irregular decomposition patterns. Furthermore, the TBS compared to ADD proved to be a good method for a PMI over 5 days. This emerging result represents an interesting new finding that still needs more research to be statistically validated and considered in real forensic practice.

On the other hand, forensic entomology has been confirmed as a well-validated approach in the field.

The study of these cases has therefore led to confirm that the PMI estimate of human remains requires multidisciplinary activity between different professional figures (anthropologists, entomologists, botanists, geologists and zoologists) aimed to integrate and interpret available data, all subjected to the medicolegal assessment of cadaveric decomposition.

Author Contributions: Conceptualization S.V., C.C., L.F.; Methodology S.V., J.P., F.T., G.G.; formal analysis and data curation L.F., J.P., F.T.; writing—original draft preparation, L.F., S.V.; writing—review and editing: L.F., S.V., G.G.; funding acquisition S.V. All authors have read and agreed to the published version of the manuscript.

Funding: This research received no external funding.

Institutional Review Board Statement: Not Applicable.

Data Availability Statement: Data are available upon request from the authors.

Conflicts of Interest: The authors declare no conflict of interest.

References

1. Henssge, C.; Madea, B. Estimation of the time since death in the early post-mortem period. *Forensic Sci. Int.* **2004**, *144*, 167–175. [CrossRef]
2. Maile, A.E.; Inoue, C.G.; Barksdale, L.E.; Carter, D.O. Toward a universal equation to estimate postmortem interval. *Forensic Sci. Int.* **2017**, *272*, 150–153. [CrossRef]
3. Gill-King, H. Chemical and ultrastructural aspects of decomposition. In *Forensic Taphonomy: The Postmortem Fate of Human Remains*; CRC Press: New York, NY, USA, 1997; pp. 93–104.
4. Vass, A.A.; Barshick, S.A.; Sega, G.; Caton, J.; Skeen, J.T.; Love, J.C.; Synstelien, J.A. Decomposition chemistry of human remains: A new methodology for determining the postmortem interval. *J. Forensic Sci.* **2002**, *47*, 542–553. [CrossRef] [PubMed]
5. Galloway, A.; Birkby, W.H.; Jones, A.M.; Henry, T.E.; Parks, B.O. Decay rates of human remains in an arid environment. *J. Forensic Sci.* **1989**, *34*, 607–616. [CrossRef]
6. Goff, M.L.; Lord, W.D. Insect as toxicological indicator and the impact of drugs and toxin on insect development. In *Forensic Entomology the Utility of Arthropods in Legal Investigations*, 2nd ed.; CRC Press: Boca Raton, FL, USA, 2010; pp. 427–434.
7. Rivers, D.B.; Dahlem, G.A. Postmortem decomposition of human remains and vertebrate carrion. In *The Science of Forensic Entomology*; Wiley Blackwell: Hoboken, NJ, USA, 2014; pp. 184–187.
8. Vass, A. The elusive universal post-mortem interval formula. *Forensic Sci. Int.* **2011**, *204*, 34–40. [CrossRef] [PubMed]
9. Campobasso, C.P.; Di Vella, G.; Introna, F. Factors affecting decomposition and Diptera colonization. *Forensic Sci. Int.* **2001**, *120*, 18–27. [CrossRef]
10. Megyesi, M.S.; Nawrocki, S.P.; Haskell, N.H. Using accumulated degree-days to estimate the postmortem interval from decomposed human remains. *J. Forensic Sci.* **2005**, *50*, 618–626. [CrossRef]
11. Kelly, J.A.; Van der Linde, T.C.; Anderson, G.S. The influence of clothing and wrapping on carcass decomposition and arthropod succession during the warmer seasons in central South Africa. *J. Forensic Sci.* **2009**, *54*, 1105–1112. [CrossRef]
12. Pakosh, C.M.; Rogers, T.L. Soft tissue decomposition of submerged, dismembered pig limbs enclosed in plastic bags. *J. Forensic Sci.* **2009**, *54*, 1223–1228. [CrossRef]
13. Simmons, T.; Adlam, R.E.; Moffat, C. Debugging decomposition data–comparative taphonomic studies and the influence of insects and carcass size on decomposition rate. *J. Forensic Sci.* **2010**, *55*, 8–13. [CrossRef] [PubMed]
14. Zhou, C.; Byard, R.W. Factors and processes causing accelerated decomposition in human cadavers–an overview. *J. Forensic Leg. Med.* **2011**, *18*, 6–9. [CrossRef]

15. Matuszewski, S.; Konwerski, S.; Frątczak, K.; Szafałowicz, M. Effect of body mass and clothing on decomposition of pig carcasses. *Int. J. Legal Med.* **2014**, *128*, 1039–1048. [CrossRef] [PubMed]

16. Hall, M.J.R.; Whitaker, A.; Richards, C. Forensic entomology. In *Forensic Ecology Handbook: From Crime Scene to Court*; Màrquez-Grant, R., Ed.; John Wiley and Sons Limited: Chichester, UK, 2012; pp. 111–140.

17. Tuccia, F.; Zurgani, E.; Bortolini, S.; Vanin, S. 2019 Experimental evaluation on the applicability of necrobiome analysis in Forensic Veterinary Science. *MicrobiologyOpen* **2019**, *12*, 828.

18. Nafte, M. *Flesh and Bone an Introduction to Forensic Anthropology*; Carolina Academic Press: Durham, NC, USA, 2000.

19. Cattaneo, C.; Gibelli, D. PMI determination of remains. In *Handbook of Forensic Medicine*; Madea, B., Ed.; Wiley Blackwell (UK): Oxford, UK, 2014; pp. 175–176.

20. Galloway, A. The process of decomposition: A model from the Arizona-Sonoran Desert. In *Forensic Taphonomy*; Haglund, H., Sorg, M., Eds.; CRC Press: Boca Raton, FL, USA, 1997; pp. 139–150.

21. Mann, R.W.; Bass, W.M.; Meadows, L. Time since death and decomposition of the human body: Variables and observations in case and experimental field studies. *J. Forensic Sci.* **1990**, *35*, 103–111. [CrossRef] [PubMed]

22. Myburgh, J.; L'Abbé, E.N.; Steyn, M.; Becker, P.J. Estimating the postmortem interval (PMI) using accumulated degree-days (ADD) in a temperate region of South Africa. *Forensic Sci. Int.* **2013**, *229*, 1–6. [CrossRef]

23. Simmons, T.; Cross, P.A.; Adlam, R.E.; Moffatt, C. The influence of insects on decomposition rate in buried and surface remains. *J. Forensic Sci.* **2010**, *55*, 889–892. [CrossRef]

24. Nawrocka, M.; Frątczak, K.; Matuszewski, S. Inter-Rater Reliability of Total Body Score-A Scale for Quantification of Corpse Decomposition. *J. Forensic Sci.* **2016**, *61*, 798–802. [CrossRef] [PubMed]

25. Dabbs, G.R.; Connor, M.; Bytheway, J.A. Interobserver Reliability of the Total Body Score System for Quantifying Human Decomposition. *J. Forensic Sci.* **2016**, *61*, 445–451. [CrossRef] [PubMed]

26. Adlam, R.E.; Simmons, T. The effect of repeated physical disturbance on soft tissue decomposition–are taphonomic studies an accurate reflection of decomposition? *J. Forensic Sci.* **2007**, *52*, 1007–1014. [CrossRef]

27. Cross, P.; Simmons, T. The influence of penetrative trauma on the rate of decomposition. *J. Forensic Sci.* **2010**, *55*, 295–301. [CrossRef]

28. De Donno, A.; Campobasso, C.P.; Santoro, V.; Leonardi, S.; Tafuri, S.; Introna, F. Bodies in sequestered and non-sequestered aquatic environments: A comparative taphonomic study using decompositional scoring system. *Sci. Justice* **2014**, *54*, 439–446. [CrossRef] [PubMed]

29. Cockle, D.L.; Bell, L.S. Human decomposition and the reliability of a 'Universal' model for postmortem interval estimations. *Forensic Sci. Int.* **2015**, *253*, 1–9. [CrossRef]

30. Suckling, J.K. A longitudinal study on human outdoor decomposition in Central Texas. *J. Forensic Sci.* **2016**, *61*, 19–25. [CrossRef]

31. Bugelli, V.; Gherardi, M.; Focardi, M.; Pinchi, V.; Vanin, S.; Campobasso, C.P. Decomposition pattern and insect colonization in two cases of suicide by hanging. *Forensic Sci. Res.* **2018**, *3*, 94–102. [CrossRef] [PubMed]

32. Gelderman, H.T.; Boer, L.; Naujocks, T.; IJzermans, A.C.M.; Duijst, W.L.J.M. The development of a post-mortem interval estimation for human remains found on land in The Netherlands. *Int. J. Leg. Med.* **2018**, *132*, 863–873. [CrossRef] [PubMed]

33. Schiel, M. Using Accumulate Degree Days for Estimating the Postmortem Interval: A Re-evaluation of Megyesi's Regression Formulae. Master's Thesis, University of Indianapolis, Indianapolis, IN, USA, 2008.

34. The Postmortem Interval: A Systematic Study of Pig Decomposition in West Central Montana. Master's Thesis, University of Montana, Missoula, MT, USA, 2009.

35. Suckling, J.K. A Longitudinal Study on the Outdoor Human Decomposition Sequence in Central Texas. Master's Thesis, University of Montana, Missoula, MT, USA, 2009.

36. Marhoff-Beard, S.J.; Forbes, S.L.; Green, H. The validation of 'universal' PMI methods for the estimation of time since death in temperate Australian climates. *Forensic Sci. Int.* **2018**, *291*, 158–166. [CrossRef]

37. Michaud, J.P.; Moreau, G. A statistical approach based on accumulated degree-days to predict decomposition-related processes in forensic studies. *J. Forensic Sci.* **2011**, *56*, 229–232. [CrossRef] [PubMed]

38. Tantawi, T.I.; Greenberg, B. The effect of killing and preservative solutions on estimates of maggot age in forensic cases. *J. Forensic Sci.* **1993**, *38*, 702–707. [CrossRef] [PubMed]

39. Wells, J.D.; Kurahashi, H. Chrysomya megacephala (Fabricius) (Diptera: Calliphoridae) development: Rate, variation and the implications for forensic entomology. *Jpn. J. Sanit. Zool.* **1994**, *45*, 303–309. [CrossRef]

40. Adams, Z.J.; Hall, M.J. Methods used for the killing and preservation of blowfly larvae, and their effect on post-mortem larval length. *Forensic Sci. Int.* **2003**, *138*, 50–61. [CrossRef]

41. Charabidze, D.; Hedouin, V. Temperature: The weak point of forensic entomology. *Int. J. Legal Med.* **2019**, *133*, 633–639. [CrossRef]

42. Amendt, J.; Richards, C.S.; Campobasso, C.P.; Zehner, R.; Hall, M.J.R. Forensic entomology: Applications and limitations. *Forensic Sci. Med. Pathol.* **2011**, 379–392. [CrossRef]

43. Introna, F.; Campobasso, C.P.; Goff, M.L. Entomotoxicology. *Forensic Sci. Int.* **2001**, *120*, 42–47. [CrossRef]

44. Campobasso, C.P.; Gherardi, M.; Caligara, M.; Sironi, L.; Introna, F. Drug analysis in blowfly larvae and in human tissues: A comparative study. *Int. J. Legal Med.* **2004**, *118*, 210–214. [CrossRef] [PubMed]

45. Campobasso, C.P.; Linville, J.G.; Wells, J.D.; Introna, F. Forensic genetic analysis of insect gut contents. *Am. J. Forensic Med. Pathol.* **2005**, *26*, 161–165.

46. Slone, D.; Gruner, S. Thermoregulation in larval aggregations of carrion-feeding blow flies (Diptera: Calliphoridae). *Med Vet. Entomol.* **2007**, *11*, 38–44. [CrossRef]

47. Kotzé, Z.; Villet, M.H.; Weldon, C.W. Heat accumulation and development rate of massed maggots of the sheep blowfly, Lucilia cuprina (Diptera: Calliphoridae). *J. Insect Physiol.* **2016**, *95*, 98–104. [CrossRef]

48. Podhorna, J.; Aubernon, C.; Borkovcova, M.; Boulay, J.; Hedouin, V.; Charabidze, D. To eat or get heat: Behavioral trade-offs between thermoregulation and feeding in gregarious necrophagous larvae. *Insect Sci.* **2018**, *25*, 883–893. [CrossRef]

49. Hart, A.J.; Hall, M.J.R.; Whitaker, A.P. The use of forensic entomology in criminal investigations: How it can be of benefit to SIOs. *J. Homicide Major Incid. Investig.* **2008**, *4*, 37–47.

50. Hofer, I.M.J.; Hart, A.J.; Martín-Vega, D.; Hall, M.J.R. Optimising crime scene temperature collection for forensic entomology casework. *Forensic Sci. Int.* **2017**, *270*, 129–138. [CrossRef]

51. Amendt, J.; Campobasso, C.P.; Gaudry, E.; Reiter, C.; LeBlanc, H.N.; Hall, M.J. European Association for Forensic Entomology. Best practice in forensic entomology–standards and guidelines. *Int. J. Legal Med.* **2007**, *121*, 90–104. [CrossRef]

52. Protocollo Nazionale per il Prelievo di Campioni Entomologici a Fini Forensi (Versione 3.0_Gief–Discusso ed Approvato dall'assemblea GIEF, 17.12.2016). In *Linee Guida Nazionali per le Autopsie a Scopo Forense in Medicina Veterinaria*; Ministero della Salute: Rome, Italy, 2018; Volume 19, pp. 60–65.

53. Giordani, G.; Grzywacz, A.; Vanin, S. Characterization and Identification of Puparia of Hydrotaea Robineau-Desvoidy, 1830 (Diptera: Muscidae) From Forensic and Archaeological Contexts. *J. Med. Entomol.* **2019**, *8*, 45–54. [CrossRef] [PubMed]

54. Szpila, K. Key for the identification of third instars of European Blowflies (Diptera: Calliphoridae) of forensic importance. In *Current Concepts in Forensic Entomology*; Amendt, J., Campobasso, C.P., Goff, M.L., Grassberger, M., Eds.; Springer Science+Business Media: Berlin, Germany, 2010; pp. 43–56.

55. Tuccia, F.; Giordani, G.; Vanin, S. A combined protocol for identification of maggots of forensic interest. *Sci. Justice* **2016**, *56*, 264–268. [CrossRef]

56. Hebert, P.D.; Ratnasingham, S.; DeWaard, J.R. Barcoding animal life: Cytochrome c oxidase subunit 1 divergences among closely related species. *Proc. Biol. Sci.* **2003**, *270*, 96–99. [CrossRef]

57. Folmer, O.; Black, M.; Hoeh, W.; Lutz, R.; Vrijenhoek, R. DNA primers for amplification of mitochondrial cytochrome c oxidase subunit I from diverse metazoan invertebrates. *Mol. Mar. Biol. Biotechnol.* **1994**, *3*, 294–299.

58. Grassberger, M.; Reiter, C. Effect of temperature on Lucilia sericata (Diptera: Calliphoridae) development with special reference to the isomegalen- and isomorphen-diagram. *Forensic Sci. Int.* **2000**, *120*, 32–36. [CrossRef]

59. Grassberger, M.; Reiter, C. Effect of temperature on development of Liopygia (= Sarcophaga) argyrostoma (Robineau-Desvoidy) (Diptera: Sarcophagidae) and its forensic implications. *J. Forensic Sci.* **2002**, *47*, 1332–1336. [CrossRef] [PubMed]

60. Grassberger, M.; Friedrich, E.; Reiter, C. The blowfly Chrysomya albiceps (Wiedemann) (Diptera: Calliphoridae) as a new forensic indicator in Central Europe. *Int. J. Legal Med.* **2003**, *117*, 75–81. [CrossRef] [PubMed]

61. Donovan, S.E.; Hall, M.J.; Turner, B.D.; Moncrieff, C.B. Larval growth rates of the blowfly, Calliphora vicina, over a range of temperatures. *Med. Vet. Entomol.* **2006**, *20*, 106–114. [CrossRef] [PubMed]

62. Cicchetti, D.V. Guidelines, criteria, and rules of thumb for evaluating normed and standardized assessment instruments in psychology. *Psychol. Assess.* **1994**, *6*, 284–290. [CrossRef]

63. Koo, T.K.; Li, M.Y. A Guideline of Selecting and Reporting Intraclass Correlation Coefficients for Reliability Research. *J. Chiropr. Med.* **2016**, *15*, 155–163. [CrossRef]

64. Rodriguez, W.C. Decomposition of buried and submerged bodies. In *Forensic Taphonomy: The Post-Mortem Fate of Human Remains*; Haglund, W.D., Sorg, M.A., Eds.; CRC Press: Boston, MA, USA, 1997; pp. 459–464.

65. Matuszewski, S.; Mądra, A. Factors affecting quality of temperature models for the pre-appearance interval of forensically useful insects. *Forensic Sci. Int.* **2015**, *247*, 28–35. [CrossRef]

66. Anderson, G.S. Comparison of decomposition rates and faunal colonization of carrion in indoor and outdoor environments. *J. Forensic Sci.* **2011**, *56*, 136–142. [CrossRef]

67. Archer, M.S. Annual variation in arrival and departure times of carrion insects at carcasses: Implications for succession studies in forensic entomology. *Aust. J. Zool.* **2004**, *51*, 569–576. [CrossRef]

68. Archer, M. Comparative analysis of insect succession data from Victoria (Australia) using summary statistics versus preceding mean ambient temperature models. *J. Forensic Sci.* **2014**, *59*, 404–412. [CrossRef] [PubMed]

69. Bhadra, P.; Hart, A.J.; Hall, M.J. Factors affecting accessibility to blowflies of bodies disposed in suitcases. *Forensic Sci. Int.* **2014**, *239*, 62–72. [CrossRef]

70. Aubernon, C.; Boulay, J.; Hédouin, V.; Charabidzé, D. Thermoregulation in gregarious dipteran larvae: Evidence of species-specific temperature selection. *Entomol. Exp. Appl.* **2016**, *160*, 101–180. [CrossRef]

71. Bugelli, V.; Forni, D.; Bassi, L.A.; Di Paolo, M.; Marra, D.; Lenzi, S.; Toni, C.; Giusiani, M.; Domenici, R.; Gherardi, M.; et al. Forensic entomology and the estimation of the minimum time since death in indoor cases. *J. Forensic Sci.* **2015**, *60*, 525–531. [CrossRef] [PubMed]

72. Mukherjee, S.; Singh, P.; Tuccia, F.; Pradelli, J.; Giordani, G.; Vanin, S. DNA characterization from gut content of larvae of *Megaselia scalaris* (Diptera, Phoridae). *Sci. Justice* **2019**, *59*, 654–659. [CrossRef]

73. Charabidze, D.; Bourel, B.; Gosset, D. Larval-mass effect: Characterisation of heat emission by necrophageous blowflies (Diptera: Calliphoridae) larval aggregates. *Forensic Sci. Int.* **2011**, *211*, 61–66. [CrossRef]

74. Parks, C.L. A study of the human decomposition sequence in central Texas. *J. Forensic Sci.* **2011**, *56*, 19–22. [CrossRef] [PubMed]

Case Report

Flies Do Not Jump to Conclusions: Estimation of the Minimum Post-Mortem Interval for a Partly Skeletonized Body Based on Larvae of *Phromia regina* (Diptera: Calliphoridae)

Senta Niederegger [1],* and Gita Mall [2]

[1] Department of Forensic Entomology, Institute of Legal Medicine at the University Hospital Jena, 07740 Jena, Germany
[2] Department of Forensic Medicine, Institute of Legal Medicine at the University Hospital Jena, 07740 Jena, Germany; gita.mall@med.uni-jena.de
* Correspondence: senta.niederegger@med.uni-jena.de

Simple Summary: Many factors can influence the appearance of human remains. In the case presented here, the remains appeared to be exposed for months, because the bones were visible. Fly maggots collected from the body, however, suggested a much shorter period of only about two weeks. The confession of the perpetrator ultimately confirmed the shorter exposure time of the remains.

Abstract: Skeletonization is often perceived as an indicator of long post-mortem intervals. The finding of feeding larvae of first colonizers, on the other hand, indicates days. We present a case in which both findings were present. Larvae of *Phormia regina*, aged 9 days, and skeletonization of the head and part of the thorax were both found on an unidentified female body. Identification of dentures eventually led to resolution of the case and a confession, which settled the seeming contradiction in favor of forensic entomology.

Keywords: thanatology; confession; post-mortem interval; forensic entomology

Citation: Niederegger, S.; Mall, G. Flies Do Not Jump to Conclusions: Estimation of the Minimum Post-Mortem Interval for a Partly Skeletonized Body Based on Larvae of *Phromia regina* (Diptera: Calliphoridae). *Insects* **2021**, *12*, 294. https://doi.org/10.3390/insects1204 0294

Academic Editors: Damien Charabidze and Daniel Martín-Vega

Received: 26 February 2021
Accepted: 26 March 2021
Published: 28 March 2021

Publisher's Note: MDPI stays neutral with regard to jurisdictional claims in published maps and institutional affiliations.

1. Introduction

The fascination with grisly crimes and the resulting supply of crime fiction in books, movies, and shows have increased the popularity of forensic entomology. If, in crime fiction, insects are recognized on a body, they predominantly lead to resolution of the case, sometimes with as little as one maggot. In casework, however, entomological evidence is oftentimes not valued as highly as other indicators. Cases in which entomological estimations are independently verified through confessions can therefore be of valuable help for forensic entomologists [1]. Unfortunately, not many case reports are published [2], and of those, even fewer contain confirmation of the estimated minimal post-mortem intervals (mPMIs) [3,4]. Oftentimes, only witness statements about when the victim was last seen alive are available [5,6].

We present a case in which entomological evidence was unappreciated because it seemingly contradicted another taphonomic indicator that suggested a far longer postmortem interval.

2. The Case

At the beginning of July, in a rural part of Germany, a dog walker discovered the body of an unidentified woman in a pit at the side of a dirty road bordering on a cornfield. The body was deposited on its right side in a fetal position. It was partly covered with soil and partly skeletonized (Figure 1); the skull and a number of ribs were visible. Police personnel collected live third instar larvae at the scene. Photographic evidence showed no presence of puparia.

requires 12 days at 15 °C and 5.5 days at 20 °C to reach a length of 16 mm. For 17.7 °C, therefore, 9–10 days' developmental time were assessed.

Accumulated degree-day (ADD) data for the emergence of adult *P. regina* from Nabity et al. [11] are 202 ADD at 15 °C and 205 ADD at 20 °C with "investigator-preferred minimum" [11] and a base temperature of 10 °C (Table 1).

Emergence of adult flies after nine days in the laboratory at 24 °C, therefore, accumulated to 126 ADD. The day of the discovery (day 0) of the body was added with 0 ADD due to 15 h in the refrigerator at 5 °C, which left 76–79 ADD prior to discovery. For each day before discovery, the respective average daily temperature was calculated from weather data (Table 1). The most likely day of egg deposition was determined to be nine days prior to the discovery of the body. Weather data showed no rainfall or cold temperature episode for this or the previous day.

Table 1. Calculation of accumulated degree-days (ADD) for *Phormia regina*, based on temperature information from the nearest weather station. bold: estimated day of body deposition

Day	Daily Average °C	Threshold °C	Degree-Days	ADD
−10	15.5	10	5.5	209.9
−9	**17.4**	**10**	**7.4**	**204.4**
−8	17.3	10	7.3	197.0
−7	19.2	10	9.2	189.7
−6	22.5	10	12.5	180.5
−5	17.2	10	7.2	168.0
−4	15.8	10	5.8	160.8
−3	16.7	10	6.7	155.0
−2	19.4	10	9.4	148.3
−1	22.9	10	12.9	138.9
Day 0	5.0	10	0	126.0
Lab1–9	24.0	10	14	126.0

With the age of *P. regina* established to have been 9 days based on both results, we were confident that this was the most likely minimum time that had elapsed since the deposition of the body.

During the TV show, however, the most probable PMI was still given as multiple weeks up to several months on grounds of the skeletonization of the head and the thorax.

2.2. Victim Identification

After the TV show, police received multiple tips from viewers. Especially, the OPG and the dentures attracted a lot of attention. Dental assistants pointed out that dental prostheses oftentimes feature serial numbers, and one was indeed detected in the artificial teeth. Investigators were able to find the dentist, and he identified one of his patients. One month after the discovery of the body, the victim was unequivocally identified through her dental prosthesis.

2.3. Suspect Identification

After the successful identification, the victim's apartment was located five weeks after the discovery of her body. Interestingly, the mailbox was empty. This prompted investigators to monitor the building's main entrance to find out who was in possession of a key to her mailbox. Only two days later, a man was observed emptying the victim's mailbox and leaving the apartment building. He was soon identified as a resident of a nearby apartment building.

During police questioning, the suspect confessed to killing the victim. He strangled her during the course of an argument over the future of their secret relationship and left the body in the bathroom. Two days later, he decided to discard the remains before suspicion would arise, and bought a large suitcase. Four days after the act, the victim's body was transported to a remote cornfield, where it was discovered nine days later, partly skeletonized and with larvae of *P. regina*.

3. Discussion

For the entomological expertise in this case, publications analyzing US populations of *P. regina* were used. This could be seen as the wrong dataset for European populations. Marchenko [12] worked with Eastern European populations of the species but calculated the pupal stage of *P. regina* to being short. His data did not correspond to our results of 7–8 days for this development period. Other publications were more comparable [11–14]. Due to its very detailed data, Nabity et al.'s study [11] was exclusively used for our calculations of accumulated degree-days.

A large number of variables affects decomposition and taphonomic processes. The interpretation of post-mortem interval determination can therefore be deceptive [15]. Skeletonization is oftentimes expected to occur only after long exposure times. This, however, might only be true in the absence of or with minimal insect activity. Studies on the decomposition of pig carcasses illustrate this phenomenon. In Matuszewski et al. [16], clothed and unclothed pig carcasses with different body masses were studied. In some cases, the onset of active decay was detected as early as two days after death. Anton et al. [17] had medium-size pig carcasses in summer, going from the fresh stage to the remains stage, with extensive skeletonization within 8 days. In this case, the larvae themselves, due to the diminished accessibility of other body parts, probably accelerated skeletonization.

In the case report presented here, entomological expertise was not valued as evidence and thus mostly disregarded during the investigation. Fortunately, the means to identify the victim were found in the serial number of the dental prosthesis. Would that not have been possible, the case might have remained unsolved. Even if a missing person report would have been filed eventually, investigators could have rejected it due to the presumption of an older age of the remains.

When human remains are found, it is therefore important to evaluate all processes and give them equal prominence in estimating plausible post-mortem intervals.

4. Conclusions

Insect evidence resulted in an estimation of time since the deposition of the body of less than two weeks and was regarded as potentially erroneous in light of the skeletonization. The time frame established by forensic entomology was later confirmed by the confession. Such results strengthen the scientific value of entomological expertise in casework and its standing in criminal investigations.

Author Contributions: Conceptualization, S.N. and G.M.; investigation, S.N.; resources, G.M.; data curation, S.N.; writing, S.N.; supervision, G.M. Both authors have read and agreed to the published version of the manuscript.

Funding: We acknowledge support by the German Research Foundation and the Open Access Publication Fund of the Thueringer Universitaets- und Landesbibliothek Jena Projekt-Nr. 433052568.

Institutional Review Board Statement: Not applicable.

Informed Consent Statement: Not applicable.

Data Availability Statement: The data presented in this study are available on request from the corresponding author.

Acknowledgments: We are grateful to the anonymous reviewers for their helpful comments.

Conflicts of Interest: The authors declare no conflict of interest.

References

1. Hall, M.J.R. The Relationship between Research and Casework in Forensic Entomology. *Insects* **2021**, *12*, 174. [CrossRef] [PubMed]
2. Lei, G.; Liu, F.; Liu, P.; Zhou, Y.; Jiao, T.; Dang, Y.H. A bibliometric analysis of forensic entomology trends and perspectives worldwide over the last two decades (1998–2017). *Forensic Sci. Int.* **2019**, *295*, 72–82. [CrossRef] [PubMed]

3. Goff, M.L.; Odom, C.B. Forensic Entomology in the Hawaiian Islands. *Am. J. Forensic Med. Pathol.* **1987**, *8*, 45–50. [CrossRef] [PubMed]

4. Grassberger, M.; Friedrich, E.; Reiter, C. The blowfly Chrysomya albiceps (Wiedemann) (Diptera: Calliphoridae) as a new forensic indicator in Central Europe. *Int. J. Leg. Med.* **2003**, *117*, 75–81. [CrossRef] [PubMed]

5. Kashyap, V.K.; Pillay, V.V. Efficacy of entomological method in estimation of postmortem interval: A comparative analysis. *Forensic Sci. Int.* **1989**, *40*, 245–250. [CrossRef]

6. Bugelli, V.; Forni, D.; Bassi, L.A.; Di Paolo, M.; Marra, D.; Lenzi, S.; Toni, C.; Giusiani, M.; Domenici, R.; Gherardi, M.; et al. Forensic Entomology and the Estimation of the Minimum Time Since Death in Indoor Cases. *J. Forensic Sci.* **2015**, *60*, 525–531. [CrossRef] [PubMed]

7. Szpila, K. Key for the identification of third instars of European blowflies (Diptera: Calliphoridae) of forensic importance. In *Current Concepts in Forensic Entomology*; Amendt, J., Campobasso, C.P., Goff, M.L., Grassberger, M., Eds.; Springer: Dordrecht, The Netherlands, 2010; pp. 43–56.

8. Niederegger, S.; Spieß, R. Muscle attachment sites of Phormia regina (Meigen). *Parasitol. Res.* **2014**, *113*, 4313–4314. [CrossRef] [PubMed]

9. Szpila, K. Key for identification of European and Mediterranean blowflies (Diptera, Calliphoridae) of medical and veterinary importance—Adult flies. In *Forensic Entomology, An Introduction*, 2nd ed.; Gennard, D., Ed.; Willey-Blackwell: Hoboken, NJ, USA, 2012; pp. 77–81.

10. Núñez-Vázquez, C.; Tomberlin, J.K.; Cantú-Sifuentes, M.; García-Martínez, O. Laboratory development and field validation of Phormia regina (Diptera: Calliphoridae). *J. Med. Èntomol.* **2013**, *50*, 252–260. [CrossRef] [PubMed]

11. Nabity, P.D.; Higley, L.G.; Heng-Moss, T.M. Effects of temperature on development of Phormia regina (Diptera: Calliphoridae) and use of developmental data in determining time intervals in forensic entomology. *J. Med. Entomol.* **2006**, *43*, 1276–1286. [CrossRef] [PubMed]

12. Marchenko, M.I. Medicolegal relevance of cadaver entomofauna for the determination of the time of death. *Forensic Sci. Int.* **2001**, *120*, 89–109. [CrossRef]

13. Byrd, J.H.; Allen, J.C. The development of the black blow fly, Phormia regina (Meigen). *Forensic Sci. Int.* **2001**, *120*, 79–88. [CrossRef]

14. Anderson, G.S. Minimum and Maximum Development Rates of Some Forensically Important Calliphoridae (Diptera). *J. Forensic Sci.* **2000**, *45*, 824–832. [CrossRef] [PubMed]

15. Haglund, W.D.; Sorg, M.H. *Forensic Taphonomy: The Postmortem Fate of Human Remains*; CRC Press: Boca Raton, FL, USA; London, UK, 1997; p. 636.

16. Matuszewski, S.; Konwerski, S.; Frątczak, K.; Szafałowicz, M. Effect of body mass and clothing on decomposition of pig carcasses. *Int. J. Leg. Med.* **2014**, *128*, 1039–1048. [CrossRef] [PubMed]

17. Anton, E.; Niederegger, S.; Beutel, R.G. Beetles and flies collected on pig carrion in an experimental setting in Thuringia and their forensic implications. *Med. Vet. Èntomol.* **2011**, *25*, 353–364. [CrossRef] [PubMed]

Review

Insect Decline—A Forensic Issue?

Jens Amendt

Institute of Legal Medicine, University Hospital Frankfurt, Goethe-University, Kennedyallee 104, D-60596 Frankfurt am Main, Germany; amendt@em.uni-frankfurt.de; Tel.: +49-69-6301-7571

Simple Summary: Numerous studies report a decline in insect biodiversity and biomass on a global scale. Since forensic entomology relies on the presence of insects, the question of whether this discipline will be or already is affected by such a decrease is not only posed to investigative authorities and the public, but also to the scientific community. While the data does indeed provide overwhelming evidence of insect decline, even if the methods of evaluation and data pooling are occasionally questioned, only a few studies deal with forensically relevant insects. These few data do hardly prove a decrease in forensically relevant insect species so far. However, one factor driving insect decline is likely to have also a strong influence on necrophagous insects in the future: climate change.

Abstract: Recent reports have shown a dramatic loss in insect species and biomass. Since forensic entomology relies on the presence of insects, the question is whether this decline effects the discipline. The present review confirms that numerous studies document insect population declines or even extinction, despite the fact that the rates of decline and the methods used to demonstrate it are still much debated. However, with regard to a decline in necrophagous insects, there is little or only anecdotal data available. A hypothetical decrease in species diversity and population density in necrophagous insects could lead to a delayed colonization of dead bodies and a modified succession pattern due to the disappearance or new occurrence of species or their altered seasonality. Climate change as one of the drivers of insect decline will probably also have an impact on necrophagous insects and forensic entomology, leading to reduced flight and oviposition activity, modified growth rates and, therefore, an over- or underestimation of a minimum postmortem interval. Global warming with increased temperature and extreme weather requires a better understanding about necrophagous insect responses to environmental variations. Here, transgeneration effects in particular should be analysed in greater depth as this will help to understand rapid adaptation and plasticity in insects of forensic importance.

Keywords: forensic entomology; climate change; global warming; development; succession; carrion

Citation: Amendt, J. Insect Decline—A Forensic Issue? *Insects* **2021**, *12*, 324. https://doi.org/10.3390/insects12040324

Academic Editors: Damien Charabidze and Daniel Martín-Vega

Received: 15 March 2021
Accepted: 1 April 2021
Published: 6 April 2021

Publisher's Note: MDPI stays neutral with regard to jurisdictional claims in published maps and institutional affil-iations.

1. Introduction

Nature seems to be losing, as a current United Nations report estimates that more than one million species are at risk of extinction in coming decades [1]. While such statements in recent years have mostly been based on summary studies of the most diverse vertebrate groups (e.g., [2,3]), insects have only become the focus of (public) interest in recent years. Despite the fact that there are more than a million described species of insects (and bearing in mind that another 4.5 to 7 million remain unnamed [4]), time-series data on e.g., ecological aspects and taxonomic nature of insect population trends are rare, compared to vertebrates, and often focused on certain groups of specialized taxa like agriculturally important species [5]. Nevertheless, recent reports have shown a dramatic loss in the biomass of flying insects over a period of less than 30 years [6,7] and sharp declines in the abundance of various insect groups [8–10]. Since forensic entomology relies on the presence and functionality of (mostly necrophagous) insects, the question is whether this massive loss of insects calls into question the applicability or flawless performance of this particular discipline. This paper summarizes the current knowledge of this topic and discusses potential issues in forensic entomology.

2. What We Know

In 2017, [6] measured total insect biomass over 27 years in 63 nature protection areas in Germany and collected a total of 53.54 kg of invertebrates within an average of 176 exposure days per location-year combination, assuming that this amount of biomass represents millions of specimens. They estimated a seasonal decline of 76%, and mid-summer decline of 82% in flying insect biomass over the 27 years of study, regardless of habitat type [6]. This study hit like a bomb and (as of 11 March 2021) has been cited more than 1500 times, triggering a flood of media coverage, conference symposia and special issues on insect decline. Follow-up studies confirmed the drastic decline of terrestrial insects e.g., [11–14] and led to newspaper headlines such as "The insect apocalypse is here" (The New York Times Magazine, 27 November 2018). However, there are conflicting critical voices about shortcomings in data selection and methodology of such studies like e.g., categorical versus continuous time series, temporal pseudoreplication, the application and comparison of different diversity metrics [15,16] or about data interpretation and communication [17,18]. There are even opposing findings, like the study by Crossley et al. which states an apparent robustness of US arthropod populations and a lack of overall increase or decline [16]. However, the majority of studies confirm a decrease and just disagree about its magnitude, and [5] recently summarises that, despite much variation across time, space, and taxonomic lineage, reported rates of annual decline in abundance frequently fall around 1% to 2%, which is a worrying number. It is pointed out by [14] that most of the studies on insect decline are restricted geographically and taxonomically. While the former can be easily verified by looking at the exemplary map of the world from the study by Sánchez-Bayo and Wyckhuys ([17] Figure 1), using total insect biomass as a proxy for biodiversity [6] or aggregating data across higher taxonomic categories and ecological groups (e.g., [16]) might help to deal with the complexity of population-level stochasticity in insects, but often overlook species-level trends [5].

It must be stated once again that the study that fuelled the whole current discussion [6] does not offer any taxonomic resolution, but merely caught and weighed "flying insects". Taxon-specific findings could hardly be derived here. Nevertheless, we can refer to some works to better assess the extent of the situation at order or family level. The majority of studies on terrestrial insects deal with the orders Coleoptera, Hemiptera, Hymenoptera, and Lepidoptera [18]. Hallmann et al. ([19]) found in a 30 year survey in the Netherlands that ground beetles (Coleoptera: Carabidae) showed a mean annual decline of 4.3% in total numbers over the period of 1985–2016 and calculate a reduction in total biomass of at least 42% for ground beetles. Interestingly, they showed an increase in carrion-beetles (Silphidae) and explained that by carrion experiments done close to the traps. This illustrates how highly context- and taxon-specific temporal changes in insect populations must be evaluated. Bell et al. [20] showed that aphid annual totals fluctuated widely in the UK, but this group was in a steady state over the long-term, with a non-significant decline of −7.6%. Such a trend may have been driven by three of the most abundant species, highlighting the need to work and understand the taxonomy of the target taxa.

Carvalheiro et al. [21] compared in three European countries four 20-year periods of rapid land-use intensification and natural habitat loss (1930–1990) with a period of increased conservation investment (post-1990) and found extensive species richness loss for bumblebees before 1990. In the same study authors found that richness of butterflies in the UK, Netherlands and Belgium declined from 1950 to 2009, and [22] showed for the tropical island state of Singapore that 32% of 413 recorded species of butterflies eradicated since 1854.

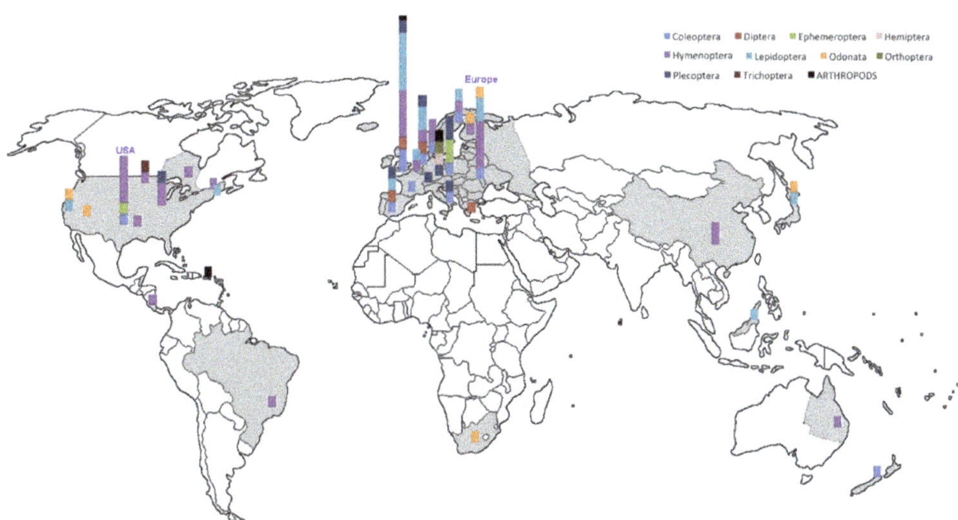

Figure 1. Geographic location of the 73 reports studied on the world map. Columns show the relative proportion of surveys for each taxa as indicated by different colours in the legend. Data for China and Queensland (Australia) refer to managed honey bees only; from [17].

These examples confirm the described trend of decline for selected groups [18]. The main causes of the aforementioned insect decline are rapid urbanisation and habitat homogenisation, industrialisation and agricultural expansion based on monocultures, and the use of pesticides like neonicotinoids. Reference [17] identify agriculture associated factors and pesticides as the main drivers in over 1/3 of the studies on insect decline (Figure 2).

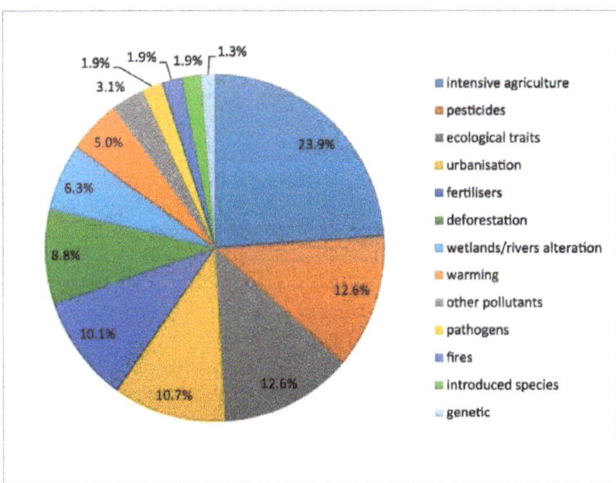

Figure 2. Main factors associated with insect declines according to reports in the literature (from [17]).

Once you have gained an overview from the literature, you will quickly realise that there are little or no data available for the families that govern forensic entomology. Here, taxa from the order Diptera and Coleoptera dominate carrion globally and thus also human cadavers [23–28]. Studies on the decline of Coleoptera mainly deals with Carabidae, saproxylic beetles, or aquatic taxa, or highly specialised species like fireflies. When it comes to carrion associated taxa, evidence for decline is rare. References [29,30] discuss the decline of some groups of dung beetles, which play an important role especially in the carrion community of tropical habitats, and [31] found in Italy a relative decrease in

Figure 1. Partly skeletonized human remains found at the edge of a cornfield.

Due to skeletonization, the doctors examining the body during autopsy the next day estimated the time of death to have been several weeks, if not months, before discovery. The buried parts of the body, however, showed only slight decomposition. Medical examiners also found conspicuous dental work, which resulted in a postmortem pantomogramm (OPG) of the victim's dentures. No additional entomological evidence was collected during autopsy.

Investigators were not able to identify the woman. No missing person's report from previous months matched the description. Interrogation of witnesses did not yield any meaningful results. About three weeks after discovery of the body, the case was therefore going to be presented on television to gain information from the public. Investigators requested a forensic entomological expertise before airing the TV show for a more accurate assessment of the possible time of deposition of the body.

2.1. Analysis of Insect Evidence

Maggots recovered from the scene reached the Institute of Legal Medicine on the day of their collection. They were divided into three groups: one was frozen, one was transferred to 70% ethanol, and one was stored in the refrigerator. The forensic entomologist was informed about it on the next day after the autopsy, and fortunately, the maggots from the refrigerator were still alive (after about 15 h at +5 °C). The 135 maggots previously stored in 70% ethanol or frozen were hot-water-fixated even after that, and the 50 largest individuals were measured. The average length of the 50 largest larvae was 16.05 mm (size range: 15.30–17.32 mm; median: 16.11 mm; standard deviation: ±0.57 mm). All the larvae collected were identified as *Phormia regina* (Meigen, 1826) (Diptera: Calliphoridae) [7,8]. The living larvae were placed on minced meat and kept at 24 °C in the laboratory. After two days in the lab, the first puparia were formed, and nine days after finding of the body, a large number of adult *P. regina* [9] emerged.

According to temperature data from a nearby weather station (8 km north from the location where the body was found), the average daily temperature for 18 days prior to the finding was 17.7 °C. Based on published data from Nunez-Vazquez et al. [10], *P. regina* frequency of roller species (an ecological subgroup of dung beetles) by 31%. Even though Staphylinidae can be present with quite a large diversity of species on carrion and often occur there as predators, data are available mainly on other ecological guilds of this family, e.g., saproxylic taxa [32]. The already mentioned increase in the number of Silphidae specimens [19] should be cited with caution due to its possible bias caused by bait and carrion experiments nearby the sampling area. However, also [33] confirm a steady state or even increase of Silphidae during two periods of sampling in New Hampshire/USA in 1973–1977 and again in 2015–2017. The state of knowledge is not much better for the most important group in forensic entomology, the Diptera. Most of the insect decline studies focus on hoverflies, which are of particular ecological and system-maintaining importance due to their role as beneficial insects in pest control and pollination [21,34,35]. Despite the fact that such approaches neglect or even ignore the importance of non-syrphid flies as pollinators [36], they lead to a surprising knowledge gap when it comes to the decline of necrophagous Diptera. The most important families due to their abundance on human bodies and often timely occurrence post mortem are the Calliphoridae, Sarcophagidae and Muscidae. While there are numerous data on their presence, seasonal occurrence and succession on carrion, there are almost no long-term studies. The only exception are studies on climatic change and warming in the Arctic. Here, the Muscidae are in focus, as they play a key role in these extreme habitats as pollinators [37]. Reference [38] showed a significant decline of Muscidae from 1996 until 2009 for high-Arctic Greenland and [39,40] confirmed such significant declines in 7 of the 14 muscid species found in five or more years between 1996 and 2014, as well as a dramatic (80%) decrease in diversity and abundance in some habitats. Beyond that, there are only avenues to speculate on potential trends. Amendt (unpublished data, [41]) and [42–44] sampled necrophagous insects in Frankfurt (Germany) and the surrounding area over a period of 20 years and could not detect any change in species diversity; moreover, the number of insect-infested corpses did not decline, on the contrary, one rather gets the impression (i.e., not statistically proven) of an increase if one looks at the number of insect-infested corpses at the Institute for Legal Medicine in Frankfurt. However, as the respective research question of these studies determined a diversity of different methods (trap and bait type, collection intervals, species selection, etc.) these data can only be used with caution for a serious evaluation of necrophagous insects decline.

In summary, there are numerous well-documented examples of insect population declines and extinctions in the context of the anthropogenic drivers of global change, and even the rate of those declines is still hotly debated, fuelled by a lack of standardised, systematically collected data. With regard to a decline in necrophagous insects or possibly opposing trends, there are little or only anecdotal data available.

3. What We Do Not Know (But Could Be ...)

Even though there are currently no reliable data on a decline in forensically relevant insects, it makes sense to address the issue. After all, it cannot be ruled out that the factors that are driving back numerous insect species and populations have at least some influence on the biological and ecological characteristics of necrophagous insects and thus ultimately on the informative value of forensic entomology.

Before discussing possible impacts of insect decline on necrophagous invertebrates, there is one major issue to emphasise: reduced resource availability, e.g., due to urbanisation or agriculture, is a very important aspect when understanding insect decline. Flower visitor declines, for example, have been linked to changes in resource availability [45]. But resource availability has always been a challenge for necrophagous insects, since carrion is an unpredictable, patchy and ephemeral resource (whether a dead mouse or a human body), and its exploitation and use as feeding and breeding habitat poses particular challenges to its necrophagous community on a spatial and temporal scale. Quantitative data on carrion biomass are lacking e.g., due to methodological issues [46–48] and it is difficult to evaluate whether carrion availability is (or will be) reduced if there is no baseline to refer to. It seems reasonable, however, not to assume any reduction; there is, for example, a clear increase in rodent populations in the US (which is thought to be linked to a rise in temperature, among other things)—and an increase in living animals will sooner or later also lead to an increased supply of carrion. Therefore, typical drivers of insect decline will probably not lead to a reduced resource availability for necrophagous insects. However, there could be a change in the composition of the carrion supply, away from large vertebrates (e.g., because of loss or modification of natural habitats) towards small species such as rodents, where at least certain pests could be the winners of climate change. This in turn would possibly have an influence on the species composition of the necrophagous community since, for example, not all fly species colonise small carcasses. In Europe, large vertebrate carrion, including human cadavers, can attract many species of Sarcophagidae (Diptera: Flesh flies), but only a few of them utilize it for their offspring, i.e., larvi- or oviposit on it [49]. Additionally, in various baiting experiments with small rodent carcasses, the author always finds the flesh fly *Sarcophaga caerulescens* (Zetterstedt, 1838) but never *S. argyrostoma* (Robineau-Desvoidy, 1830), the only flesh fly species of forensic relevance in Central Europe. The situation is similar for the blow fly *Protophormia terraenovae* (Robineau-Desvoidy, 1830)—this species is of certain forensic relevance due to its common presence on human bodies, but seems to ignore small carrion. Hence, a shift towards more small carrion might favour certain taxa and modify, therefore, the species composition and abundance. However, a reduced species diversity would not necessarily have to be a disadvantage from a forensic-entomological point of view: 4–5 species might be easier to deal with than 24–25 species. At the end of the day, it could even be an advantage to work with a small number of species, since one would have to study and know a smaller set of species in terms of their biology and ecology Less reference data on growth and less ecological data on phenology need to be gathered, identification becomes easier due to fewer possibilities for confusion and error, and even the appropriate sampling and further rearing of insect evidence could be easier. However, in the worst case, species relevant for the assessment of the succession interval become rarer or will even disappear, and thus the templates and patterns defined on the basis of previous studies can no longer be applied in certain cases or simply lead to less accurate estimates. A decline in species leading to only one fly species instead of three or four species developing on the corpse after e.g., a postmortem interval (PMI) of 8 days could impact the estimation of an accurate date based on the age of those specimens for the same reason, as data based on e.g., three or four different species and their development rates provide more certainty in the assessment. However, one should not lose sight of the fact that there is already now no high species diversity on one and the same body, especially during the first days or even weeks postmortem [43]. Reference [42], for example, found 13 species of flesh flies attracted to baited traps, but on human corpses just one of them, *S. argyrostoma*, is representing the family in central Europe almost exclusively. Hence, a small number of species on dead bodies does not necessarily have a negative effect when it comes to PMI estimation as forensic entomology is already used too.

Reduced species diversity and insect abundance could lead to delayed colonisation, but it is not proven so far that a smaller number of species means a lower number of colonised corpses and a higher number of insect-free bodies. Instead, the low number of species which succeed the competition could be much more present than before. Typical "winners" could be highly adaptive and competitive native species such as *Lucilia sericata* (Meigen, 1826), or invasive taxa like several species of the genus *Chrysomya* (Robineau-Desvoidy, 1830, e.g., *Ch. albiceps* (Wiedemann, 1819) or *Ch. megacephala* (Fabricius, 1784).

One driving factor of insect decline is climate change, and global warming and extreme weather (e.g., heavy rain, storms and droughts) are one of its most important effects. It can be expected that they will also have an impact on necrophagous insects, whereby one must distinguish between two targets: The "adult level" and the "juvenile level" (Figure 3).

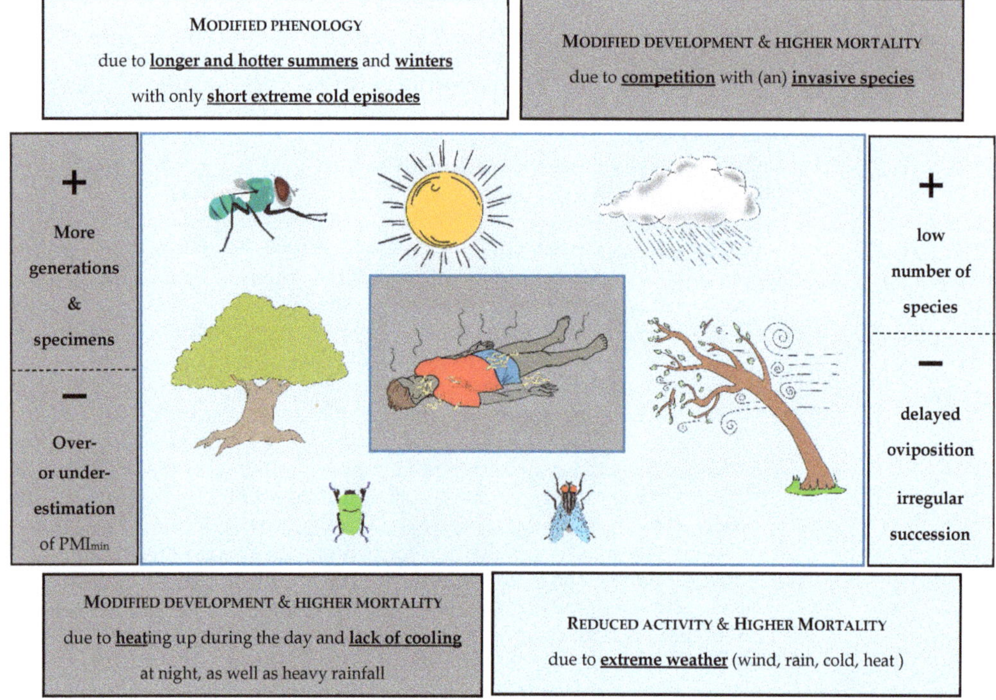

Figure 3. Global warming as a driver of insect decline and its possible impact (+ and −) on the necrophagous fauna and forensic entomology (grey background: juvenile stages, light blue: adult stages).

Increased (average) temperature and frequency of extreme weather events will impact the activity and behaviour of adult insects, but the direction of such an effect is not clear. Heavy wind or rain might decrease the rate of detection of carrion due to reduced flight activity and the scattering of odours that are relevant for cadaver discovery. Due to milder winters and longer summers, adult activity will start earlier and end later over the course of the year. Low temperatures are generally a key factor limiting the range of insect species, and even a small increase in winter temperatures enables survival in areas that were previously inaccessible. The adult phenology of single species will, therefore, change. Beside this shift (or extension) of activity, species will be able to access more habitats, e.g., be found in higher altitudes as insects move uphill to escape warming temperatures. All this could increase the timely detection and colonization of dead bodies in more habitats and for a longer period of the year. A shift in temperature profiles could enable larger range expansion due to higher chances of winter survival, as it is typical for invasive species like *Ch. albiceps* [50], *Ch. megacephala* [51] or *L. cuprina* (Wiedemann, 1830) [52]. The more common presence could affect the local fauna due to interspecific interactions and competition and there might be negative effects of invasive species (see also above "What we know"). But even though this is regrettable for many reasons, it is not proven whether there will be a negative effect for routine work in forensic entomology. Of course, a lack of awareness about the possible presence of invasive species in the native region could lead to misidentifications and incorrect age determinations [52]. Moreover, there is no doubt that certain biological characteristics like the predatory behaviour of the larvae of that invasive species have a negative impact on the presence, abundance and maybe even development of native species. But I also put forward the hypothesis that a small number of species (the "winners") might be more to hand and useful for the expert than several dozens of different taxa. Nevertheless, a low number of species could lead to a decreased population and, due to e.g., hot weather and high wind speeds, climate change might lead to a reduced flight and oviposition activity and as a consequence to a delayed colonization of dead bodies. Moreover, existing succession templates may need to be revalidated or adapted.

The juvenile stages on a dead body in the field could experience increased mortality or possibly aberrant growth behaviour due to periods of extreme heat or heavy rainfall. Temperatures can also impact the transition to the next developmental stage in summer. The effects of mild winter temperatures and extreme summer temperature fluctuations on the development of the local necrophagous fauna have hardly been investigated so far. New invasive taxa could eliminate native species through cannibalism and cause altered growth through competitive situations. Such alteration or modification might lead to over- or underestimations of a PMI_{min} and, therefore, impact the outcome of the entomological report. An exciting interface between the adult and the juvenile stages are the so-called transgenerational effects. Phenotypic plasticity across generations is an important mechanism for organisms to cope with climate variability at different temporal scales [53]. In the context of fast climate change and increase in temperature fluctuation and unpredictability, a better knowledge about response to environmental variations through such transgeneration effect may help to understand the ability of rapid adaptation and plasticity [53,54]—an issue which so far has been neglected in forensic entomology.

4. Conclusions

Up until now there has been no evidence for a decline in necrophagous insects. If there is any indication, it has no relevance for forensic entomology so far. This need not always be the case, and it is quite possible that single drivers of insect decline will sooner or later also affect forensically important taxa. Here, in particular, climate change and global warming can be named. It cannot really be expected that these factors will result in a decline, but they will have a strong impact on the biology and ecology of necrophagous insects. In this respect, therefore, it is not the decline in insects but climate change that will have a strong effect on forensic entomology and which research will have to address in the future. Here, the focus should be on transgenerational plasticity, because climate change will pose major challenges to the short-term adaptive abilities of insects. Existing associations such as the European Association for Forensic Entomology or the North American Forensic Entomology Association should establish a network of baseline data on the occurrence and activity of necrophagous insects, backed up with the relevant local climatic data and next-generation effects. Such cooperation and interaction will help to better understand key species and to better target the impact of climate change on forensic entomology, for example, by identifying specific factors such as e.g., droughts or the presence of heat islands in urban environments, as key players in triggering transgenerational effects and associated modified development.

Funding: This research received no external funding.

Institutional Review Board Statement: Not applicable.

Data Availability Statement: Not applicable.

Acknowledgments: Lena Lutz designed the single elements in Figure 3.

Conflicts of Interest: The author declares no conflict of interest.

References

1. United Nations Report. *Nature's Dangerous Decline 'Unprecedented'; Species Extinction Rates 'Accelerating'*; United Nations: New York, NY, USA, 2019.
2. Ceballos, G.; Ehrlich, P.R.; Dirzo, R. Biological annihilation via the ongoing sixth mass extinction signaled by vertebrate population losses and declines. *Proc. Natl. Acad. Sci. USA* **2017**, *114*, E6089–E6096. [CrossRef] [PubMed]
3. González-del-Pliego, P.; Freckleton, R.P.; Edwards, D.P.; Koo, M.S.; Scheffers, B.R.; Pyron, R.A.; Jetz, W. Phylogenetic and Trait-Based Prediction of Extinction Risk for Data-Deficient Amphibians. *Curr. Biol.* **2019**, *29*, 1557–1563.e3. [CrossRef] [PubMed]
4. Stork, N.E. How Many Species of Insects and Other Terrestrial Arthropods Are There on Earth? *Annu. Rev. Entomol.* **2018**, *63*, 31–45. [CrossRef] [PubMed]

5. Wagner, D.L.; Grames, E.M.; Forister, M.L.; Berenbaum, M.R.; Stopak, D. Insect decline in the Anthropocene: Death by a thousand cuts. *Proc. Natl. Acad. Sci. USA* **2021**, *118*, 1–10. [CrossRef]

6. Hallmann, C.A.; Sorg, M.; Jongejans, E.; Siepel, H.; Hofland, N.; Schwan, H.; Stenmans, W.; Müller, A.; Sumser, H.; Hörren, T.; et al. More than 75 percent decline over 27 years in total flying insect biomass in protected areas. *PLoS ONE* **2017**, *12*. [CrossRef]

7. Wagner, D.L. Insect declines in the anthropocene. *Annu. Rev. Entomol.* **2020**, *65*, 457–480. [CrossRef]

8. Conrad, K.F.; Warren, M.S.; Fox, R.; Parsons, M.S.; Woiwod, I.P. Rapid declines of common, widespread British moths provide evidence of an insect biodiversity crisis. *Biol. Conserv.* **2006**, *132*, 279–291. [CrossRef]

9. Hallmann, C.A.; Ssymank, A.; Sorg, M.; de Kroon, H.; Jongejans, E. Insect biomass decline scaled to species diversity: General patterns derived from a hoverfly community. *Proc. Natl. Acad. Sci. USA* **2021**, *118*, 1–8. [CrossRef]

10. Lister, B.C.; Garcia, A. Climate-driven declines in arthropod abundance restructure a rainforest food web. *Proc. Natl. Acad. Sci. USA* **2018**, *115*, E10397–E10406. [CrossRef] [PubMed]

11. Van Klink, R.; Bowler, D.E.; Gongalsky, K.B.; Swengel, A.B.; Gentile, A.; Chase, J.M. Meta-analysis reveals declines in terrestrial but increases in freshwater insect abundances. *Science* **2020**, *368*, 417–420. [CrossRef]

12. Daskalova, G.N.; Phillimore, A.B.; Myers-Smith, I.H. Accounting for year effects and sampling error in temporal analyses of invertebrate population and biodiversity change: A comment on Seibold et al. 2019. *Insect Conserv. Divers.* **2021**, *14*, 149–154. [CrossRef]

13. Desquilbet, M.; Gaume, L.; Grippa, M.; Céréghino, R.; Humbert, J.F.; Bonmatin, J.M.; Cornillon, P.A.; Maes, D.; Van Dyck, H.; Goulson, D. Comment on "meta-analysis reveals declines in terrestrial but increases in freshwater insect abundances". *Science* **2021**, *370*, 1–3. [CrossRef]

14. Saunders, M.E.; Janes, J.K.; O'Hanlon, J.C. Moving on from the Insect Apocalypse Narrative: Engaging with Evidence-Based Insect Conservation. *Bioscience* **2020**, *70*, 80–89. [CrossRef]

15. Saunders, M.E.; Janes, J.K.; O'Hanlon, J.C. Semantics of the insect decline narrative: Recommendations for communicating insect conservation to peer and public audiences. *Insect Conserv. Divers.* **2020**, *13*. [CrossRef]

16. Crossley, M.S.; Meier, A.R.; Baldwin, E.M.; Berry, L.L.; Crenshaw, L.C.; Hartman, G.L.; Lagos-Kutz, D.; Nichols, D.H.; Patel, K.; Varriano, S.; et al. No net insect abundance and diversity declines across US Long Term Ecological Research sites. *Nat. Ecol. Evol.* **2020**, *4*, 1368–1376. [CrossRef] [PubMed]

17. Sánchez-Bayo, F.; Wyckhuys, K.A.G. Worldwide decline of the entomofauna: A review of its drivers. *Biol. Conserv.* **2019**, *232*, 8–27. [CrossRef]

18. Sánchez-Bayo, F.; Wyckhuys, K.A.G. Further evidence for a global decline of the entomofauna. *Austral Entomol.* **2021**, *60*, 9–26. [CrossRef]

19. Hallmann, C.A.; Zeegers, T.; Klink, R.; Vermeulen, R.; Wielink, P.; Spijkers, H.; Deijk, J.; Steenis, W.; Jongejans, E. Declining abundance of beetles, moths and caddisflies in the Netherlands. *Insect Conserv. Divers.* **2020**, *13*, 127–139. [CrossRef]

20. Bell, J.R.; Blumgart, D.; Shortall, C.R. Are insects declining and at what rate? An analysis of standardised, systematic catches of aphid and moth abundances across Great Britain. *Insect Conserv. Divers.* **2020**, *13*, 115–126. [CrossRef] [PubMed]

21. Carvalheiro, L.G.; Kunin, W.E.; Keil, P.; Aguirre-Gutiérrez, J.; Ellis, W.N.; Fox, R.; Groom, Q.; Hennekens, S.; Landuyt, W.; Maes, D.; et al. Species richness declines and biotic homogenisation have slowed down for NW-European pollinators and plants. *Ecol. Lett. Ecol. Lett.* **2013**, *16*, 870–878. [CrossRef]

22. Theng, M.; Jusoh, W.F.A.; Jain, A.; Huertas, B.; Tan, D.J.X.; Tan, H.Z.; Kristensen, N.P.; Meier, R.; Chisholm, R.A. A comprehensive assessment of diversity loss in a well-documented tropical insect fauna: Almost half of Singapore's butterfly species extirpated in 160 years. *Biol. Conserv.* **2020**, *242*. [CrossRef]

23. Tabor, K.L.; Brewster, C.C.; Fell, R.D. Analysis of the Successional patterns of insects on carrion in southwest Virginia. *J. Med. Entomol.* **2004**, *41*, 785–795. [CrossRef]

24. Sharanowski, B.J.; Walker, E.G.; Anderson, G.S. Insect succession and decomposition patterns on shaded and sunlit carrion in Saskatchewan in three different seasons. *Forensic Sci. Int.* **2008**, *179*, 219–240. [CrossRef] [PubMed]

25. Matuszewski, S.; Bajerlein, D.; Konwerski, S.; Szpila, K. Insect succession and carrion decomposition in selected forests of Central Europe. Part 3: Succession of carrion fauna. *Forensic Sci. Int.* **2011**, *207*, 150–163. [CrossRef]

26. Wang, Y.; Ma, M.Y.; Jiang, X.Y.; Wang, J.F.; Li, L.L.; Yin, X.J.; Wang, M.; Lai, Y.; Tao, L.Y. Insect succession on remains of human and animals in Shenzhen, China. *Forensic Sci. Int.* **2017**, *271*, 75–86. [CrossRef]

27. Meira, L.M.R.; Barbosa, T.M.; Jales, J.T.; Santos, A.N.; Gama, R.A. Insects Associated to Crime Scenes in the Northeast of Brazil: Consolidation of Collaboration Between Entomologists and Criminal Investigation Institutes. *J. Med. Entomol.* **2020**, *57*, 1012–1020. [CrossRef]

28. Griffiths, K.; Krosch, M.N.; Wright, K. Variation in decomposition stages and carrion insect succession in a dry tropical climate and its effect on estimating postmortem interval. *Forensic Sci. Res.* **2020**, *5*, 327–335. [CrossRef] [PubMed]

29. Lobo, J.M. Decline of roller dung beetle (Scarabaeinae) populations in the Iberian peninsula during the 20th century. *Biol. Conserv.* **2001**, *97*, 43–50. [CrossRef]

30. Cuesta, E.; Lobo, J.M. A comparison of dung beetle assemblages (Coleoptera, Scarabaeoidea) collected 34 years apart in an Iberian mountain locality. *J. Insect Conserv.* **2019**, *23*, 101–110. [CrossRef]

31. Carpaneto, G.M.; Mazziotta, A.; Valerio, L. Inferring species decline from collection records: Roller dung beetles in Italy (Coleoptera, Scarabaeidae). *Divers. Distrib.* **2007**, *13*, 903–919. [CrossRef]

32. Seibold, S.; Brandl, R.; Buse, J.; Hothorn, T.; Schmidl, J.; Thorn, S.; Müller, J. Association of extinction risk of saproxylic beetles with ecological degradation of forests in Europe. *Conserv. Biol.* **2015**, *29*, 382–390. [CrossRef]
33. Harris, J.E.; Rodenhouse, N.L.; Holmes, R.T. Decline in beetle abundance and diversity in an intact temperate forest linked to climate warming. *Biol. Conserv.* **2019**, *240*, 108219. [CrossRef]
34. Powney, G.D.; Carvell, C.; Edwards, M.; Morris, R.K.A.; Roy, H.E.; Woodcock, B.A.; Isaac, N.J.B. Widespread losses of pollinating insects in Britain. *Nat. Commun.* **2019**, *10*, 1–6. [CrossRef]
35. Outhwaite, C.L.; Gregory, R.D.; Chandler, R.E.; Collen, B.; Isaac, N.J.B. Complex long-term biodiversity change among invertebrates, bryophytes and lichens. *Nat. Ecol. Evol.* **2020**, *4*, 384–392. [CrossRef]
36. Orford, K.A.; Vaughan, I.P.; Memmott, J. The forgotten flies: The importance of non-syrphid Diptera as pollinators. *Proc. R. Soc. B Biol. Sci.* **2015**, *282*. [CrossRef]
37. Tiusanen, M.; Hebert, P.D.N.; Schmidt, N.M.; Roslin, T. One fly to rule them all—Muscid flies are the key pollinators in the arctic. *Proc. R. Soc. B Biol. Sci.* **2016**, *283*. [CrossRef]
38. Høye, T.T.; Post, E.; Schmidt, N.M.; Trøjelsgaard, K.; Forchhammer, M.C. Shorter flowering seasons and declining abundance of flower visitors in a warmer Arctic. *Nat. Clim. Chang.* **2013**, *3*, 759–763. [CrossRef]
39. Loboda, S.; Savage, J.; Buddle, C.M.; Schmidt, N.M.; Høye, T.T. Declining diversity and abundance of High Arctic fly assemblages over two decades of rapid climate warming. *Ecography* **2018**, *41*, 265–277. [CrossRef]
40. Gillespie, M.A.K.; Alfredsson, M.; Barrio, I.C.; Bowden, J.J.; Convey, P.; Culler, L.E.; Coulson, S.J.; Krogh, P.H.; Koltz, A.M.; Koponen, S.; et al. Status and trends of terrestrial arthropod abundance and diversity in the North Atlantic region of the Arctic. *Ambio* **2020**, *49*, 718–731. [CrossRef]
41. Amendt, J. *Insect Survey 2004*; Institute Legal Medicine: Frankfurt am Main, Germany, 2006.
42. Fremdt, H.; Amendt, J. Species composition of forensically important blow flies (Diptera: Calliphoridae) and flesh flies (Diptera: Sarcophagidae) through space and time. *Forensic Sci. Int.* **2014**, *236*. [CrossRef]
43. Bernhardt, V.; Bálint, M.; Verhoff, M.A.; Amendt, J. Species diversity and tissue specific dispersal of necrophagous Diptera on human bodies. *Forensic Sci. Med. Pathol.* **2018**, *14*, 76–84. [CrossRef]
44. Lutz, L.; Verhoff, M.A.; Amendt, J. Environmental factors influencing flight activity of forensically important female blow flies in Central Europe. *Int. J. Legal Med.* **2019**, *133*, 1267–1278. [CrossRef]
45. Potts, S.G.; Biesmeijer, J.C.; Kremen, C.; Neumann, P.; Schweiger, O.; Kunin, W.E. Global pollinator declines: Trends, impacts and drivers. *Trends Ecol. Evol.* **2010**, *25*, 345–353. [CrossRef]
46. Barton, P.S.; Evans, M.J.; Foster, C.N.; Pechal, J.L.; Bump, J.K.; Quaggiotto, M.M.; Benbow, M.E. Towards Quantifying Carrion Biomass in Ecosystems. *Trends Ecol. Evol.* **2019**, *34*, 950–961. [CrossRef] [PubMed]
47. Moleón, M.; Selva, N.; Quaggiotto, M.M.; Bailey, D.M.; Cortés-Avizanda, A.; DeVault, T.L. *Carrion Availability in Space and Time*; Springer: Cham, Switzerland, 2019; pp. 23–44.
48. Moleón, M.; Selva, N.; Sánchez-Zapata, J.A. The Components and Spatiotemporal Dimension of Carrion Biomass Quantification. *Trends Ecol. Evol.* **2020**, *35*, 91–92. [CrossRef] [PubMed]
49. Szpila, K.; Mądra, A.; Jarmusz, M.; Matuszewski, S. Flesh flies (Diptera: Sarcophagidae) colonising large carcasses in Central Europe. *Parasitol. Res.* **2015**, *114*, 2341–2348. [CrossRef] [PubMed]
50. Szpila, K.; Matuszewski, S.; Bajerlein, D.; Konwerski, S. *Chrysomya albiceps* (Wiedemenn, 1819), a forensically important blowfy (Diptera: Calliphoridae) new for the Polish fauna. *Polish J. Entomol.* **2008**, *77*, 351–355.
51. Picard, C.J. First record of *Chrysomya megacephala* in Indiana, U.S.A. *Proc. Entomol. Soc. Washingt.* **2013**, *115*, 265–267. [CrossRef]
52. Owings, C.G.; Picard, C.J. New distribution record for *Lucilia cuprina* (Diptera: Calliphoridae) in Indiana, United States. *J. Insect Sci.* **2018**, *18*, 1–6. [CrossRef]
53. Tougeron, K.; Devogel, M.; van Baaren, J.; Le Lann, C.; Hance, T. Transgenerational effects on diapause and life-history-traits of an aphid parasitoid. *J. Insect Physiol.* **2020**, *121*, 104001. [CrossRef]
54. Donelson, J.M.; Salinas, S.; Munday, P.L.; Shama, L.N.S. Transgenerational plasticity and climate change experiments: Where do we go from here? *Glob. Chang. Biol.* **2017**. [CrossRef] [PubMed]

MDPI

St. Alban-Anlage 66

4052 Basel

Switzerland

Tel. +41 61 683 77 34

Fax +41 61 302 89 18

www.mdpi.com

Insects Editorial Office

E-mail: insects@mdpi.com

www.mdpi.com/journal/insects